Recent Advances in Creep and Fracture of Engineering Materials and Structures

Recent Advances in Creep and Fracture of Engineering Materials and Structures

Edited by:

B. Wilshire
Department of Metallurgy and Materials Technology,
University College, Swansea.

D. R. J. Owen
Department of Civil Engineering, University College, Swansea.

PINERIDGE PRESS

Swansea, U.K.

First published, 1982 by
Pineridge Press Limited
91 West Cross Lane, West Cross, Swansea, U.K.

ISBN 0-906674-18-2

British Library Cataloguing in Publication Data
Recent Advances in Creep and Fracture of Engineering
Materials and Structures

1. Materials—Creep
2. Fracture mechanics
I. Wilshire, B. II. Owen, D.R.J.
620.1′123 TA418.38

ISBN 0-906674-18-2

The opinions expressed in each chapter are entirely those of the authors. Although the computer programs listed in the text have been extensively verified, the publishers cannot accept responsibility for their performance in practice.

Printed and bound in Great Britain by
Redwood Burn Limited,
Trowbridge, Wiltshire

PREFACE

Rapid progress is now being made on many different aspects of high temperature creep and fracture. Yet, the information available on any individual topic is usually distributed widely throughout the scientific literature. This Recent Advances series aims to combine the advantages of a 'review style' with significant new developments in appropriate areas. Self-contained articles by leading authorities in the field are therefore designed to provide both a general overview and reference for the present position on individual topics as well as a detailed coverage of the authors' current activities.

The specialist articles in this initial volume were selected to emphasize some of the major new developments which were emerging at the International Conference on 'Creep and Fracture of Engineering Materials and Structures' held at the University College of Swansea during 23rd to 27th March 1981. Consequently, the themes included range from detailed mechanistic considerations to engineering applications, while the materials covered extend from ceramics to commercial creep-resistant alloys.

Subsequent volumes in this Series will also be composed of definitive articles from internationally recognized authorities and will continue the approach of combining full coverage of new concepts with a comprehensive review of the relevant background. However, it is intended that even the publication philosophy itself should possess the flexibility needed to respond to the changing needs of the field as challenging new areas of theoretical and practical importance become apparent. With this attitude in mind, the set of articles constituting each future volume will be chosen to offer an integrated coverage of a specific theme. It has therefore been decided that the next volume should be devoted to the critical area of 'Engineering Approaches to High Temperature Design'. In this way, the present volume is the first in an evolving Recent Advances series which aims to provide a valuable and distinctive contribution to the literature on 'Creep and Fracture of Engineering Materials and Structures'.

B. Wilshire
D.R.J. Owen

Swansea, June 1982.

CONTENTS

Chapter 1

MECHANISMS AND MECHANICS OF FRACTURE IN CREEPING ALLOYS

A.S. Argon

Department of Mechanical Engineering
Massachusetts Institute of Technology
Cambridge, Massachusetts 02139, U.S.A.

SUMMARY

Important advances in the understanding of the mechanics and mechanisms of creep fracture in polycrystalline alloys registered in the past decade are presented. Much progress has been made both by the materials scientist in investigating directly in detail the damage mechanisms due to creep, and by the applied mechanician in modeling the micro-structural and macro-structural aspects of this damage. The most impressive nature of this advance is, the close coupling that has resulted between the two approaches of materials science and structural design. The review furnishes many examples of this heartening development, but also points out areas where more progress is necessary.

1. INTRODUCTION

High temperature service often combines extreme conditions of aggressive environments, and high stresses having fluctuating components, in addition to the high temperatures. Fracture under these conditions is usually the performance limiting phenomenon and is of primary concern to the designer. In many cases the loading is stationary and components can be protected from the aggressive environments by the application of protective coatings so that the problem becomes often one of accumulation of creep damage. In the past, engineering design against fracture in creep has been handled satisfactorily largely by empirical approaches, while the metallurgist and materials scientist has concerned himself mostly with the study of relatively simple alloys under simple tension. The demands of the nuclear power industry, and the general strong trend toward more efficient energy systems of all types has, however, brought about a remarkable synthesis in these two separate studies of creep fracture within the past decade. On one side, this has resulted in the large scale application of mechanics to the solution of both macro-scale problems of notches, cracks, and the like, and also micro-scale problems of sliding grain-boundaries and the growth of grain-boundary cavities. On the other side, the synthesis has come about by the precise and direct studies of damage mechanisms in both simple and complex alloys by the powerful new tools of the materials scientist. As a result creeping alloys have ceased to be "black boxes" to the design engineer, and the materials scientist has begun to appreciate the complex nature of load shedding and stress redistribution in inhomogeneously stressed parts. This fortunate trend has already been perceived and partially aided by the review of McLean, Dyson and Taplin [1] at the 4th International Fracture Conference.

Here we will pick up the mechanistic thread and review the important recent advances in creep fracture occurring in the past decade and, wherever possible, relate them to engineering practice.

2. TYPES OF HIGH TEMPERATURE FAILURE

Ashby, Gandhi and Taplin [2] have catalogued the phenomenologically different types of fracture in a number of important engineering alloys in the creep range. The information is presented typically in to complementary forms as shown in Figs. 1a and 1b for the case of Nimonic 80A. In Fig. 1a the isochronal fracture data is presented on a field of the normalized tensile stress and homologous temperature conditions under which the tests were conducted. The strains to fracture in each case are indicated by the number next to the data point. This form of presentation, when taken together with the corresponding deformation mechanism map for the same material, relates the fracture information to the deformation process that has preceded it. The fields outlined by the tie-contours characterize the regions in which one of the specific phenomenologically different forms of fracture depicted in Fig. 1c dominates. Figure 1b shows the information in terms of isothermal contours on a field of normalized tensile stress and time to fracture, familiar to the design engineer. The tie-contours again delineate the regions in which phenomenologically different forms of fracture dominate. Since ductile fracture is relatively time independent, and rupture occurs at very low stresses and very high temperatures, often associated with dynamic re-crystallization, Fig. 1b represents almost entirely the various forms of fracture encountered in creeping alloys. As the figure demonstrates clearly, the type of fracture in a typical commercial alloy such as Nimonic 80A at high stress and high strain rates, leading to short times to fracture, is transgranular and occurs by micro-void coalescence -- with features almost identical to those in

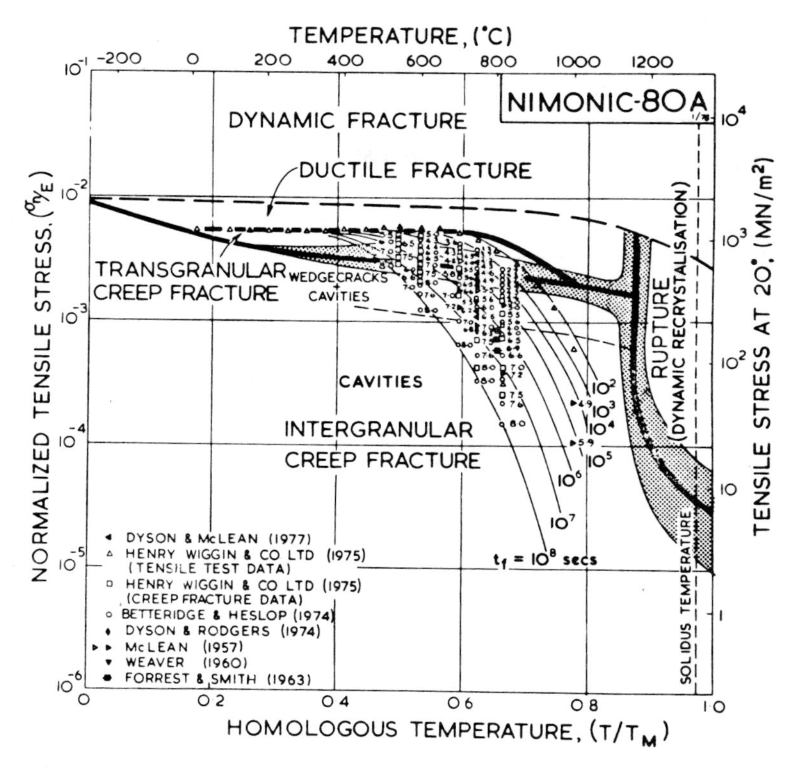

Fig. 1a. Map of isochronal fracture data for Nimonic 80A (from Ashby, Gandhi, and Taplin [2]; courtesy of Pergamon Press); for sources of data see Ref. [2].

ductile fracture at low temprature. As the stress is decreased to increase the fracture time, fracture becomes intergranular and is apparently dominated by wedge cracks leading to the accumulation of grain facet fractures which then collectively lead to final fracture by eventual interaction and linkage. At lower stresses leading to long fracture times, typically encountered in service, the fracture is dominated by the coalescence of individual grain-boundary cavities. It is these intergranular fracture processes, that are dominant in service conditions, that will be the primary subject of our discussion.

The transition of fracture from a transgranular form of micro-void coalescence to intergranular cavitation is influenced, more than anything else, by grain-boundary sliding. In the range of stresses and temperatures where the strain-rate, stress response of the grain matrix is by power-law creep, the response of grain-boundary sliding to the shear stress acting across the grain-boundary is linear. This results in nearly complete relaxation of shear stress across grain-boundaries in a polycrystal at low stresses and strain rates, and in no significant relaxation of such

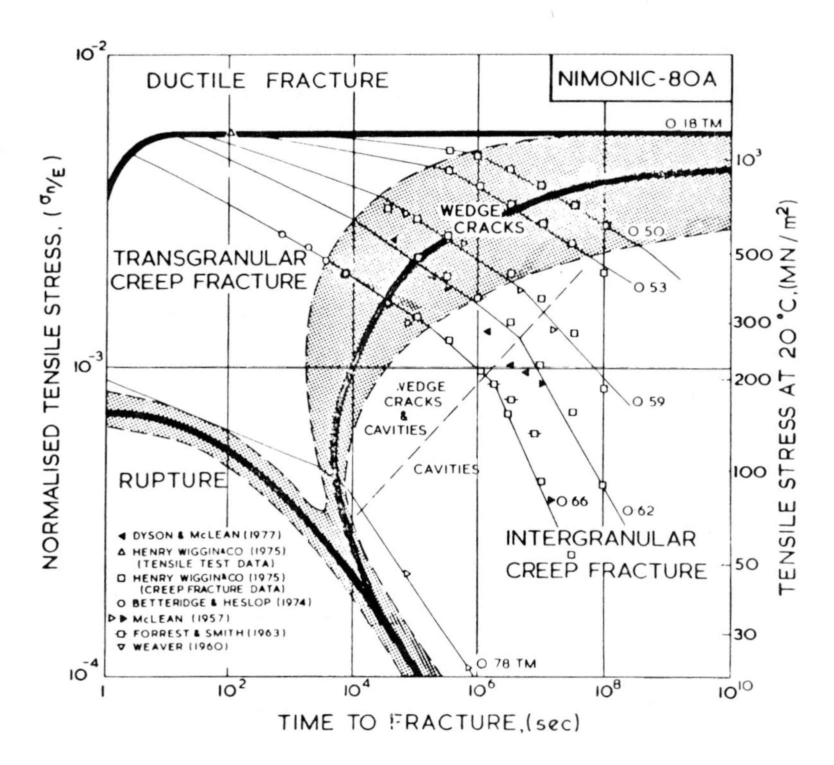

Fig. 1b. Map of isothermal fracture data for Nimonic 80A (from Ashby, Gandhi, and Taplin [2]; courtesy of Pergamon Press); for sources of data see Ref. [2].

stresses at high applied stress and strain rates [3-5]. The problem has been treated in a very general way by Chen and Argon [5] for space filling polyhedral grains of size d, creeping according to a power law with a stress exponent of m, i.e.

$$\dot{\gamma} = A(\tau/G)^m \quad (1)$$

where τ is the shear stress, $\dot{\gamma}$ shear strain rate, G the shear modulus, and A a constant. In this treatment the grain-boundaries are of thickness δ obeying a linear viscous response given by

$$\dot{\gamma}_{gb} = \tau/\mu_{gb} , \quad (2)$$

where $\dot{\gamma}_{gb}$ is the shear strain rate in the grain boundary material and μ_{gb} is the grain boundary shear viscosity. The application of the well known self consistent method widely used for computation of composite properties, and the use of a slip line field analysis for terminal plastic behavior of grains ($m \to \infty$), has resulted in the typical response given in Fig. 2 for columnar grains with surface filling polygonal

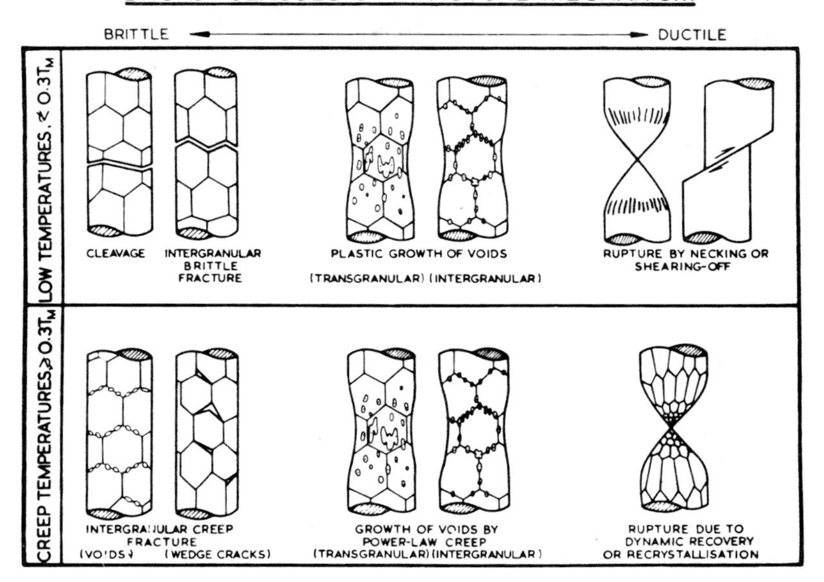

Fig. 1c. Pictorial classification of fracture mechanisms. The lower row are observed at high temperatures (from Ashby, Gandhi and Taplin [2]; courtesy of Pergamon Press).

cross section. The figure shows for a specific case of m = 5, $A = 10^{18}s^{-1}$, $G/\mu^*_{gb} - 10s^{-1}$ (where $\mu^*_{gb} = (4/3\pi)(d/\delta)\mu_{gb}$ is a defined "grain equivalent" boundary viscosity) the average computed response of the polycrystal over a wide stress range. At large applied stresses the grain-boundaries transmit full shear tractions, and the polycrystal is unrelaxed, with a uniform stress distribution of τ. At very low stresses the grain-boundaries transmit no such tractions, the stress distribution in the polycrystal is highly non-uniform but the average strain rate, $\dot{\gamma}$, exhibits the same power dependence on the average shear stress $\bar{\tau}$ as in the unrelaxed case. In the latter case the response can be best characterized, at a given strain rate, as a reduced average shear resistance that varies between 3/4 and 2/3 as m goes from 1 to ∞. The point of transition in behavior between relaxed and unrelaxed behavior occurs when the power law creep response of the matrix equals the "grain equivalent" flow response of the grain-boundary material, as shown in Fig. 2 or where

$$\dot{\gamma}_{gb} = \frac{\tau}{G}\left(\frac{G}{\mu_{gb}}\right) = A\left(\frac{\tau}{G}\right)^m , \tag{3}$$

giving a transition at $\tau/G = 5.623 \times 10^{-5}$, for the specific case which is, of course, grain size dependent. Thus, the

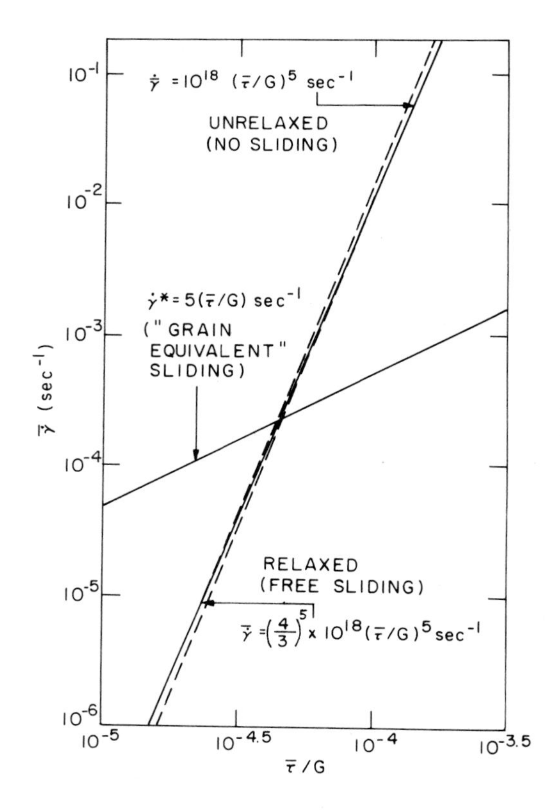

Fig. 2. Transition between relaxed and unrelaxed behavior of grain-boundaries with increasing stress in a plane strain geometry in a polycrystal of parallel columnar grains transversely deformed (from Chen and Argon [5]; courtesy of Pergamon Press).

effect of the shear relaxation across sliding grain-boundaries is a relatively minor reduction in the shear resistance of the assembly that is almost unaffected by the power exponent m and is independent of grain size. The consequence of the redistribution of stress resulting from the shear relaxation is, however, of vital importance in the initiation of grain-boundary cavities.

At high strain rates where grain-boundaries are unrelaxed and the stress distribution is uniform, the deformation of the polycrystal approximates that of a continuum and only non-deformable particles perturb the deformation, regardless of their location. If the resulting interface tractions are high enough, or the particle strength itself is reached, cavities form much like in ductile fracture at low temperature [6,7] and grow by power-law creep to coalescence [8,9] and eventual transgranular fracture. We discuss such transgranular fractures by power-law creep in Section 3.

At lower stresses the tractions on matrix particles become insufficient to cavitate their interfaces, but much larger stress concentrations begin to arise at particles along sliding grain-boundaries as the shear tractions on the boundary between such particles are relaxed. This results in the nucleation of grain-boundary cavities which we discuss in some detail in Section 4.2.

The grain-boundary cavities grow under the local prevailing normal boundary tractions by a combination of diffusional flow and power-law creep until they link together to result in intergranular fracture. This we discuss in Section 4.3.

As the applied stress decreases, power-law creep in the matrix is sharply reduced, and the strain to fracture drops to a limiting low value of roughly the inter-cavity spacing along the grain-boundaries divided by the grain size. In the limit this cavitational strain becomes the most important contribution to the total inelastic strain. We discuss in Section 5 the developments to calculate the overall time to fracture under tensile and multi-axial stresses by these processes leading to the forms of data used widely in design engineering.

In the absence of sharp notches and more major sources of strain concentration even the levels of the above mentioned cavitational strain is sufficient to equalize the short wave length variations in the stress distribution made initially non-homogeneous by the sliding of grain-boundaries. This results in quasi-homogeneous cavitation and grain-boundary facet cracking over large regions under relatively uniform stress, on a scale larger than the grain size. At intermediate stresses where the added matrix creep increases the final strain to fracture, well above the cavitational strain, the stress leveling is more complete. This capacity for compensating against moderate stress non-uniformities of relatively short range by the formation of cavitational wedges and by creep renders many, except the most acute, notches and cracks in a creeping alloy ineffective. We discuss this behavior of stress redistribution at notches and the development of intergranular fracture under such conditions in Section 6.

At very acute notches and at the tips of sharp cracks the developing concentrations of strain can not be completely eliminated by the stress leveling action of cavitation and creep strain. In such inhomogeneous fields the limiting strain to fracture is reached locally before the concentration of stress can be smoothed out, and gradually accelerating creep fracture extends the crack. These processes of crack extension by the focusing of creep damage, even in the stress field extensively modified by the cavitation itself, are not yet fully understood quantitatively. We will discuss in Section 7 the most recent developments in this field.

3. TRANSGRANULAR CREEP FRACTURE

The growth of cavities in deforming continua subjected to a general stress state has been of interest for some time to explain the strain to fracture in ductile metals undergoing a micro-void growth and coalescence process. In that context the problem has been considered by McClintock [10] for periodic cylindrical cavities in power-law hardening plastic continua, and by Rice and Tracy [11] for periodic spherical cavities. The specific application of these developments to creeping solids has been considered first by Hancock [8]. A more general analysis of expansion of cavities of initially spherical shape, in inelastic continua obeying a power-law, has recently been provided by Budiansky, Hutchinson and Slutsky [9]. They have considered how initially spherical cavities reach asymptotic shapes in the deforming continuum, and how such shapes of cavities eventually grow in a general axi-symmetric stress field (i.e., σ_{zz}, $\sigma_{rr} = \sigma_{\theta\theta}$). An important finding in their analysis is that unless the triaxiality of stress $\sigma_m/\sigma_e \gtrsim 1.2$, the initially spherical cavities deform into elongated cylinders, aligned parallel to the principal extension axis. This is regardless of the value m of the power-law creep exponent in the general creep relation of the material, given by

$$\frac{\dot{\varepsilon}_e}{\dot{\varepsilon}_0} = \left(\frac{\sigma_e}{\sigma_0}\right)^m \quad (4)$$

Here the triaxiality is defined as the ratio of the negative pressure on σ_m $(=(\sigma_1 + \sigma_2 + \sigma_3)/3)$ to the Mises equivalent stress σ_e in the distant field, and $\dot{\varepsilon}_e$ stands for the Mises equivalent strain rate, σ_0 is a constant having the dimensions of stress and representing a reference tensile deformation resistance of the solid, $\dot{\varepsilon}_0$ is the reference tensile strain rate at the latter stress, and m the power-law exponent, valid for $\sigma_e < \sigma_0$. For triaxialities $(\sigma_m/\sigma_e) \gtrsim 1.2$, Budiansky et al find that initially, spherical cavities reach asymptotic forms of expansion that are oblate or prolate pseudo-spheres. Such large triaxialities are rarely attainable in creeping solids in all but the most acute notches or crack tips. Therefore, we consider the case of expanding parallel cylinders as the most widely applicable asymptotic form of interest for the purpose of obtaining an expression for the time to fracture t_f by coalescence of cavities in a creeping solid. for this geometry Budiansky et al [9] find the normalized volumetric expansion rates for two limiting cases of the power exponent $m = 1$ and $m = \infty$ as follows,

$$\frac{\dot{V}}{\dot{\varepsilon}V} = \frac{3\sigma_m}{\sigma_e} - 1 \qquad (m = 1) \tag{5}$$

$$\frac{\dot{V}}{\dot{\varepsilon}V} = \sqrt{3}\,\sinh\frac{1}{\sqrt{3}}\left(\frac{3\sigma_m}{\sigma_e} - 1\right) \qquad (m = \infty) \tag{6}$$

which for $(\sigma_m/\sigma_e) < 0.6$ differ by less than 1.5%, and hence, cover the behavior over the entire range of non-linearity.

The critical equivalent strain to fracture ε_{cr} is stated by Budiansky et al [9] to be,

$$\varepsilon_{cr} = \frac{\ln(\overset{*}{A}_{cr}/\overset{*}{A}_0)}{(\dot{V}/\dot{\varepsilon}V)} \tag{7}$$

where $\overset{*}{A}_0$ is the initial areal porosity along a plane perpendicular to the principal direction of extension and $\overset{*}{A}_{cr}$ the critical value of this porosity (usually about 0.4) where cavities have a high touching probability. The total time to fracture t_f under constant loading conditions then becomes,

$$t_f = \varepsilon_{cr}/\dot{\varepsilon} \ . \tag{8}$$

This, upon use of Eqns (5), (6) and (7) gives

$$\{\ln(\overset{*}{A}_{cr}/\overset{*}{A}_0)/\sqrt{3}\sinh[\frac{1}{\sqrt{3}}(\frac{3\sigma_m}{\sigma_e} - 1)]\} < \dot{\varepsilon}t_f$$

$$< \{\ln(\overset{*}{A}_{cr}/\overset{*}{A}_0)(\frac{3\sigma_m}{\sigma_e} - 1)\} \tag{9}$$

which brackets the entire range of non-linear behavior from $m = 1$ to $m \to \infty$. For a specific case of $\sigma_m/\sigma_e = 0.6$, that is typical for the center of a necked, creeping bar, $\overset{*}{A}_{cr} = 0.4$ and $\overset{*}{A}_0 = 0.01$ we find

$$4.45 < \dot{\varepsilon}t_f < 4.61 \ . \tag{10}$$

This corresponds to a local reduction of area of nearly 99% and is far in excess of the usual reductions of area of around 60% that are common for transgranular creep fracture. A similar overestimation by the theoretical model of cavity growth also exists for ductile fracture [12] and is usually attributed to early cavity interactions and strain localization effects. Transgranular fracture in creeping alloys is of interest in high temperature forming processes where high stresses and strain rates are present, and where little relaxation of grain-boundary shear tractions occurs.

4. INTERGRANULAR FRACTURE

4.1 Phenomenology

As stated in Section 2 above, under service conditions where stresses and strain rates are low, the unsupported portions of grain-boundaries can undergo sliding to relax the shear stresses. The loss of shear support is made up for short periods of time by concentrated stresses at interfaces of grain-boundary particles and on faces of grain-boundary ledges that now become overloaded. This results in the fracture becoming intergranular and in a marked decrease in overall strain to fracture due to the decrease of the creep deformation inside grains. The early investigators viewing the development of intergranular fracture distinguished between two apparently distinct forms of damage: wedge cracks at triple grain junctions occurring predominantly at higher stresses, and round cavities along all boundaries that are not parallel to the tensile axis, occurring under lower stresses [13, 14]. Fracture was found to result from such damage by the linkage of growing cavities and by extension of wedge cracks producing grain facet cracks that eventually were bridged together by cavity coalescence along the slanted bridging grain-boundaries. Examination of the intergranular fracture surfaces after final fracture generally reveals a uniform coverage with fine scale dimples corresponding roughly to the mean spacing along the boundaries of grain-boundary particles. This and the fact that intergranular fracture can be completely suppressed when grain-boundary particles are eliminated [15], demonstrates clearly that this fracture is closely linked to the presence of such particles.

Once nucleated, cavities can grow either by a continuum mode of creep deformation as discussed in Section 3 or by a flow of vacancies into the cavities along the grain-boundaries. As Eqns (5) and (6) show, the volumetric growth rate per unit cavity volume by continuum creep deformation is independent of the size of the cavity, while it turns out that the corresponding growth rate by flow of vacancies is inversely proportional to the cavity volume itself. Therefore, it is now well established that very small cavities grow preferentially by diffusional flow of vacancies until they become quite large where finally growth by creep deformation takes over.

In numerous engineering studies of creep fracture in tension it has been established by Monkman and Grant [16] that the overall time to fracture t_f in long term creep tests is inversely proportional to a power function of the minimum creep rate $\dot{\varepsilon}_{min}$, i.e.,

$$t_f = C/(\dot{\varepsilon}_{min})^n , \tag{11}$$

where the exponent n is found to range from a low of 0.77 for Nimonic 80 to a high of 0.93 for austenitic stainless steels, and where the constant C ranges from about 1.8 for Nimonic 80 to 15 for aluminum. The proper explanation of this quite general observation by detailed mechanistic approaches will be given in Section 5 below.

4.2 The Nucleation of Grain Boundary Cavities

4.2.1 Nucleation theory

Grant and co-workers [14, 15] have demonstrated experimentally that cavity or crack formation along grain-boundaries in creeping alloys requires both the presence of grain-boundary particles and grain-boundary sliding. Since such cavitation occurs at high temperatures where point defect concentrations and mobilities are both high, a classical nucleation process of a vacant phase by the agglomeration of vacancies on stressed interfaces must be suspected. This possibility has been investigated in great detail by Raj and Ashby [17], Raj [18], and Argon, Chen and Lau [19].

Consider a stressed, non-coherent interface such as a grain-boundary or the boundary between a particle and the surrounding matrix as shown in Fig. 3. the shear stresses acting across the boundary are assumed to be relieved by boundary sliding, giving rise to only a normal stress σ_n, which may be related to the distant stress by an appropriate stress concentration factor k. Formation of a cavity at the

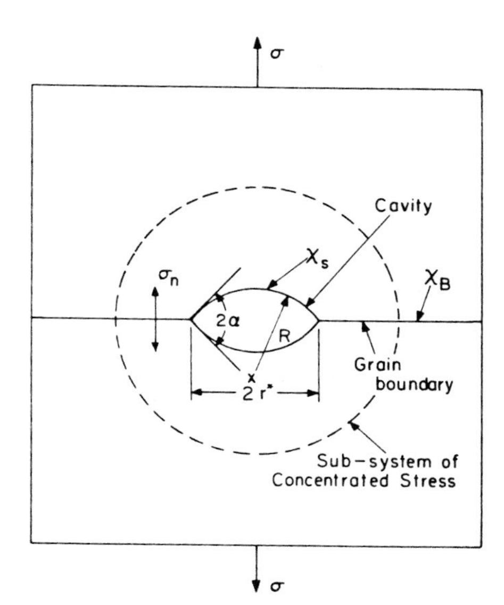

Fig. 3. Nucleation of a cavity on a stressed interface (from Argon, Chen, and Lau [19]; courtesy of AIME).

interface under the action of the normal stress σ_n raises the energy of the local sub-system. Three contributions to the free energy change must be considered: a) the production of free surface for the cavity and elimination of a corresponding portion of interface energy; b) elimination of elastic strain energy in the volume to be replaced with a cavity; and c) the work done by the applied normal stress acting across the boundaries of the local sub-system due to the formation of the cavity. It can be shown that even under the highest stresses, bordering on athermal cavitation, the contribution of the released elastic strain energy is of smaller order than the work done by the local concentrated stress across the boundary, i.e. (b) << (c). On the basis of this, determination of the critical activation configuration of a cavity is obtained in the usual way by maximizing the free energy, under constant conditions, to give a critical cavity radius r* in the plane of the boundary and a free energy ΔG* given as,

$$r^* = 2\alpha^2\chi_s/\sigma_n \tag{12}$$

$$\Delta G^* = \frac{4}{3}\frac{\alpha^2\chi_s^3}{\sigma_n^2}F_v(\alpha) \tag{13}$$

$$\text{where } \alpha = \cos^{-1}(\chi_B/2\chi_s) \tag{14}$$

is the half dihedral angle, χ_s the surface free energy, χ_B the boundary free energy, and $F_v(\alpha)$ [17, 18] is a numerical constant affected by the shape of the cavity depending upon whether it is along a flat boundary, curved boundary, or at a corner.

The reciprocal time for cavity nucleation, $\dot{\rho}$, is then given by the usual considerations of obtaining the flux of the constituent making up the new phase (vacancies into the cavity) across the interface of the critical configuration (the equatorial periphery of the cavity in the plane of the boundary having a height δ). This gives,

$$\dot{\rho} = \dot{\rho}_0\exp(-\Delta G^*/kT) \tag{15}$$

$$\dot{\rho}_0 = (2\pi r^* D_b\delta/\Omega^{4/3})\exp(\sigma_n\Omega/kT) , \tag{16}$$

where D_b is the diffusion constant along grain-boundaries (we assume that the diffusion constant along an incoherent phase boundary is not much different), δ the effective grain boundary thickness, and Ω the atomic volume. The numerical evaluation of the critical cavity size r*, the activation free energy ΔG*, and the reciprocal time $\dot{\rho}$, for cavity nucleation as a function of stress σ_n is instructive for a well defined case. All constants except $F_v(\alpha)$ are well established for a given metal. The shape constant $F_v(\alpha)$ is

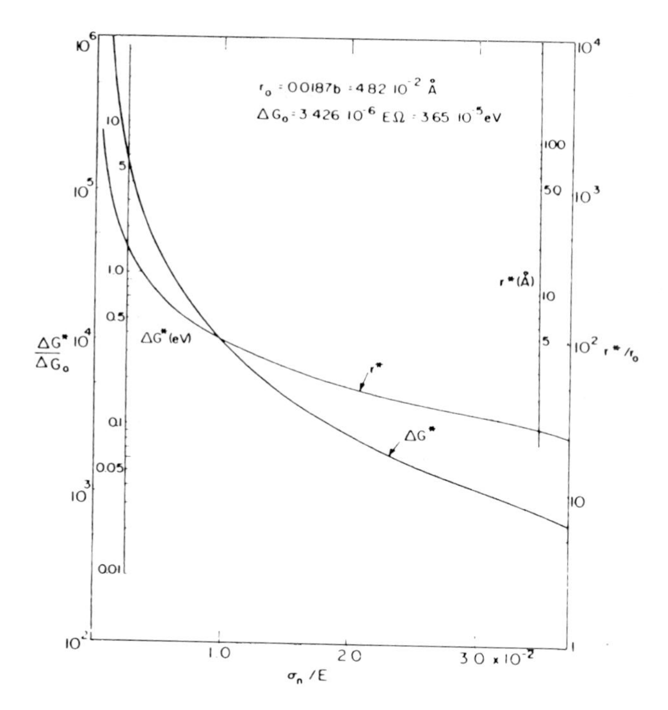

Fig. 4. Dependence of the free-energy for nucleation ΔG* and the critical cavity radius r* on the local normal stress σ_n for γ-iron (from Argon, Chen and Lau [19]; courtesy of AIME).

unity for a flat boundary but can be considerably smaller for re-entrant corners at cavity apexes or grain corners as Raj and Ashby [17] have shown. A reasonable range for this constant is probably between 0.1 and unity, with the low values being in re-entrant corners where the interface stress may not be very high or maybe even reversing in sign. Argon, Chen and Lau [19] have considered only the case for $F_v(\alpha) = 1$, and evaluated the expressions in Eqns (12), (13), and (16) for the following sets of physical constants typical for γ iron (or stainless steel), for which some properties, like the surface free energy χ_s, and the ideal cohesive strength σ_i, were determined from a Lennard-Jones potential fitted to the elastic properties of iron [19]: $\sigma_i = 0.0374E$; $\chi_s = 9.35 \cdot 10^{-3}Eb$; $E = 1.4 \times 10^2 GPa$; $\Omega = 1.21 \times 10^{-29} m^3$; $\delta D_{bo} = 7.5 \times 10^{-14} m^3/s$; $Q_b = 159$ kJ/mole; $b = 2.5 \times 10^{-10} m$; $\alpha = 1$ rad; and where $\delta D_b = \delta D_{bo} \exp(-Q_b/kT)$. The computed dependence of the critical cavity radius r* and the free energy ΔG* on the stress to modulus ratio σ_n/E is plotted in Fig. 4. In Fig. 5 the calculated dependence of the reciprocal nucleation time of a critical size cavity, on

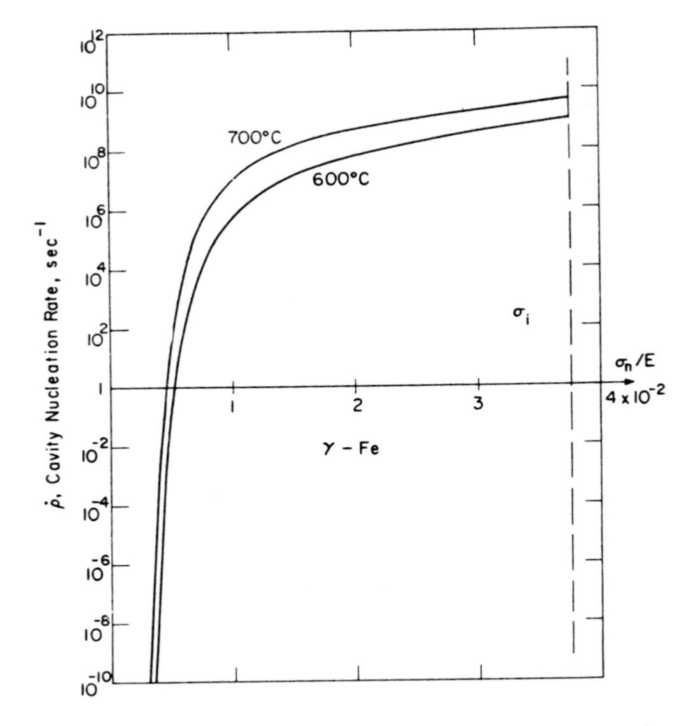

Fig. 5. Dependence of the cavity nucleation rate $\dot{\rho}$ on the local normal stress at 600 and 700°C in γ-iron (from Argon, Chen, and Lau [19]; courtesy of AIME).

the ratio σ_n/E, is shown, for two temperatures of 600 and 700 ^{o}C where stainless steel is used in high temperature service. It is clear from these figures that the activaion free energy and the critical cavity size rise to unattainably high levels at σ_n/E becoming less than 5 x 10^{-3}. As observed by Argon et al [19] this is only a factor of 2 or so below the level of the interface cavitation strengths measured in low temperature ductile fracture [7]. On the basis of this it must be concluded that cavitation at interfaces at elevated temperature or nucleation of cavities at stressed interfaces is only marginally easier than at low temperature -- a conclusion reached by McLean [20] much earlier on equivalent by more elementary analysis. When the geometrical shape factor $F_v(\alpha)$ is smaller than unity the apparent nucleation threshold shown in Fig. 5 drifts down somewhat lower on the σ_n/E axis. Since the normal levels of service stresses are a factor lf 10 to 30 below this level, it is necessary to conclude that nucleation of cavities during service can not be a result of the normal service stresses but that stress concentrations of the order of 10 to 30 are required.

4.2.2 Stress concentrations

That the above mentioned stress concentrations must be the result of incompatibilities produced by sliding boundaries has been suspected for a long time. Their treatment by early investigators [21, 22], however, had been based on linear elastic approximations utilizing dislocation pile-ups or shear cracks, which are clearly inapplicable once the matrix can undergo creep deformation.

To obtain better solutions for stress concentrations that may arise from grain-boundary sliding, Argon, Chen and Lau [19] have modeled in detail the two basic forms of grain-boundary decohesion reported by all creep investigators: (I) hard grain-boundary particles giving rise to round grain-boundary cavities; and (II) triple point stress concentrations giving rise to wedge cracks as shown in Fig. 6. In such a computation it is important to note that because of the different time dependent properties of grain-boundaries and grain matrix, important local stress re-distributions will occur, and that in addition short range stress concentrations can be wiped out very

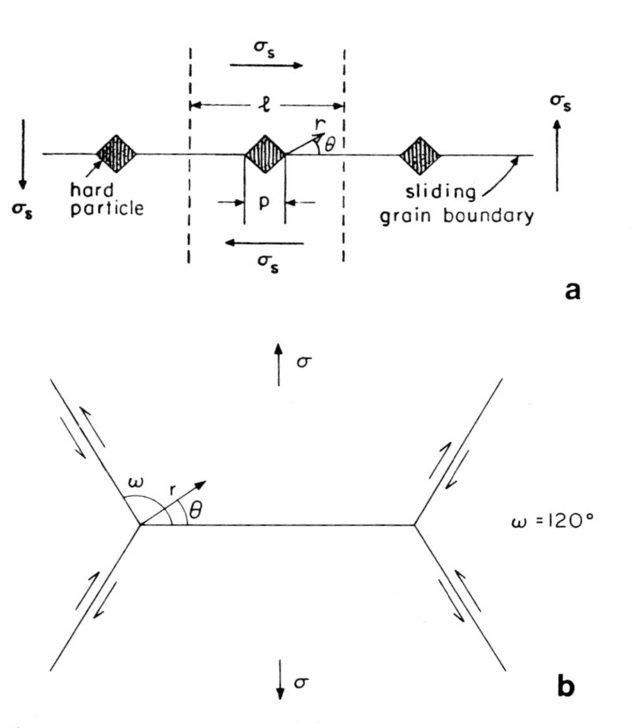

Fig. 6. (a) Idealized square particles on a sliding grain-boundary; and (b) triple grain junctions on sliding grain-boundaries.

effectively by local diffusional flow. It can be shown readily [19] that the large stress concentrations that arise around grain-boundary particles as they interrupt the free sliding of grains over each other, occur only during transients. These occur during start up or during sudden re-distributions of stress due to a number of stochastic processes, such as sudden migration of a boundary released from pinning points, separation or fracture of previously load carrying particles on neighboring boundaries, etc. On the other hand triple point stress concentrations occur only when all load support by grain-boundary particles is lost. Furthermore, it is necessary to note that the severity of the stress concentrations around particles is governed by the range of diffusional flow given by the critical diffusion length Λ defined as

$$\Lambda = \left(\frac{\sigma D_b \Omega \sigma}{kT\dot{\varepsilon}}\right)^{1/3} \tag{17}$$

Thus, if the size p of the grain-boundary particles are smaller than Λ, the stress concentrations that might arise on particle interfaces can be eliminated quite effectively by diffusional flow [19]. When $p \gg \Lambda$, then the incompatibilities pressed onto the particles by the sliding boundary can only be counteracted by creep flow around the particles. Since sizes, d of grains are very much larger than grain-boundary particles, $d \gg \Lambda$ is almost always true also, and diffusional flow is ineffective in counteracting the mounting incompatibilities arising from boundary sliding. By a number of specialized techniques involving the use of stress functions, and finite element modeling Lau and Argon [23], and Lau, Argon, and McClintock [24] have calculated the local singularities both around the apexes of grain-boundary particles and those at grain-boundary triple junctions, in material creeping according to a general power-law of the type given by Eqn (4). They found that when diffusional flow was not present, the stress and strain rate distributions obey a relationship of the form

$$\sigma_{ij} = (K/r^{\lambda})\hat{\sigma}_{ij}(\theta) \tag{18}$$

$$\dot{\varepsilon}_{ij} = (AK^{m}/r^{m\lambda})\hat{\varepsilon}_{ij}(\theta) \tag{19}$$

around the particle apex on the sliding boundary and around the triple point, for the coordinate systems given in Fig. 6. Here, K is a generalized stress intensity factor given below, λ a range exponent, that is a function of m and the dihedral angle of the particle, $\hat{\sigma}_{ij}(\theta)$ and $\hat{\varepsilon}_{ij}(\theta)$ are angle dependent functions without a singularity, and A is a temperature dependent constant, related to the specific creep law. The dependence of λ on m for two prominent exponents of $m = 3$ and $m = 5$, appropriate for certain solid solution alloys and sub-grain forming metals and alloys are given in Table I for the two problems being discussed.

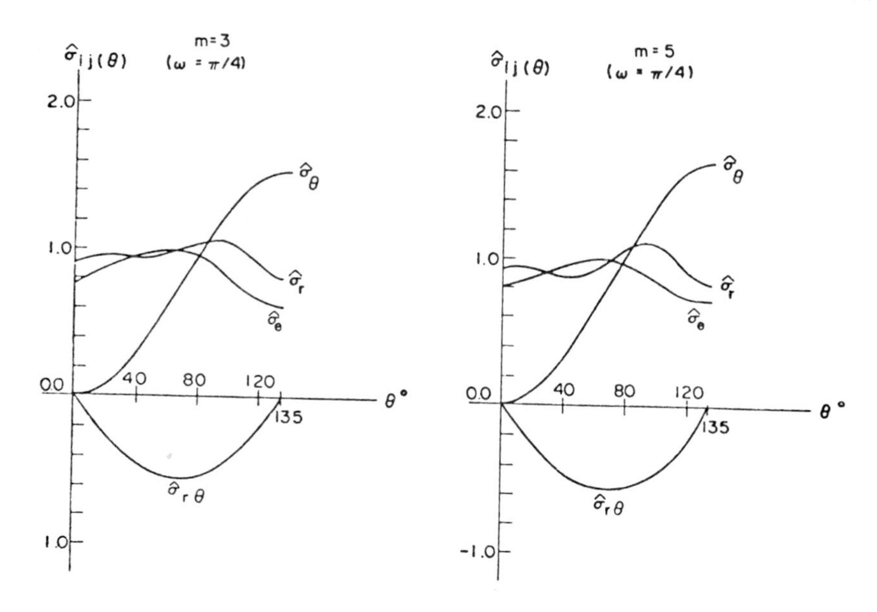

Fig. 7a. Angular distribution of stresses at the apex of a hard, square particle on a sliding grain-boundary, for m = 3 and m = 5 (from Lau, Argon, and McClintock [24]; courtesy of ASTM).

Table I. Values of $\lambda(m)$ for Singularities at the Apexes of Square Particles and at Triple Grain Junctions

	Square Particle		Triple Junction	
m	$\lambda(m)$	$C(m)$	$\lambda(m)$	$C(m)$
3	0.225	0.689	0.231	0.613
5	0.152	0.767	0.150	0.633

Values of λ for many other integer values of m have been calculated by Lau [25]. The dependence of $\hat{\sigma}_{ij}(\theta)$ on θ for the two problems for the same two exponents given above are given in Figs. 7a and 7b. The computed generalized stress intensity factors K for the two problems are:

$$K = C(m)\sigma_s\left(\frac{\ell}{p}\right)\frac{(1-\lambda)(p/\sqrt{2})^{\lambda}}{\hat{\sigma}_{\theta\theta}(3\pi/4)} \tag{20}$$

for a square grain-boundary particle, and

$$K = \frac{3}{2}C(m)\sigma_n\frac{(1-\lambda)(d/2\sqrt{3})^{\lambda}}{\hat{\sigma}_{\theta\theta}(0)} \tag{21}$$

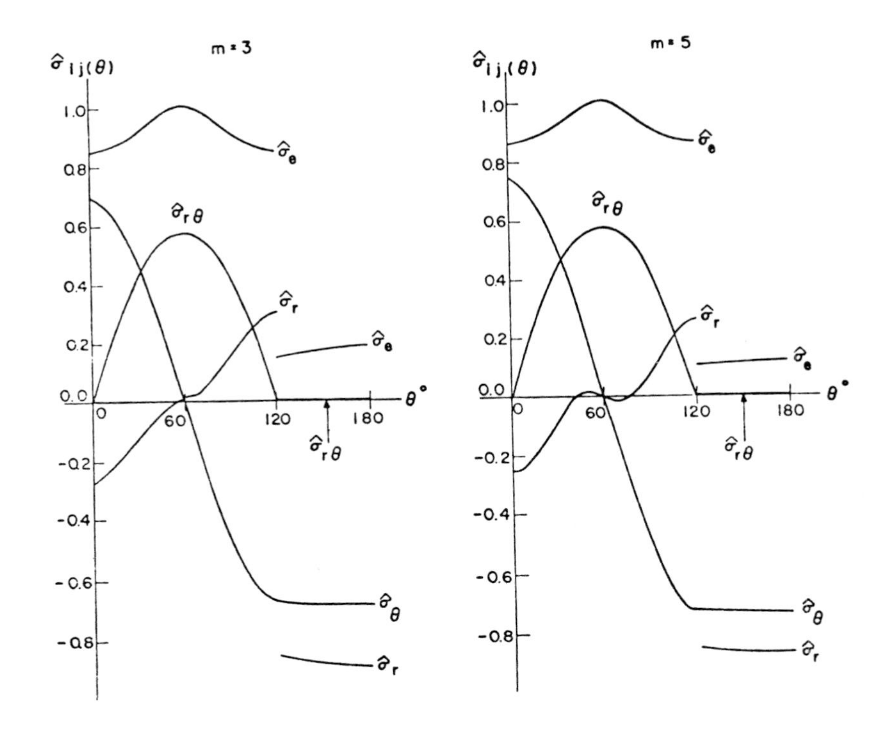

Fig. 7b. Angular distribution of stresses at a triple junction of sliding grain-boundaries, for m = 3 and m = 5 (from Lau, Argon, and McClintock [24]; courtesy of ASTM).

for the triple grain junction, where (p/ℓ) is the ratio of particle size to particle spacing (to be replaced with the area fraction of boundary covered with particles in a three dimensional problem), d the grain size, σ_s and σ_n the distant shear stress and normal stress in the two problems, $\hat{\sigma}_{\theta\theta}(3\pi/4)$ and $\hat{\sigma}(0)$ are the specific values of these functions on the particle interface and grain-boundary surface (given in Figs. 7a and 7b), and C(m) are specific numerical constants listed in Table I).

If a real material diffusional flow modifies these stress concentrations by eliminating the singularity at the origin by relaxing the stresses there within $r = \Lambda$ to very low values (assumed to be zero here), and by jacking up the stresses outside this range slightly to preserve overall equilibrium. This leaves the maximum stress at $r = \Lambda$, and permits computation of specific stress concentration factors k_p and k_{tp} for the particle and triple point problems respectively [19, 24] as follows,

$$k_p = C(m)\left(\frac{\ell}{p}\right)\frac{(1-\lambda)}{(1-(\sqrt{2}\Lambda/p)^{1-\lambda})}\left(\frac{p}{\sqrt{2}\Lambda}\right)^{\lambda} \tag{22}$$

$$k_{tp} = \frac{3}{2}C(m)\frac{(1-\lambda)}{(1-(2\sqrt{3}\Lambda/d)^{1-\lambda})}\left(\frac{d}{2\sqrt{3}\Lambda}\right)^{\lambda} . \tag{23}$$

It is instructive to calculate actual stress concentrations for a specific case such as e.g. gamma iron at some representative stress levels and temperatures. These calculated values are given in Tables II-a and II-b respectively for two problems.

Table II-a. Stress Concentration Factors k_p at Square Grain Boundary particles, and Threshold Stresses for No-Cavitation along Boundaries

p(m)	L(m)	Λ(m)	k_p	σ_{TH}/G
	$T = 600^oC$, $\sigma_\infty/G = 3 \times 10^{-3}$, $m = 3$			
4×10^{-7}	8×10^{-6}	3.24×10^{-9}	30.14	8.77×10^{-6}
4×10^{-7}	4×10^{-6}	5.14×10^{-9}	13.77	1.75×10^{-5}
	$T = 600^oC$, $\sigma_\infty/G = 3 \times 10^{-3}$, $m = 5$			
4×10^{-7}	8×10^{-6}	2.10×10^{-8}	21.71	6.62×10^{-4}
4×10^{-7}	4×10^{-6}	5.32×10^{-8}	11.07	1.32×10^{-3}
	$T = 700^oC$, $\sigma_\infty/G = 1.5 \times 10^{-3}$, $m = 3$			
4×10^{-7}	8×10^{-6}	3.04×10^{-9}	30.52	4.00×10^{-6}
4×10^{-7}	4×10^{-6}	4.83×10^{-9}	13.94	8.00×10^{-6}
	$T = 700^oC$, $\sigma_\infty/G = 1.5 \times 10^{-3}$, $m = 5$			
4×10^{-7}	8×10^{-6}	3.15×10^{-8}	21.50	4.47×10^{-4}
4×10^{-7}	6×10^{-6}	7.94×10^{-8}	11.96	8.94×10^{-4}

Table II-b. Stress Concentration Factors k_{tp}, at Triple Grain Junctions of Hexagonal Grains

m	Λ(m)	k_{tp}
	$T = 600^oC$, $\sigma_\infty/G = 3 \times 10^{-3}$, $d = 5 \times 10^{-5}m$	
3	1.99×10^{-8}	3.66
5	1.29×10^{-6}	1.61
	$T = 700^oC$, $\sigma_\infty/G = 1.5 \times 10^{-3}$, $d = 5 \times 10^{-5}m$	
3	1.85×10^{-8}	3.73
5	2.14×10^{-6}	1.63

Inspection of the stress concentration factors in Tables II-a and II-b show that for particles they are of the right magnitude to bridge the gap between the applied stress and the required high stresses for cavitation of the interfaces. It is useful to note that the main contribution to the high stress concentration at particles comes from the term (ℓ/p), or the reciprocal of the area fraction of the boundary covered with particles, and indicates further that in non-regular coverage, portions of the boundary where (ℓ/p) is large will result in large stress concentrations at the bordering particles. We note in comparison that the stress concentrations for triple junctions are far too low to bridge the gap, and should present no problem if particles were absent on the boundaries. That this must be so was established by Grant and co-workers [14, 15] in tests on super pure aluminum which always failed by rupture by thinning down to a pin-point without undergoing intergranular fracture. We note further that the stress concentrations are higher for small values of m and lower for larger values of m -- well in keeping with the near disappearance of stress concentrations for ideally plastic behavior at $m \to \infty$.

4.2.3 Formation of stable cavities

Such models for development of high stress concentrations at grain-boundaries indicate that cavitation could occur at particles during transient periods of rapid sliding when the local stresses are high. But, this is only a necessary condition. The actual cavity formation must be a very complex process. Since the locally concentrated stresses are of relatively short range, reversing sign from one apex of the particle to the other, the opening up of the cavity will largely relieve the concentrated local stress and subject the cavity to a stress that is somewhere in between the initially high value and the relatively low background level. This will make many cavities sub-critical in size and allow them to sinter shut. Two processes counteract shrinkage. First, the sliding action of the adjoining grain over the particle interface will tend to spread the cavity out and enlarge it. Second, any impurity near the interface will segregate rapidly to the free surfaces of the cavity and thereby significantly lower the surface enregy and tend to stabilize an otherwise sub-critical cavity. Seah [26], and Hondros and Seah [27] have evaluated this effect and find the segregation coefficient Γ_s to the cavity surface and its effect on the surface free energy χ_s of the cavity to be given by

$$\Gamma_s = C[(4D_i t/\pi)^{1/2} + (D_i t/r)] \tag{24}$$

$$\chi_s = \chi_{so} + \Gamma_{so} RT\ell n(1 - \Gamma_s/\Gamma_{so}) \tag{25}$$

where Γ_s is the segregation level in moles per square meter on the cavity surface, C the background impurity concentration in moles per cubic meter, D_i the bulk diffusivity of the impurity in the solvent matrix, Γ_{so} the saturation coverage of the impurity on the surface in moles per square meter, χ_{so} the surface free energy of the clean surface, χ_s the surface energy of the partially contaminated surface, and t the time in seconds. Seah and Hondros find that although most substitutional impurities such as P, Sn, Sb, As and Cu produce only a relatively weak effect in iron, interstitial impurities such as N and B can potentially reduce the stable cavity size in a stress field by about a factor of 3. In any event we expect that cavity nucleation at particles is not automatically assured with the presence of an initial high stress concentration of short range. Many cavities must form and collapse before some stabilized cavities can grow to maturity. This, plus the recurring nature of grain-boundary sliding transients, observed directly by Chang and Grant [28] in aluminum and indirectly by Chen and Argon [29] in 304 stainless steel, explains why nucleation of cavities continues throughout the creeping process, almost in parallel with the creep strain.

It has been stated that cavities can also be formed by impingement of dislocation pile-ups on grain-boundaries, grain-boundary ledges acting as stress concentration sites, and cooperative action of sliding grain-boundaries and slip bands cutting through them. While the dislocation pile-ups can in principle produce the needed stress concentrations they still require grain-boundary particles as hard obstacles. This appears to be the case in some super alloys, such as Nimonic 80A investigated by Dyson and coworkers [31]. On the other hand grain-boundary ledges are not good candidates because they are readily straightened out under the local stress concentrations, while the cooperative action of sliding of a boundary and a slip band cutting through it should, if anything, produce material interference instead of cavitation [32]. This leaves grain-boundary particles as the only serious possibility of general applicability.

Since cavity nucleation is a key step in creep fracture, the suppression of cavitation or its postponement should be an attractive goal. Based on the understanding developed here, several possibilities should be noted. Elimination of all grain-boundary particles should eliminate the problem, but in relatively non-linearly creeping alloys with large creep exponents, this would increase the creep rate significantly by the stress redistribution produced by the freely sliding boundaries. According to Chen and Argon's [5] model for boundary sliding, discussed in Section 2, a creep rate increase by $(1.4)^m$ should be expected, which for $m \simeq 8$ would be nearly 15 fold, and very undesirable. A more promising possibility is to produce a uniform and high coverage (20 - 30%) of grain-boundaries with relatively coarse (0.5 - 1.0 mm) grain-boundary particles. This will reduce the stress concentrations by a factor of roughly 3 - 6 from those given in Table II-a, and make them too small for cavitation, but still effectively inhibit grain-boundary sliding. A final alternative for any particle size is to reduce the service stress to such a level where $\Lambda > p$, where the diffusional smoothing should be far reaching, over particle dimensions. Then, as discussed by Koeller and Raj [33] little stress concentration should remain. Such theshold stresses were calculated for the specific cases considered in connection with Table II-a and are listed in the last column of that table. They are quite low for small creep exponents but are meaningfully large in the more interesting cases of large creep exponents. Thus, with careful control of grain-boundary particle uniformity, and monitoring the applied stress levels to maintain $\Lambda \gtrsim p$, it should become possible to delay or even suppress cavitation.

4.2.4 Experimental measurements of cavity nucleation

The experimental verification of the rate of nucleation as given by Eqn (15) is an exceedingly difficult task. As indicated in this equation and shown in Fig. 5, the nucleation rate is a very strong function of stress. Because of the relatively wide distribution of particle sizes and spacings along boundaries, and the local variation of grain shapes, a wide distribution of stress concentrations is expected in a given alloy. Hence, unless special measures are taken to control these variables, it is nearly impossible to measure a rate that can be related to the predictions of the theory. An equally difficult task is the observation of the cavities near their critical activatiion configurations of only 10 Å radius. As discussed by Argon et al [19] the usual techniques of direct observation utilizing conventional electron microscopy and measurements of density change are incapable of detecting such small cavities. Although special techniques of high resolution electron microscopy [30] can in principle resolve cavities of this size, application of them should prove to be very difficult because of the rarity of the event. The most promising current techniques suitable for the task are small angle X-ray diffraction and small angle neutron scattering (SANS). The latter has recently been applied on both crept and cyclically deformed samples and looks promising [34,35].

Most investigators who have measured the evolution of cavities in creep experiments have detected relatively large cavities of a diameter of 0.2 μm, or larger, and have reported the increase in the numbers as a function of strain. Cane and Greenwood [36] and Cane [37] have studied in considerable detail nucleation of grain-boundary cavities

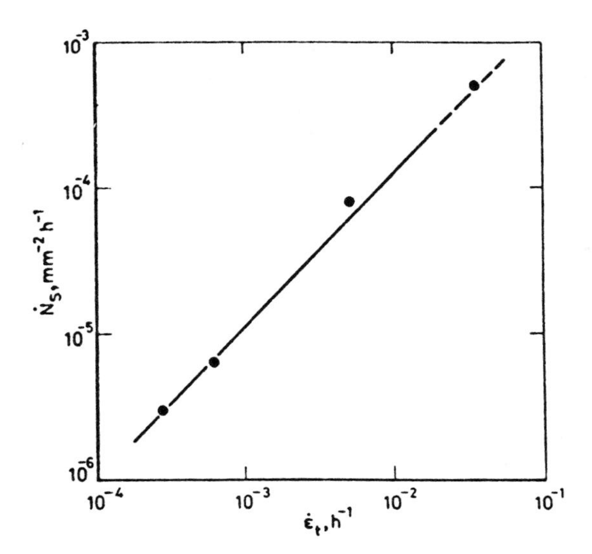

Fig. 8. Measured dependence of cavity nucleation rate on steady state strain rate in iron (from Cane [37]; courtesy of the Metals Society).

in iron with a carbon content of 60 ppm. By means of a very elegant technique of intergranularly separating at low temperature (*) the crept specimens, they were able, not

---

(*) This separation technique, also used by others, e.g., Chen and Argon [29], is a result of boundary embrittlement due to segregation of trace elements, to the boundaries during creep, particularly S and P [38].

---

only to obtain number counts of cavities, but also observe the evolution of the cavity shapes as a function of time. As Fig. 8 shows, Cane [37] has observed, for small strains, a linear relationship between the nucleation rate and the strain rate, that is expected from the stochastic sliding model of grain-boundaries discussed in Section 4.2.3 above, where repeated trials are necessary to bring a cavity through the sub-critical range into a super critical size. Since grain-boundary sliding must be synchronized with matrix creep for reasons of compatibility, a proportionality between $\dot{N}$ and $\dot{\varepsilon}$ and between N and ε must follow. Cane and Greenwood also studied the distribution of cavity size and found this distribution to gradually widen because of the continued nucleation of cavities.

Dyson and McLean [39] measured metallographically the evolving cavity density as a function of creep strain in tension and torsion in Nimonic 80A. Their results obtained

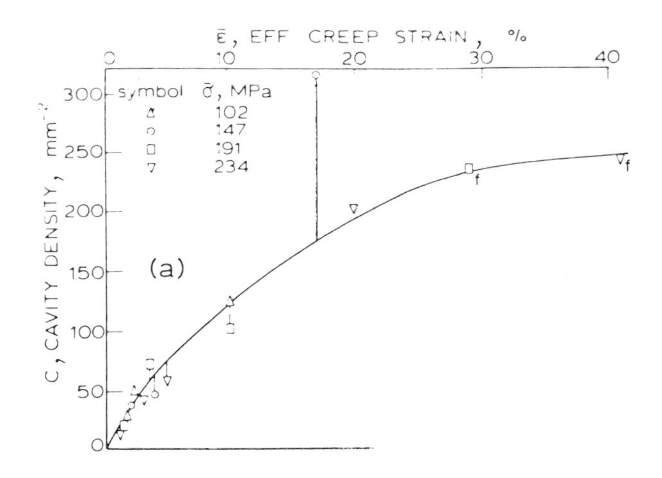

Fig. 9. Measured dependence of cavity density on effective creep strain in Nimonic 80A (from Dyson and McLean [39]; courtesy of the Metals Society).

in tension experiments is shown in Fig. 9 as a function of strain. As in the experiments of Cane and Greenwood [36] a nearly linear relationship between cavity density and strain was found for small strains, while for large strains a distinct saturation effect is discernable. Furthermore, all measurements made at different stress levels are correlated by the unique curve indicating that the total cavity density correlates with the total amount of boundary sliding --

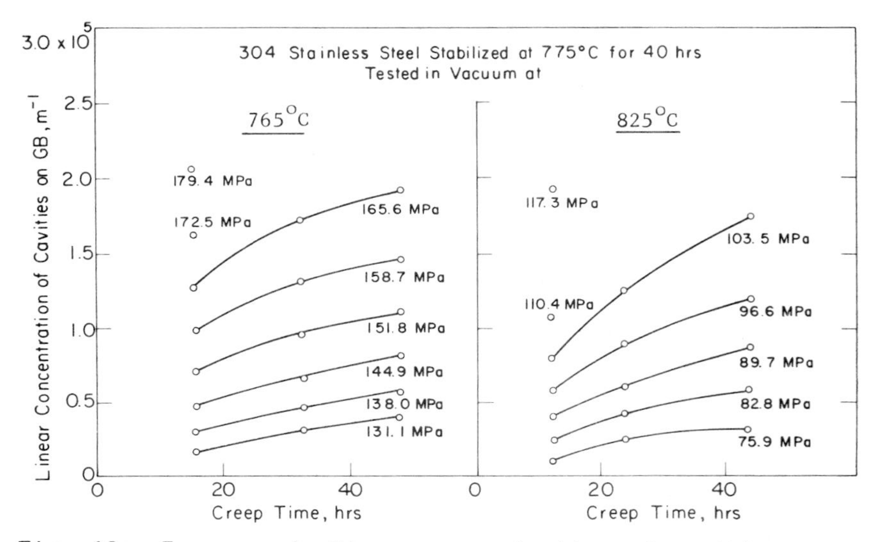

Fig. 10a. Increase in linear concentration of cavities on grain-boundaries with creep time in 304 stainless steel (from Chen and Argon [29]; courtesy of Pergamon Press).

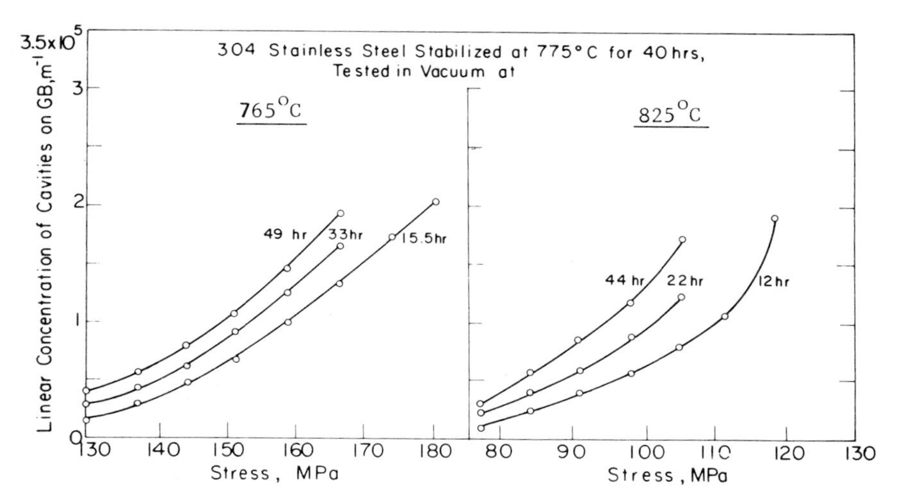

Fig. 10b. Increase in linear concentration of cavities on grain boundaries with tensile stress in 304 stainless steel (from Chen and Argon [29]; courtesy of Pergamon Press).

higher stresses accomplishing the end result faster.

Chen and Argon [29] have made cavity nucleation measurements on type 304 stainless steel specimens of contoured shape permitting the determination of stress dependence of the phenomenon in an efficient way. By a combination of surface removal and slight re-straining followed by scanning electron microscopy the results shown in Figs. 10a and 10b were obtained (*).

(*) In Figs 10a and 10b the temperatures have been corrected from previously published values by Chen and Argon, when it was discovered later that a faulty temperature measurement had been made.

Those figures show that cavity nucleation continues as a function of time (and roughly parallel to strain) at all stress levels, but appears to show a saturation effect as that observed by Dyson and McLean [39], discussed above. Figure 10b shows, furthermore, an apparent stress threshold for cavitation which is roughly in the range of the prediction given in Table II-a. A more revealing measurement of Chen and Argon is given in Fig. 11 in which the relative nucleation rates are plotted as a function of inclination of the cavitated grain-boundaries. The observed curve fits well a $\cos^2\theta$ dependence of the boundary inclination with the tensile direction, indicating that the stress dependence of the observed cavitation is linear and not at all the strong dependence expected from the

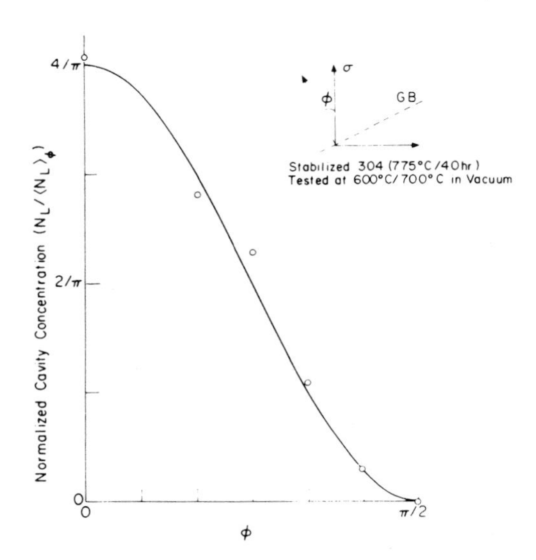

Fig. 11. Dependence of normalized cavity density along boundaries on the inclination ϕ of the boundary with the tensile axis (from Chen and Argon [29]; courtesy of Pergamon Press).

theoretical expression given in Eqn. (15). This perhaps, more than anything else, indicates that the cavitation phenomenon on this relatively coarse scale, differing from that of the activation configuration by at least two orders of magnitude, does not reflect the kinetics of cavity formation but rather cavity growth by diffusional flow as will be discussed in Section 4.3 below. Using the same technique of intergranular brittle separation at low temperature introduced by Cane and Greenwood, Chen and Argon [29] were able to find a very close correlation of cavitation with grain-boundary carbides, and practically no cavitation on grain-boundaries containing no carbides.

Less complete cavitation measurements, or observations, have been made by numerous other workers. Of these the following are noteworthy: Fleck, Taplin and Beevers [40], have used high voltage electron microscopy to observe cavities in copper; and Raj [41] has studied cavitation in copper bi-crystals, from copper oxide particles.

The discussion of cavitation along grain-boundaries from second phase particles given above pre-supposes that no other sources but vacancies initiate the cavity formation under stress. This should apply to situation in vacuum or in neutral environments. There are, showever, many other cases where supersaturation of gases are present in the metal producing cavitation by condensation of such gases along grain-boundaries, even when no stresses are applied. Such cavitation requires no second phase and is important both industrially and often very useful as a powerful tool for fundamental experiments. The penetration of oxygen along grain-boundaries in nickel upon exposure to air at high temperatures in both unstrained and creeping samples has been studied by Bricknell and Woodford [42, 43]. They have shown extensive "bubbling" on boundaries of near-surface grains, often penetrating to depths of several grain diameters. Goods and Nix [44,45] have intergranularly pre-cavitated silver polycrystals with water vapor bubbles by a combination of internal oxidation followed by annealing in hydrogen, to study cavity growth, that will be discussed in Section 4.2 below. Similar procedures were applied by Nieh and Nix [46,47], Hanna and Greenwood [48], and by Stanzl, Argon, and Tschegg [49] on polycrystalline copper all for the purpose of making precision cavity growth measurements, unhampered by much additional cavity nucleation from second phase particles.

4.3 The Growth of Grain Boundary Cavities

4.3.1 Theoretical models

Balluffi and Seigle [50] appear to have been first in pointing out the possibility of rapid growth under stress of creep cavities by a drainage of vacancies into them along the surrounding portion of the grain-boundary. A theoretical model of the process has been advanced first by Hull and Rimmer [51] for the growth of a cavity in an equilibrium, spherical-caps, shape. This model has been improved by a number of investigators, and finally by Speight and Beere [52] who have added a number of necessary kinematical refinements. Further developments have come from Chuang and Rice [53], and Chuang, Kagawa, Rice, and Sills [54] who have investigated the entire range of cavity growth by diffusional flow ranging from the quasi-equilibrium spherical-caps shape considered by Hull and Rimmer to non-equilibrium shapes resembling crack like cavities. Additional important developments were made in recognizing that when stresses become large or cavities grow to large sizes, their growth must be influenced by the creep deformation of the surrounding matrix to finally link to the creep expansion developments discussed above in Section 3 under transgranular creep fracture. This additional problem was discussed by Beere and Speight [55] and more thoroughly by Needleman and Rice [56]. The latter have pointed out that the expansion of the cavity by diffusional flow must not be viewed to occur in a "rigid" or purely elastic background, but in a background of a creeping solid which removes the need of growing neighboring cavities from diffusionally communicating with each other. The full mathematical details of the diffusional growth of grain-boundary cavities in a creeping background, leading to various asymptotic shapes are too complex to give here in

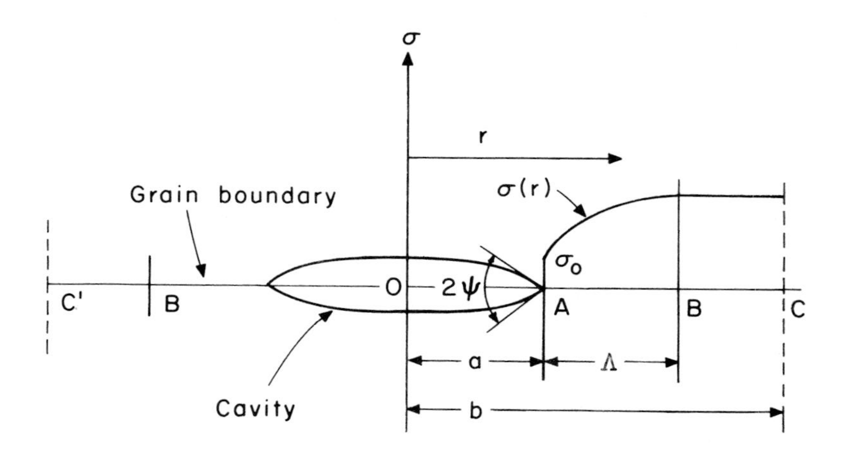

Fig. 12. The distribution of stress on a grain-boundary outside the cavity.

any detail. We will instead merely outline the various requirements for the solution of the problem and discuss the final results on the basis of a very useful approximate treatment suggested by Chen and Argon [57]. The reader interested in the considerable mathematical refinements of diffusional cavity growth in a rigid or in a creeping background is referred to Chuang et al [54] and to Needleman and Rice [56] for a remarkably thorough discussion of this detail.

Consider as in Fig. 12 a principal cross section of a cylindrically symmetric cavity of radius a, in a region of the grain-boundary plane of radius b, corresponding to the half distance between neigboring cavities in the plane. In relation to the matrix creep rate an effective range Λ of the diffusional transport from the cavity tip is defined, given by Eqn. (17) above. Within a region $a \leq r \leq \Lambda$ diffusional "thickening" of the grain-boundary by the plating out of atoms draining into the grain-boundary from the cavity dominates the rate of separation of any pair of symmetrically placed material points on either side of the boundary. In the region $\Lambda \leq r \leq b$ creep flow is the dominant mechanism. The effective diffusional transport range expands with reduced stress as given by Eqn (17) and the power-law creep expression of Eqn (4). In the usual approximations, even in most of the very detailed developments, elastic strains are neglected everywhere and the material is considered as a rigid, creeping solid accompanied by boundary diffusional flow.

Inside the cavity the distribution of the chemical potential is governed by the steady state curvature, and the gradient of this chemical potential drives a surface flux J_s of atoms foward the tip of the cavity. The condition of conservation of matter and flux continuity at every point of the cavity surface governs the shape of the cavity for a given level of flux, J_{s-tip}, at the cavity tip. In the region of the boundary $a \leq r \leq \Lambda$, a uniform thickening of the boundary (required by the assumption of a rigid surrounding matrix) prescribes a definite variation of grain boundary flux J_b from the cavity tip to the periphery of $r = \Lambda$ where the gradient of the chemical potential, and the flux are taken to be zero. At the cavity tip a certain conservation relation exists between surface flux J_{s-tip}, on the cavity side, J_{b-tip} on the boundary side, and a kinematically prescribed virtual "siphoning flux" due to the jacking up of the cavity by the matter plating rate along the boundary between $r = a$ and $r = \Lambda$ [57]. The required variation of the boundary flux J_b between $r = a$ and $r = \Lambda$ is compatible with a definite gradient of chemical potential. This gradient, in turn, evokes a distribution of normal stress acting across the boundary, starting from a level σ_0 to equilibrate the surface curvature forces on the cavity side of the tip. The consistent solution of these conditions and equilibration of the stress distribution, rising from σ_0 at $r = a$ to a constant level at $r = \Lambda$, and remaining constant over the region $\Lambda \leq r \leq b$, relate finally the ligament stress distribution to the applied uniform stress σ_∞ at a distance [54, 57]. When surface diffusivity D_s is very high in relation to the boundary diffusivity D_b the cavity grows in a fully equilibrated spherical-caps shape as originally assumed by Hull and Rimmer. When the surface diffusivity is restricted, however, or when boundary diffusivity is enhanced by various causes, then the cavity can not equilibrate its shape as rapidly as the tip moves away and the resulting cavity shape is crack-like as analyzed in great detail by Chuang [58]. In addition to these quasi-steady state growth shapes, several important transient processes of growth have been considered by numerical or asymptotic analysis techniques. These will be discussed later below.

As mentioned above, the theoretical models for growth of cavities in a creeping background for the quasi-equilibrium spherical-caps shape has been synthesized in a simple, approximate form by Chen and Argon [57].

The volumetric growth rate, dv/dt, of the cavity is governed by the net outflow of matter across the cavity tip giving

$$\frac{dv}{dt} = 4\pi a \Omega J_{s-tip} \,. \qquad (26)$$

According to the theoretical model of cavity growth of Chuang et al [54] described above, the tip flux is given as a function of the rate of advance da/dt of the tip for the two limiting models of quasi-equilibrium (spherical-caps) and the non-equilibrium (crack-like) growth modes are:

$$J_{s\text{-}tip} = \frac{h(\psi)}{\Omega}\, a\, \frac{da}{dt} \qquad \text{(quasi-equilibrium)} \tag{27}$$

$$J_{s\text{-}tip} = 2\sin(\psi/2)\left(\frac{D_s\delta_s\chi_s}{kT}\right) x \left(\frac{kT(da/dt)}{D_s\delta_s\chi_s\Omega}\right) \qquad \text{(crack-like)} \tag{28}$$

where

$$h(\psi) = [1/(1 + \cos\psi) - (\cos\psi)/2]/\sin\psi \tag{29}$$

is a geometrical shape constant involving the dihedral half angle ψ of the cavity tip.

Inspecting the developments of Beere and Speight [52] and that of Needleman and Rice [56], Chen and Argon [29] have observed that for the case where $(a + \Lambda) < b$ diffusional transport need to be only to $r = (a + \Lambda)$ while the region outside, $\Lambda \leq r \leq b$, can deform compatibly by power-law creep deformation and needs no special consideration. In that case it should be possible to consider the cavity growth according to the Hull and Rimmer model as modified by Speight and Beere and treat the dimension $a + \Lambda$ as if it were the half cavity separation in the model of Speight and Beere. This leads to the volumetric draining rate dv/dt through the cavity tip [29] of,

$$\frac{dV}{dt} = (a^3\dot{\varepsilon}_\infty)2\pi(\Lambda/a)^3\left[\ln\left(\frac{a + \Lambda}{a}\right) + \left(\frac{a}{a + \Lambda}\right)^2\left(1 - \frac{1}{4}\left(\frac{a}{a + \Lambda}\right)^2\right) - \frac{3}{4}\right]^{-1} . \tag{30}$$

Substitution of this relation into Eqn (26) and the use of the specific flux expressions at the cavity tip for the quasi-equilibrium shaped cavity and the crack-like cavity, given by Eqns (27) and (28) results in cavity tip advance rates da/dt for the two limiting modes of growth, for $\Lambda < b$, which are:

$$\left(\frac{da}{dt}\right)_{\text{quasi-eq.}} = \frac{(a\dot{\varepsilon}_\infty)}{2h(\psi)}\left(\frac{\Lambda}{a}\right)^3\left[\ln\left(\frac{a + \Lambda}{a}\right) + \left(\frac{a}{a + \Lambda}\right)^2\left(1 - \frac{1}{4}\left(\frac{a}{a + \Lambda}\right)^2\right) - \frac{3}{4}\right]^{-1} , \tag{31}$$

$$\left(\frac{da}{dt}\right)_{\text{crack-like}} = \frac{1}{(4\sin\psi/2)^{3/2}}\left(\frac{D_b\delta_b}{D_s\delta_s}\right)^{1/2}\left(\frac{D_b\delta_b\Omega\chi_s}{a^3kT}\right)\left(\frac{\sigma_\infty}{\chi_s/a}\right)^{3/2} x \left[\qquad\right]^{-3/2} \tag{32}$$

where the last term in brackets in Eqn (32) is the same as that in Eqn (30). When $\Lambda > b$, and the cavities diffusionally communicate $(a + \Lambda)$ is to be replaced with b in Eqn (32). For the crack-like growth case the asymptotic solution for the cavity tip velocity as worked out by Chuang

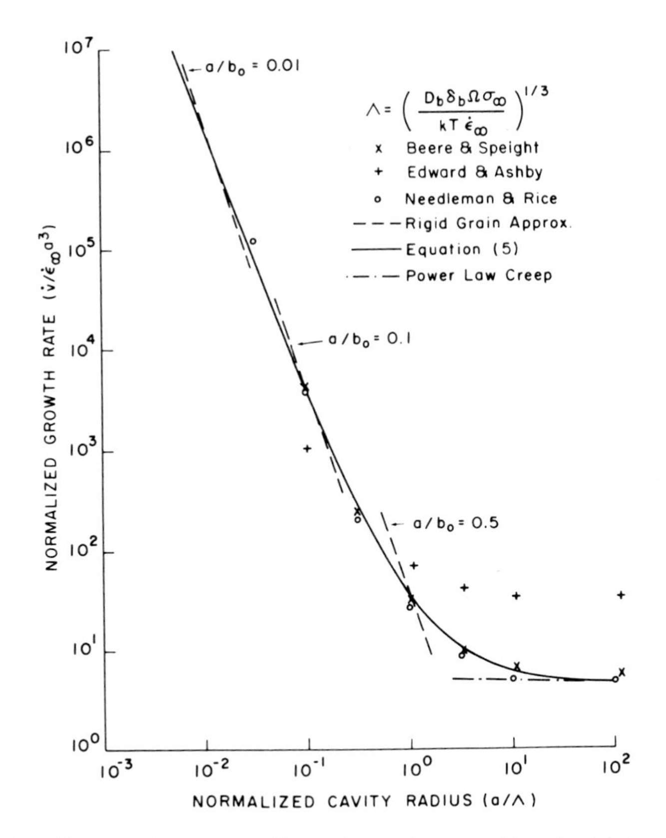

Fig. 13. Normalized growth rate of a cavity in the quasi-equilibrium mode (from Chen and Argon [57]; courtesy of Pergamon Press).

et al [54] is

$$\left(\frac{da}{dt}\right)_{cracklike} = \frac{3}{(2\sin\psi/2)^3}\left(\frac{D_s\delta_s\Omega\chi_s}{b^3kT}\right)\left(\frac{\sigma_\infty}{\chi_s/b}\right)^3\frac{1}{(1-(a/b)^2)^3} \ . \quad (33)$$

In Fig. 13 the simplified cavity growth expression is compared with the exact development of Needleman and Rice [56] and the model of Beere and Speight [55]. The agreement of the simplified model with the more exact one is remarkably good. The dotted lines in the figure give the growth relation of Hull and Rimmer which considers the background as rigid. Thus, at small cavity sizes a rigid grain approximation is quite satisfactory. Inspection of the crack-like growth expression for diffusionally communicating cavities indicates that the growth is governed by surface diffusion in the cavity. This results in a third power stress dependence.

Pharr and Nix [59] and Martinez and Nix [60] have analyzed by a numerical finite difference method the evolution of initially lenticular cavities into crack-like cavities for a case where the surface diffusivity is rate controlling as given in Eqn. (33). Their computations have verified that the asymptotic behavior when the cavity develops crack-like "shoots" is that given by Chuang et al [54} in Eqn. (33), but that a third power stress dependence requires the boundary diffusivity to be 10^4 times that of the surface diffusivity. There is no evidence that such a large disparity is possible between these diffusivities.

4.3.2. Experimental measurements

Cane and Greenwood [36] furnished the first reliable measurements of cavity growth rates in a creeping alloy by measuring the principal axis dimensions in the plane of the boundary of small cavities revealed by intergranularly fracturing, at low temperatures, previously crept iron samples. Fig. 14 shows the measurements made at 700 °C of the cavity width as a function of stress and time, exhibiting a very good linear relationship between the cavity width and the square root of the product of the third power of stress with time. This indicates that the cavity velocity at any given time is a 3/2 power of the applied stress but that the cavity velocity is decreasing with increasing time. Since the observed cavities have been far apart and most likely did not diffusionally communicate (i.e. $\Lambda < b$), the observed stress exponent suggests a crack-like growth model in a creeping background as given by Eqn. (32). The decelerating nature of the cavity growth,

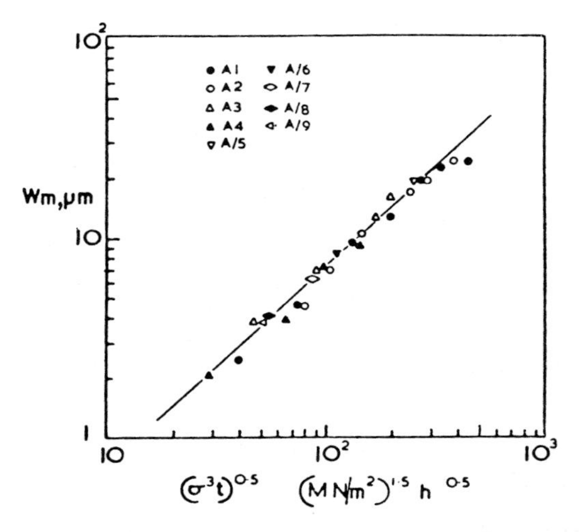

Fig. 14. Linear relationship between the cavity width and $(\sigma^3 t)^{1/2}$ observed in iron (from Cane and Greenwood [36]; courtesy of the Metals Society).

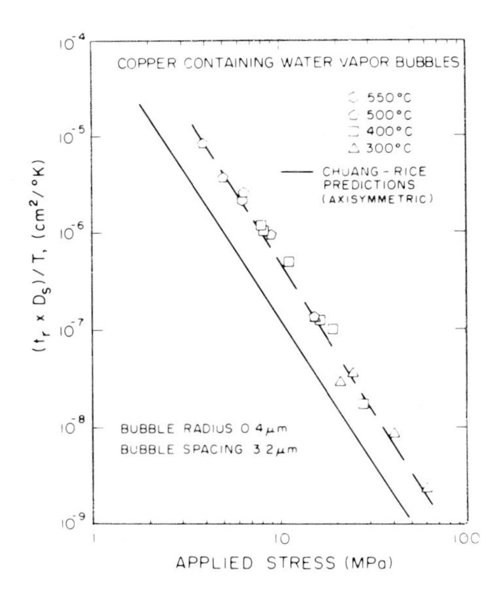

Fig. 15. Dependence of the creep fracture time in pre-cavitated copper, on the applied stress (from Nieh and Nix [47]; courtesy of Pergamon Press).

however, is puzzling and indicates either a transient process of the type modeled by Pharr and Nix, or more likely a tranport limitation to the cavity tip of a species initially segregated to the embryonic cavity or the surface of the particle from which the cavity was nucleated.

The cavity growth models have been tested in a series of elegant experiments by Nix and co-workers [44-47] who have initially intergranularly pre-cavitated both silver and copper by first internally oxidizing these metals and subsequently annealing them in hydrogen under no stress to precipitate a uniform collection of water vapor bubbles on all grain-boundaries that could then be used as initial cavities in creep fracture experiments. The results on copper which are typical for the two metals are shown in Fig. 15 [45]. The creep fracture time normalized by the surface diffusion constant of copper and the temperature is a function of the reciprocal third power of the applied stress. This is predicted by the crack-like cavity growth model of Chuang et al [54] which results in an expression of

$$t_f = 3.66\left(\frac{b^4 kT}{D_s \delta_s \chi_s \Omega}\right)\left(\frac{\chi_s/b}{\sigma_\infty}\right)^3 (\sin\psi/2)^3 H((a/b)_o) \qquad (34)$$

that is obtained by integrating the expression given in Eqn (33). In Eqn (34) $H((a/b)_o)$ stands for a function of the initial cavity concentration which in the experiments on copper for an initial cavity diameter of 0.8 μm and cavity spacing of 3.2 μm is equal to 0.56. Nieh and Nix [46,47] have found from the temperature dependence of the fracture time an activation energy of 92 kJ/mol that is in the range of expectations for the activation energy of surface self diffusion. Clearly in this case the regular and relatively close spacing of the cavities has made $\Lambda > b$, and the crack-like growth model applicable.

In an attempt to test the cavity growth models Dyson and Rodgers [62] have pre-cavitated Nimonic 80A by straining at low temperature followed by annealing. This produces cavities on boundaries parallel to the principal direction of extension due to preferential ductile cavitation on interfaces of grain-boundary particles which round-out under the heat treatment procedure. In subsequent creep experiments, with stress applied normal to the cavitated boundaries, Dyson and Rodgers have found that the time to fracture followed neither an inverse first power nor an inverse third power stress relationship but was instead proportional to the reciprocal 0.93 power of the minimum creep strain rate. This is in accord with the creep fracture observations of Monkman and Grant [16] on numerous industrial alloys crept without initial pre-cavitation -- even though the strain rate exponent reported by Monkman and Grant for Nimonic 80 is 0.77. This observation indicates that the fracture in these experiments -- even though by cavity growth -- was actually governed by creep deformation. If the initial size of the cavities had been very large so that $\Lambda << a$ then fracture could have been explained by expansion of cavities by pure creep flow of the type discussed in Section 3 above on transgranular fracture. The initial cavities in the Dyson and Rodgers experiments were, however, only of 0.25 μm diameter, requiring a totally different explanation that is discussed in Section 4.3.3 below.

The growth model of Needleman and Rice [56] based on the combined effect of power-law creep deformation and diffusional flow has been tested by Stanzl, Argon, and Tschegg [49] in copper with initially implanted water vapor bubbles on all grain-boundaries to act as initial cavities as in the experiments of Nix and co-workers [44-47], discussed above. On the assumption that cavity growth by the creep flow mechanism should respond strongly to the presence of a negative pressure, σ_m, as is clear from Eqns (5) or (6), its contribution to the overall cavitation rate was tested in a series of tension and torsion experiments at 500°C where the tensile stress in the tension experiment and the shear stress in the matching torsion experiment were kept equal and varied togeter. Since this produces the same maximum principal tensile stress across the appropriate boundaries but adds an additional negative pressure to the tension experiments, any possible strong influence of the mode of cavity growth by power-law creep should manifest itself in shorter times to fracture in the tension

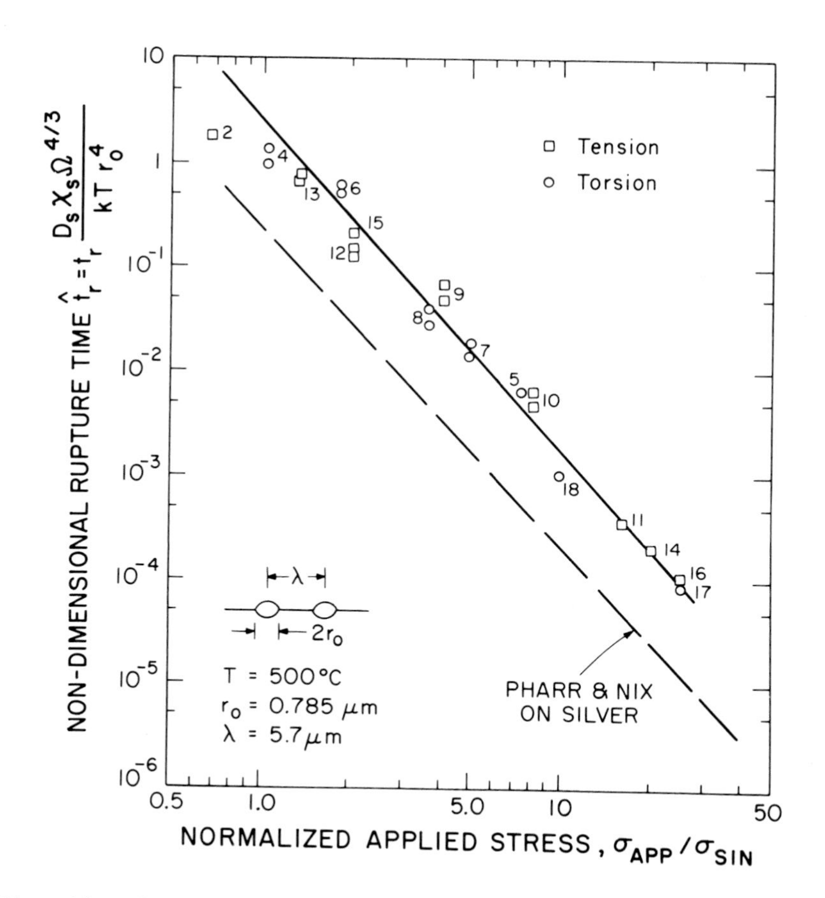

Fig. 16. Dependence of the creep fracture time in tension and torsion of pre-cavitated copper on the applied stress.

experiments. The measured fracture times for all experiments are plotted in Fig. 16 as a function of the maximum principle tensile stress in both the tension and torsion experiments. In both cases the time to fracture shows a reciprocal third power stress dependence and fits the crack-like growth model of Chuang et al [54] for rigid grains. From this, it must be concluded that the contribution to cavity growth by power-law creep comes too late in the fracture life to have an appreciable influence. This is in keeping with the information in Fig. 13.

4.3.3. Constrained diffusional growth of cavities

On the basis of the experiments of Dyson and Rodgers [62] discussed above, Dyson [63,64] has pointed out that when cavities grow on only some boundaries, the thickening of these boundaries would not, in general, make the

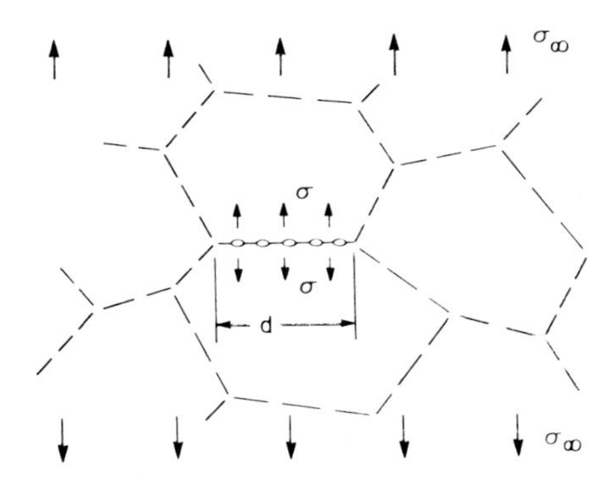

Fig. 17. Inhomogeneous cavitation of a boundary surrounded by non-cavitating boundaries leading to load shedding and constrained growth (after Rice [65]; courtesy of Pergamon Press).

polyhedral grains fit together compatibly. Thus, additional grain deformations are necessary to satisfy compatability between grains. This is clear to see in the sketch of Fig. 17 where cavitation occurs along an isolated boundary normal to the applied stress. The thickening of the boundary due to the expansion of cavities acts as an extra sheet of material that produces some unloading over the cavitating grain-boundary and overloading in the immediate vicinity. This "load shedding" toward the sides continues until when fully cavitated the boundary transmits no normal tractions at all. Thus, the degree at which cavitation can continue depends on the actual tractions maintained on the boundary by the creeping surroundings. This problem has been analyzed by Rice [65] in an approximate manner by matching the thickening rate of a cavitating sheet of extent d along the grain-boundary under an unknown traction, to the opening rate of a penny shapped crack of length d in the creeping surrounding, carrying the same unknown traction on the crack faces. The unknown traction determined from this compatibility match is then used in the cavity growth law to determine the actual cavity growth rate, constrained by the surrounding creeping medium, as,

$$\frac{da}{dt} = \frac{1}{a^2 h(\psi)} \frac{(D_b \delta_b \Omega \sigma_\infty / kT)}{[(D_b \delta_b \Omega \sigma_\infty / kT)(4/\alpha \dot{\varepsilon} b^2 d) + Q(a/b)]} , \qquad (35)$$

where d is the grain facet diameter, α a numerical constant slightly less than unity (0.9 for a stress exponent of 5) and Q(a/b) stands for

$$Q(a/b) = \ell n(b/a) + \left(\frac{a}{b}\right)^2 \left(1 - \frac{1}{4}\left(\frac{a}{b}\right)^2\right) - \frac{3}{4} . \qquad (36)$$

Clearly when the matrix strain rate is small, the load shedding relaxation is so strong that the first term in the denominator of Eqn (35) dominates over the second term, leading to an asymptotic growth rate of cavities, and a corresponding separation time of the facet as follows.

$$\left(\frac{da}{dt}\right)_{asympt} = \frac{\alpha}{4h(\psi)}(b/a)^2 d\dot{\varepsilon} \quad , \qquad (37)$$

$$t_{fac} = (4h(\psi)/3\alpha)(b/a)(1 - (a_i/b)^3(1/\dot{\varepsilon}) \qquad (38)$$

where a_i is the initial cavity radius. If it can be assumed (as Rice has) that isolated facet fracture is synonymous with overall fracture or that a scaling exists between the two then Eqn (38) explains the experiment of Dyson and Rodgers [62]. Parenthetically it is interesting here to note that the load shedding phenomenon discussed here is also responsible for much of the final acceleration of the tertiary creep rate (far above what can be expected from the cavity density according to an upper bound argument) as has been shown by Dyson, Verma, and Szkopiak [66].

Finally, it is necessary to return to the experiments of Nix and coworkers [44-47] discussed in Section 4.3.2 and enquire why they did not reflect a constraint effect of the matrix. inspection of the constrained growth equation (35) indicates that when the strain rate in the matrix is large, little or no constraint exists. Nieh and Nix [47] found, however, no effect on the fracture behavior with incorporation of MgO particles in silver to slow the matrix strain rate down by an order of magnitude. Another possibility would appear to be an absence of constraint when all boundaries are equally pre-cavitated. It is easy to see, however, that a constraint must always exist with space filling polyhedral grains because the required separation across inclined boundaries varies with $\cos\theta$ of the angle of inclination of the boundary while what can be provided by the resolved normal component of stress across the boundary varies with $\cos^2\theta$. This produces always added tensions across inclined boundaries or the load shedding that was discussed above. Although grain-boundary sliding could help rectify the balance, it is not clear how this can be fully successful in irregular polyhedral grain assemblies. This, together with the very large disparities between surface diffusion and grain-boundary diffusion (a factor of 10^4 in the modeling study of Martinez and Nix [60] that is required, makes it useful to search also for alternative explanations for the third power reciprocal stress dependence. An explanation relying on the growth of asymmetrically shaped cavities due to grain-boundary sliding has been proposed and analyzed by Chen [67] which needs, however, additional scrutiny.

5. TIME TO FRACTURE BY INTERGRANULAR CAVITATION

In commercial alloys intergranular creep fracture involves both a continued nucleation of cavities and the growth under stress of the existing cavities by combined diffusional flow and power-law matrix creep. This sequence of processes has been discussed by Raj and Ashby [17], Skelton [68], and Lonsdale and Flewitt [69]. None of these models, however, include the important effect of constrained cavity growth under load shedding. We incorporate this necessary addition into these developments below.

As demonstrated by Cane [37] the number density N of successfully produced mature cavities ready to grow, is proportional to the equivalent strain $\dot{\varepsilon}_e$ through the stochastic processes of boundary sliding outlined in Section 4.2.3 above, i.e.,

$$\dot{N} = \beta\dot{\varepsilon}_e \;, \tag{39}$$

where β is an experimentally determinable coefficient having dimensions of reciprocal area. If no cavities existed initially and the creep strain rate remained constant (*)

---

* In transient creep the overall creep strain rate decreases. This, however, is due to strain hardening of the grain matrix. The nucleation rate, being dependent on grain-boundary sliding, probably continues to scale with the steady state component of creep.

---

the cavity density should increase linearly with time as,

$$N = \beta\dot{\varepsilon}_e t \;. \tag{40}$$

According to the constrained growth considerations the rate of increase of area fraction $\overset{*}{A} = (a/b)^2$ of the boundaries that are cavitated is given by,

$$\frac{d\overset{*}{A}}{dt} = \frac{2a}{b^2}\frac{da}{dt} = \frac{2}{b^3 h(\psi)}\,\frac{(D_b\delta_b\Omega\sigma_\infty/kT)}{(a/b)[(D_b\delta_b\Omega\sigma_\infty/kT)(4/\alpha\dot{\varepsilon}_e b^2 d) + Q(a/b)]} \tag{41}$$

where the continued decrease of b with added nucleation is to be taken as

$$b = (\pi N)^{-1/2} = (\pi\beta\dot{\varepsilon}t)^{-1/2} \tag{42}$$

where we interpreted $\dot{\varepsilon}$ in Eqn (35) as the equivalent strain rate.

Two limiting conditions are of interest: at very small strain rates where load shedding is fully effective,

$$(\overset{*}{A})^{1/2} \frac{d\overset{*}{A}}{dt} = \frac{\alpha}{2} \frac{(d/b)}{h(\psi)} \dot{\varepsilon}_e = \frac{\alpha d(\pi\beta)^{1/2}(\dot{\varepsilon}_e)^{3/2}}{2h(\psi)} t^{1/2} \; ; \qquad (43)$$

and at larger strain rates where little or no load shedding occurs ($\dot{\varepsilon} >> (4/\alpha b^2 dQ(a/b))(D_b\sigma_b\Omega\sigma/kT)$);

$$\frac{d\overset{*}{A}}{dt} = \frac{2}{b^3 h(\psi)}\left(\frac{D_b\delta_b\Omega\sigma}{kT}\right)((b/a)[Q(a/b)]^{-1}) \qquad (44a)$$

$$= \frac{2K(a/b)}{h(\psi)}\left(\frac{D_b\delta_b\Omega\sigma}{kT}\right)(\pi\beta\dot{\varepsilon}_e)^{3/2} t^{3/2} \; , \qquad (44b)$$

where the constant $K(a/b) = (b/a)[Q(a/b)]^{-1}$ is in the range of about 10 for the range of interest in a/b, from 0.05 to 0.5.

Integration of the growth rates in Eqns (43) and (44b) from $\overset{*}{A} = 0$ to $\overset{*}{A} \simeq \overset{*}{A}_{crit}$ ($\simeq 0.4$) where final separation should occur, gives,

$$t_f = \frac{\overset{*}{A}_{crit.}}{\dot{\varepsilon}_e}\left(\frac{2h(\psi)}{\alpha(\pi\beta)^{1/2} d}\right)^{2/3} \qquad (45)$$

for fully constrained growth at small strain rates, and

$$t_f = \left[\frac{5h(\psi)\overset{*}{A}_{crit} kT}{4KD_b\delta_b\Omega(\pi\beta)^{3/2}}\right]^{2/5} \left(\frac{\sigma_0^m}{\dot{\varepsilon}_0}\right)^{3/5} \frac{1}{\sigma_e^{3m/5}\sigma_1^{2/5}} \qquad (46a)$$

$$t_f = \left[\frac{5h(\psi)\overset{*}{A}_{crit} kT}{4KD_b\delta_b\Omega(\pi\beta)^{3/2}}\right]^{2/5} \left(\frac{\dot{\varepsilon}_0^{1/m}}{\sigma_0}\right) \frac{1}{(\dot{\varepsilon}_e)^{(2+3m)/5m}} \qquad (46b)$$

for unconstrained growth at larger strain rates.

Equation (45) for small strain rates should apply under service conditions and is exactly the idealized form of the Monkman and Grant [16] law. Within the spirit of the nucleation relation given by Eqns (39) and (40) the fracture time according to Eqn (45) should be relatively insensitive to triaxial stresses. Under large susperimposed compression it would be expected that the coefficient β in Eqn (39) would become pressure dependent. At higher strain rates the fracture time given by Eqns (46a) shows a specific dependence on equivalent stress and on the maximum principal tensile stress σ_1 ($\sigma_\infty = \sigma_1$) in the distant field. In the form given in Eqn (46b) the fracture time shows a Monkman and Grant strain rate exponent of $(2 + 3m)/5m$ which for the normal stress exponents of 5 - 8 gives a value 0.68 - 0.60. This is considerably smaller than the average Monkman-Grant exponent of about 0.85. Clearly the real situation will be somewhere in between the two limits, leading to an exponent of somewhat less than unity as reported for most alloys by Monkman and Grant [16]. The form of the equations in (46a) and (46b) are, to within some numerical constants, identical with that given by Lonsdale and Flewitt [69]. An accurate quantitative test of the relations given by Eqns (45) or (46b), particularly in relation to the grain size dependence, has not yet been made.

On the basis that the stress dependence of the creep strain rate is often an exponential function of stress (for a compilation of evidence see Bassani [70]), given from fundamental considerations to be

$$\dot{\varepsilon}_e = \dot{\varepsilon}_o \exp(-\Delta H(\sigma_e)/kT) \ , \tag{47}$$

the relation for fracture time given by Eqn (45) can be converted into

$$\theta \equiv T[\log(\dot{\varepsilon}_0/B) + \log t_f] = \frac{0.43}{k}\Delta H(\sigma_e) \tag{48}$$

where B is the collection of terms in Eqn (45) other than t_f and $\dot{\varepsilon}_e$, and $0.43 = \log_{10} e$. In Eqn (48) a combined fracture time, temperature term, θ, is defined and the structure of the equation implies that θ is only a function of the equivalent stress σ_e. The first introduction of such a combined time temperature (or strain rate temperature) parameter to scale inelastic deformation processes is due to MacGregor and Fisher [71], and its application to the correlation of creep fracture times at different temperatures against stress is due to Larson and Miller

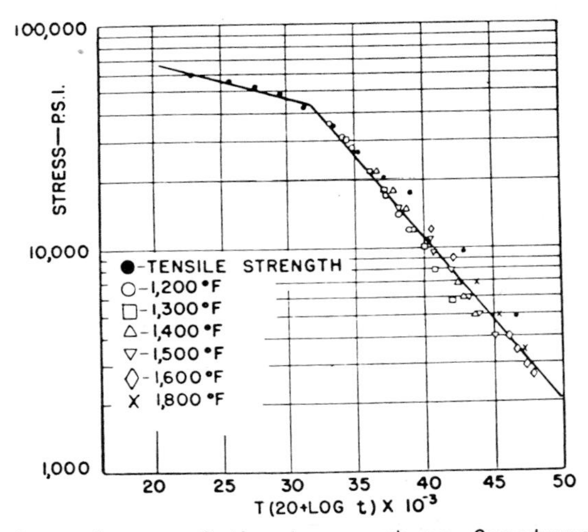

Fig. 18. Dependence of the temperature-fracture time parameter for 18-8 stainless steel on the applied stress (from Larson and Miller [72]; courtesy of ASME).

[72]. The procedure is usually quite successful as is shown by the typical plot for 18-8 stainless steel in Fig. 18. The figure indicates that in many cases the value of the term $\log(\dot{\varepsilon}_o/B)$ is 20 for time measured in hours. Cocks and Ashby [73] have considered a different mechanistic interpretation of this relationship on the basis of an approximate development for the growth of cavities under combined power law creep and diffusional flow, with, however, no allowance being made for continued cavity nucleation nor for constraint effects resulting from load shedding.

6. CREEP CAVITATION AND FRACTURE EVOLUTION IN NOTCHES

Through the relatively large distortional and cavitational strains that creeping materials can exhibit, that results in an important capacity to relax and re-distribute concentrated stresses, these materials can absorb considerable damage and render ineffective most notches. The actual process is complicated and involves both the notch stresses, and the strain to fracture of the creeping alloy.

One of the earliest investigations of this effect is that of Davis and Manjoine [74] whose extensive experiments on notch effects that have been summarized by McLean, Dyson and Taplin [1] are shown in Fig. 19. The figure shows for four different alloys with strains to fracture ranging from 3.5% to 36%, the ratio of the 1000 hour creep strengths of notched and un-notched bars, plotted as a function of the increasing notch acuity. If no fracture occurred, the creep

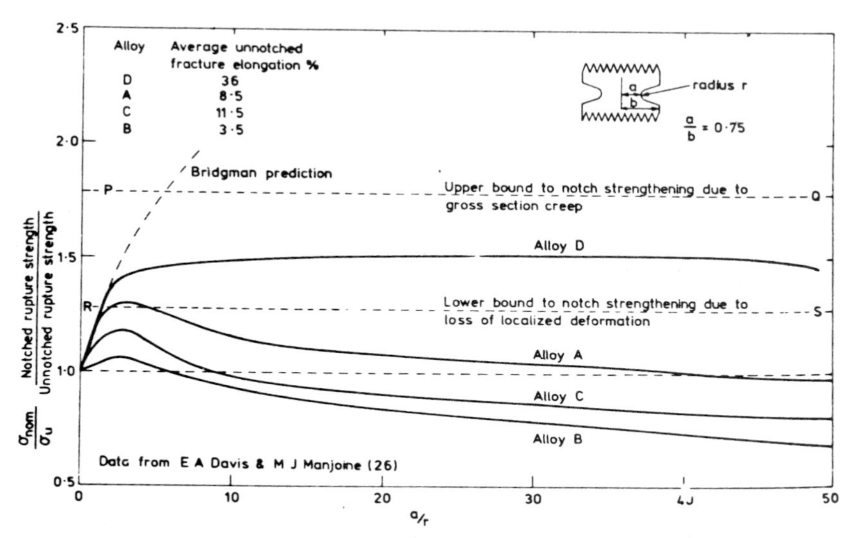

Fig. 19. Notch strengthening at different notch acuity for four alloys with different strains to fracture (from Davis and Manjoine [74]; courtesy of ASTM).

deformation in the notch would produce a stress redistribution and creep constraint that should increase the load carrying capacity of the notched bar over the unnotched bar having the same neck radius. This effect can be predicted according to Bridgman's [75] well known development and Fig. 19 shows that all four alloys follow this pattern initially with increasing a/r. Increasing the notch acuity, at a fixed ratio of shoulder area to neck area, produces eventually general deformation in the shoulder region. This limits the notch strengthening at PQ. On the other hand below a stress ratio given by RS, deformation across the neck is no longer possible. The data of Davis and Manjoine show that the most ductile alloy (D) maintains fully the plastic constraint over all levels of notch acuity by successfully redistributing the stress before creep fracture is completed. Alloys A, C, and B, with less ductility, are capable to re-distribute the stress successfully for only small notch acuity, while with increasing notch acuity creep fracture begins before the non-uniform creep deformation can develop the creep constraint that gives notch strengthening.

Hayhurst and Leckie [76,77] have investigated the development of fracture in creeping copper and aluminum plates with circular holes and slits having the same lateral dimension as the holes but with much smaller radii of curvature at the tip. They appear to have been first in the creep field to point attention to the important stress redistribution that arises from cavitational deformation. In more detailed experiments Hayhurst and Leckie [78] have investigated the evolution of creep damage in round bars

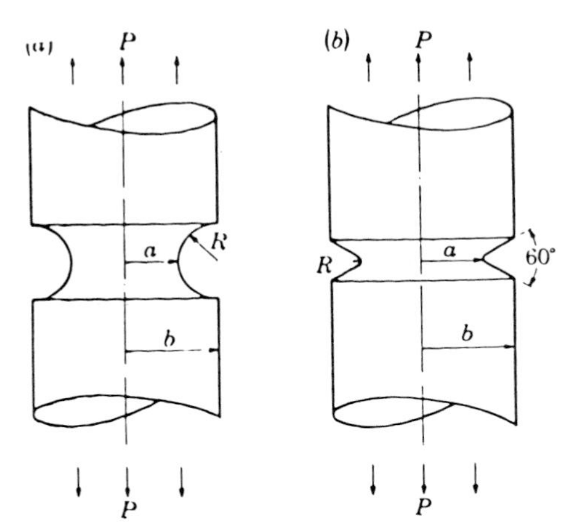

Fig. 20. Circumferentially notched specimens with: (a) a round notch; and (b) a British Standard notch used by Hayhurst and Leckie to study the evolution of creep fracture (from Hayhurst, Leckie, and Morrison [78]; courtesy of the Royal Society of London).

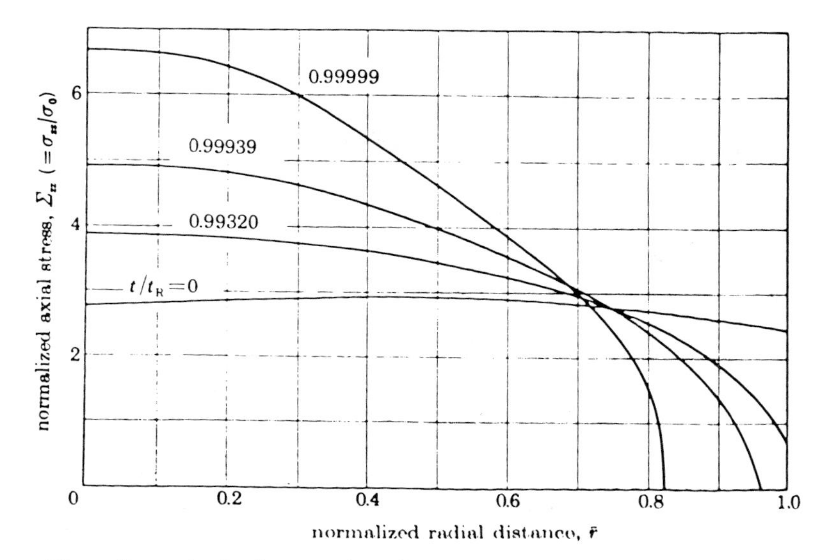

Fig. 21a. Computed change in stress distribution with time in a round notch (from Hayhurst, Leckie, and Morrison [78]; courtesy of the Royal Society of London).

with: a) circular circumferential notches, and b) British standard notches having the geometry shown in Fig. 20. These experiments were guided and interpreted by detailed finite element computations of stress redistribution in bars of these same geometries, carried out by Hayhurst and Henderson [79] and Hayhurst, Leckie, and Henderson [80]. The numerical computations have incorporated not only the usual distortional steady state creep relationships but also the accelerating creep due to increased cavitation in the tertiary creep range, without, however, (apparently) incorporating specifically any dilatation to relieve negative pressures. Two sample computations of axial stress distribution σ_{zz} in the circularly notched bar and the bar with the British Standard notch are shown in Figs. 21a and 21b for a stress exponent of m = 5.9 in the steady state creep law, and a damage exponent of $\nu = 5.6$ (both considered to be typical for copper by Hayhurst and Leckie) in a damage production law of

$$\dot{\Psi} = -(\sigma_1/\Psi\sigma_0)^{\nu} \qquad (49)$$

where σ_1 is the maximum principal tensile stress, σ_0 has the same meaning as in Eqn. (4), and the damage function $\Psi = 1$ for undamaged material (with zero cavitation) and $\Psi = 0$ for the fully cavitated material at fracture (100% cavitation). Figure 21a shows the change in the distribution of the axial stress σ_{zz} across the bar from a flat distribution at the beginning of steady state deformation (after a transient that has eliminated the initial elastic distribution) to the

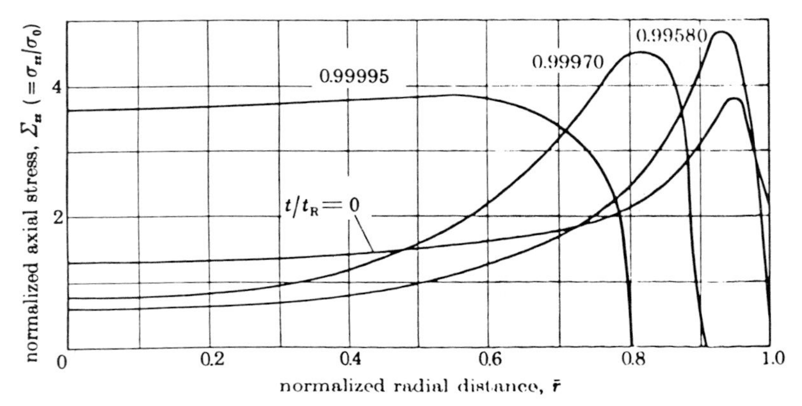

Fig. 21b. Computed change in stress distribution with time in a British Standard notch (from Hayhurst, Leckie, and Morrison [78]; courtesy of the Royal Society of London).

redistributions shown at very nearly the end of life when fracture begins at the outer periphery at t/t_f between 0.99320 and 0.99939. Figure 21b, on the other hand, shows the corresponding case of the evolution of fracture in the much more acute British standard notch. Here at $t/t_f = 0$ the initial steady state distribution of stress shows a prominent peak underneath the notch root, typical of blunted notches, and fracture begins again at the notch tip at nearly the end of the creep life at $t/t_f = 0.9958$. It then propagates across the section rapidly through material that too has nearly fully exhausted its creep ductility.

These developments of Hayhurst and Leckie, discussed above, are dramatic presentation of the nature of the creep fracture process where the material locally comes apart when, through cavitation and associated load shedding, the traction drops to zero. The same mechanism should apply in the growth of creep cracks where, however, the crack should focus the damage more sharply than in the notches.

7. STRESS REDISTRIBUTION AT CRACKS IN CREEPING ALLOYS

As in the more benign situation of evaluation of creep damage in notches the problem of primary initial concern for cracks in creeping alloys is the redistribution of stress that produces the damage and couples with it to affect the distribution. Unfortunately, as we will see below the loop has not yet been closed in modeling creep fracture at cracks where all except two linear analyses have concentrated in the computation of the stress distribution in non-cavitating and non-accelerating material and have not bothered to enquire about how this distribution is altered by the damage that the distribution evokes.

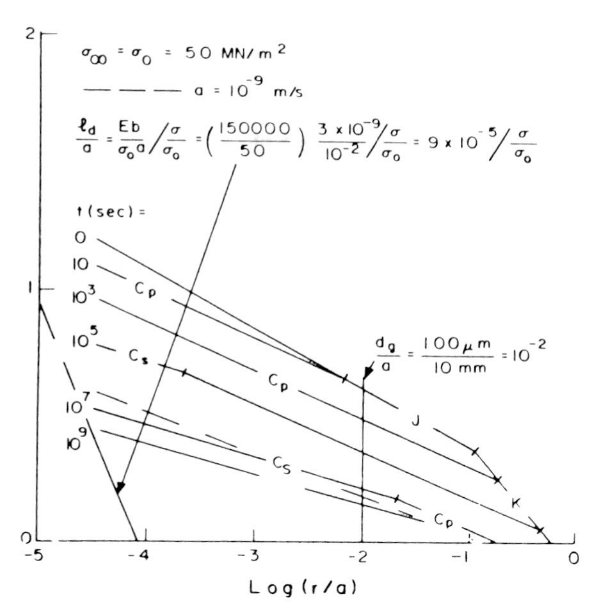

Fig. 22. Time dependent change in the computed dominance of different singularities at the tip of a very sharp crack in a creeping alloy (from McClintock and Bassani [84]; courtesy of North Holland).

The developments in the computation of stresses around stationary cracks in creeping alloys have been catalogued by Bassani [81], making it unnecessary for us to repeat such a presentation here beyond giving a simple summary.

Riedel and Rice [82] and Riedel [83] have derived the asymptotic stress singularity for a sharp crack in a model material exhibiting both elasticity and power-law creep [82], and additional primary creep together with the former two responses. When time independent plastic response according to a power-law hardening model is included the crack tip shows different responses at different times. This has been discussed by McClintock and Bassani [84]. In all instances, since cracks have not been allowed to blunt the stresses are singular at the crack tip. this singularity, however, changes in nature with the passage of time. In Fig. 22, due to McClintock and Bassani, the changes in the dominant singular field at the crack tip is shown for a specific case of 304 stainless steel subjected to a service stress of 50 MPa at 600°C. The figure shows the distribution on logarithmic coordinates of the stress $\sigma_{\theta\theta}$ as a function of radial distance from the crack tip in units of crack length. At time zero the stress distribution near the crack tip is dominated by a time independent, plastic field of the HRR (Hutchinson-Rice-Rosengren) type of singularity shown by J and the distant field is characterized by the typical elastic field, shown by K. As time advances the J field is rapidly superceded by the lower grade singularity of the primary creep strain analysed by Riedel [83]: this is shown by C_p. At around 10^5 seconds the modification due to the secondary creep singularity analysed by Riedel and Rice [82] has arrived at the crack tip region of interest. Its effect is shown by the line C_s. At very long times of 10^9 seconds the secondary creep field singularity due to Riedel and Rice [82] dominates the entire region. The additional dotted line in the figure represents an asymptotic crack tip singular field calculated by Hui and Riedel [85] for a crack moving with a constant velocity (10^{-9} m/s, in this case).

These asymptotic stress distributions for infinitely sharp and stationary cracks represent what is known among investigators in applied mechanics as "well posed problems." As in the case of linear elastic fracture mechanics they may be useful as characterizing parameters for the distant field for long cracks in very large specimens, provided that they are valid distant field parameters. For mechanistic modeling of the crack tip damage processes, however, their utility is quite small. As the computations of the notch tip stress redistribution discussed in Section 6 above have demonstrated dramatically, the crack tip blunting and the cavitational creep damage must radically alter the stress field near the crack tip to at least several radii of the blunted crack tip that the singular distributions are certain to be completely wiped out. To what extent this is the case for creep cracks is at the present unknown -- still awaiting the application of interactive numerical computations that consider blunting and cavitation in the stress redistribution for an answer.

Two recent attempts in the direction of the required interactive computations are noteworthy. Raj and Baik [86] have modeled the crack tip stress relaxation in an elastic material where the singular distribution is modified by the introduction of a cavitational wedge resulting from cavity growth. Even without any inelastic strain, significant stress redistribution at stationary cracks is concentrated showing the beginning of a reversal in the crack tip stress gradient, and an approach of the crack tip stress to zero that is necessary for eventual growth. A second application is that of Evans and Hsueh [87] for a cavitating ceramic where the development of an incipient crack by the interaction of a central core of cavities has been modeled giving the relaxation of the stress at the central region to zero and a gradual spreading of a cavitational zone extending the initial crack. Clearly computations of this type applied to creeping solids are much needed to model and guide the many experimental measurements of creep crack growth.

Acknowledgement: The author's research on fracture in creeping alloys is supported by the U.S. Department of Energy under Contract No. DE-AC02-77ER04461. It is a pleasure to acknowledge discussions with many colleagues in the creep field many of whom are directly quoted in this paper, but particularly with my colleagues at M.I.T., Professors F.A. McClintock, I-W. Chen, J.L. Bassani (now at the University of Pennslyvania) and Dr. C.W. Lau.

REFERENCES

1. McLEAN, D., DYSON, B.F., and TAPLIN, D.M.R., 'The Prediction of Creep Fracture in Engineering Alloys,' in Fracture 1977, Ed. Taplin, D.M.R., University of Waterloo Press: Waterloo, Canada, 1977, vol. 1 p. 325.
2. ASHBY, M.F., GANDHI, C., and TAPLIN, D.M.R., 'Fracture Mechanism Maps and their Construction for F.C.C. Metals and Alloys,' Acta Met., 1979, 27, 699.
3. CROSSMAN, F.W., and ASHBY, M.F., 'The Non-uniform Flow of Polycrystals by Grain-Boundary sliding Accomodated by Power-Law Creep,' Acta Met., 1975, 23, 425.
4. GHAHREMANI, F., 'Effect of Grain-Boundary Sliding on Steady State Creep of Polycrystals,' Int. J. Solids Structures, 1980, 16, 847.
5. CHEN, I-W, and ARGON, A.S., 'Grain-Boundary and Interphase Boundary Sliding in Power Law Creep, 'Acta Met., 1979, 27, 749.
6. ARGON, A.S., 'Formation of Cavities from Nondeformable Second-Phase Particles in Low Temperature Ductile Fracture,' J. Eng. Mat. Tech., 1976, 98, 60.
7. ARGON, A.S., IM, J., and SAFOGLU, R., 'Cavity Formation from Inclusions in Ductile Fracture,' Met. Trans. 1975, 6A, 825.
8. HANCOCK, J.W., 'Creep Cavitation without a Vacancy Flux,' Metal Sci., 1976, 10, 319.
9. BUDIANSKY, B., HUTCHINSON, J.W., and SLUTSKY, S., 'Void Growth and Collapse in Viscous Solids,' in Mechanics of Solids, Eds. Hopkins, H.G., and Sewell, M.J., Pergamon press: Oxford, 1981, p. 13.
10. McCLINTOCK, F.A., 'A Criterion for Ductile Fracture by Growth of Holes,' J. Appl. Mech., 1968, 35, 363.
11. RICE, J.R., and TRACY, D.M., 'On the Ductile Enlargement of Voids in Triaxial Stress Fields,' J. Mech. Phys. Solids, 1969, 17, 201.
12. HANCOCK, J.W., and MacKENZIE, A.C., 'On the Mechanisms of Ductile Failure in High-Strength Steels subjected to Multi-axial Stress States,' J. Mech. Phys. Solids, 1976, 24, 147.
13. GREENWOOD, J.N., MILLER, D.R., and SUITER, J.W., 'Intergranular Cavitation in Stressed Metals,' Acta Met., 1954, 2, 250.

14. GRANT, N.J., and MULLENDORE, A.W., Deformation and Fracture at Elevated Temperatures, M.I.T. Press: Cambridge, Mass. 1965.

15. SERVI, I., and GRANT, N.J., 'Creep and Stress Rupture Behavior of Aluminum as a Function of Purity,' Trans. AIME, 1951, 191, 909.

16. MONKMAN, F.C., and GRANT, N.J., 'An Empirical Relationship Between Rupture Life and Minimum Creep Rate in Creep-Rupture Tests,' Proc. ASTM, 1956, 56, 593.

17. RAJ, R., and ASHBY, M.F., 'Intergranular Fracture at Elevated Temperature,'Acta Met., 1975, 23, 653.

18. RAJ, R., 'Nucleation of Cavities at Second Phase Particles in Grain Boundaries,' Acta Met., 1978, 26, 995.

19. ARGON, A.S., CHEN, I-W., and LAU, C.W., 'Intergranular Cavitation in Creep: Theory and Experiments,' in Creep-Fatigue-Environment Interactions, Eds. Pelloux, R.M., and Stoloff, N.S., AIME: New York, 1980, p. 46.

20. McLEAN, D., 'Point Defects and the Mechanical Properties of Metals and Alloys at High Temperatures,' in Vacancies and Other Point Defects in Metals and Alloys, The Institute of Metals: London, 1958, p. 187.

21. SMITH, E., and BARNBY, J.T., 'Nucleation of Grain-Boundary Cavities during High-Temperature Creep,' Metal Sci., 1967, 1, 1.

22. BRUNNER, H., and GRANT, N.J., 'Calculation of the Contribution Made by Grain-Boundary Sliding to Total Tensile Elongation,' J. Inst. Met., 1956, 85, 77.

23. LAU, C.W., and ARGON, A.S., 'Stress Concentrations Caused by Grain-Boundary Sliding in Metals Undergoing Power-Law Creep,: in Fracture 1977, Ed. Taplin, D.M.R., University of Waterloo Press: Waterloo, Canada 1977, vol. 2, 595.

24. LAU, C.W., ARGON, A.S., and McCLINTOCK, F.A., 'Stress Concentrations due to Sliding Grain-Boundaries in Creeping Alloys,' in Elastic-Plastic Fracture Mechanics, ASTM: Philadelphia, in the press.

25. LAU, C.W., Dominant Singularity and Finite Element Analysis of Plane-Strain Stress Fields in Creeping Alloys with Sliding Grain Boundaries, Ph.D. Thesis Department of Mechanical Engineering, M.I.T., 1981.

26. SEAH, M.P., 'Impurities, Segregation, and Creep Embrittlement,' Phil. Trans. Roy. Soc., 1980, A295, 27. HONDROS, E.D., and SEAH, M.P., 'The Theory of Grain-Boundary Segregation in Terms of Surface Adsorption Analogues,' Met. Trans., 1977, A8, 1363.

27. HONDROS, E.D., and SEAH, M.P., 'The Theory of Grain-Boundary Segregation in Terms of Surface Adsorption Analogues,' Met. Trans., 1977, A8, 1363.

28. CHANG, H.C., and GRANT, N.J., 'Inhomogeneity in Creep Deformation of Coarse Grained High Purity Aluminum,' Trans. AIME, J. Metals, 1953, 5, 1175.

29. CHEN, I-W., and ARGON, A.S., 'Creep Cavitation in 304 Stainless Steel,' Acta Met., 1981, 29, 1321.
30. RUHLE, M., and WILKENS, M., 'Defocusing Contrast of Cavities I. Theory,' Crystal Lattice Defects, 1975, 6, 129.
31. DYSON, B.F., LOVEDAY, M.S., and RODGERS, M.J., 'Grain-Boundary Cavitation under various States of Applied Stress,' Proc. Roy. Soc., 1976, A349, 245.
32. COTTRELL, A.H., 'The Intersection of Gliding Screw Dislocations,' in Dislocations and Mechanical Properties of Crystals, Eds. Fisher, J.C., Johnston, W.G., Thomson, R., Vreeland, T., Jr., Wiley: New York, 1957, p. 509.
33. KOELLER, R.C., and RAJ, R., 'Diffusional Relaxation of Stress Concentration at Second Phase Particles,' Acta Met., 1978, 26, 1551.
34. SHIOZAWA, K., and WEERTMAN, J.R., 'Nucleation of Grain Boundary Voids in Astroloy during High Temperature Creep,' Scripta Met., in the press.
35. PAGE, R., WEERTMAN, J.R., and ROTH, M., 'Investigation of Fatigue-Induced Grain-Boundary Cavitation by Small Angle Neutron Scattering,' Scripta Met., 1980, 14, 773.
36. CANE, B.G., and GREENWOOD, G.W., 'The Nucleation and Growth of Cavities in Iron during Deformation at Elevated Temperatures,' Metal Sci., 1975, 9, 55.
37. CANE, B.J., 'Deformation-Induced Intergranular Creep Cavitation in Alpha-Iron,' Metal Sci., 1978, 12, 102.
38. WHITE, C.L., and LIU, C.T., 'The Effect of Phosphorous Segregation to Grain-Boundaries in Ir + 0.3 wt% W Alloys on High Temperature Ductility,' Scripta Met., 1978, 12, 727.
39. DYSON B.F., and McLEAN, D., 'Creep of Nimonic 80A in Torsion and Tension,' Metal Sci., 1977, 11, 37.
40. FLECK, R.G., TAPLIN, D.M.R., and BEEVERS, C.J., 'An Investigation of the Nucleation of Creep Cavities by 1 MV Electron Microscopy,' Acta Met., 1975, 23, 415.
41. RAJ, R., 'Intergranular Fracture in Bi-crystals,' Acta Met., 1978, 26, 341.
42. BRICKNELL, R.H., and WOODFORD, D.A., 'Cavitation in Nickel During Oxidation and Creep,' in Creep and Fracture of Engineering Materials and Structures, Eds. Wilshire, B., and Owen, D.R.J., Pineridge Press: Swansea, U.K., 1981, p. 249.
43. BRICKNELL, R.H., and WOODFORD, D.A., 'The Mechanism of Cavity Formation During High Temperature Oxidation of Nickel,' Acta Met., 1982, 30, 257.
44. GOODS, S.H., and NIX, W.D., 'The Kinetics of Cavity Growth and Creep Fracture in Silver Containing Implanted Grain-Boundary Cavities,' Acta Met., 1978, 26, 739.
45. GOODS, S.H., and NIX, W.D., 'The Coalescense of Large Grain Boundary Cavities in Silver during Tension Creep,' Acta Met., 1978, 26, 753.

46. NIEH, T.G., and NIX, W.D., 'A Study of Intergranular Cavity Growth in Ag + 0.1% MgO at Elevated Temperatures,' Acta Met., 1979, 27, 1097.

47. NIEH, T.G., and NIX, W.D., 'The Formation of Water Vapor Bubbles in Copper and Their Effect on Intergranular Creep Fracture,' Acta Met., 1980, 28, 557.

48. HANNA, M.D., and GREENWOOD, G.W., 'Cavity Growth and Creep in Copper at Low Stresses,' Acta Met., 1982, 30,

49. STANZL, S., ARGON, A.S., and TSCHEGG, E., 'Intergranular Cavity Growth in Copper in Tension and Torsion,' to be published.

50. BALLUFFI, R.`W., and SEIGLE, L.I., 'Growth of Voids in Metals During Diffusion and Creep,' Acta Met., 1957, 5,

51. HULL, D., and RIMMER, D.E., 'The Growth of Grain-Boundary Voids Under Stress,' Phil. Mag., 1959, 4, 673.

52. SPEIGHT, M.V., and BEERE, W., 'Vacancy Potential and Void Growth on Grain-Boundaries,' Metal Sci., 1975, 9,

53. CHUANG, T-J., and RICE, J.R., 'The Shape of Intergranular Creep Cracks Growing by Surface Diffusion,' Acta Met., 1973, 21, 1625.

54. CHUANG, T-J., KAGAWA, K.I., RICE, J.R., and SILLS, L.B., 'Non-Equilibrium Models for Diffusive Cavitation of Grain Interfaces,: Acta Met., 1979, 27, 265.

55. BEERE, W., and SPEIGHT, M.V., 'Creep Cavitation by Vacancy Diffusion in Plastically Deforming Solid,' Metal Sci., 1978, 12, 172.

56. NEEDLEMAN, A., and RICE, J.R., 'Plastic Creep Flow Effects in the Diffusive Cavitation of Grain-Boundaries,' Acta Met., 1980, 28, 1315.

57. CHEN, I-W., and ARGON, A.S., 'Diffusive Growth of Grain-Boundary Cavities,' Acta Met, 1981, 29, 1759.

58. CHUANG, T-J., Models of Intergranular Creep Crack Growth by Coupled Crack Surface and Grain-Boundary Diffusion, Ph.D. Thesis, Brown University, 1975.

59. PHARR, G.M., and NIX, W.D., 'A Numerical Study of Cavity Growth Controlled by Surface Diffusion,' Acta Met., 1979, 27, 1615.

60. MARTINEZ, L., and NIX, W.D., 'A Numerical Study of Cavity Growth Controlled by Coupled Surface and Grain-Boundary Diffusion,' Met. Trans., 1982, 13A, 427.

61. HANNA, M.D., and GREENWOOD, G.W., 'Cavity Growth and Creep in Copper at Low Stresses,' Acta Met., 1982, 30,

62. DYSON, B.F., and RODGERS, M.J., 'Intergranular Creep Fracture in Nimonic 80A Under Conditions of Constant Cavity Density,' in Fracture 1977, Ed. Taplin D.M.R., University of Waterloo Press: Waterloo, Canada, 1977, vol. 2, p. 621.

63. DYSON, B.F., 'Constraints on Diffusional Cavity Growth Rates,' Metal Sci., 1976, 10, 349.

64. DYSON, B.F., 'Constrained Cavity Growth, its use in Quantifying Recent Creep Fracture Results,' Can. Metal Quart., 1979, 18, 31.
65. RICE, J.R., 'Constraints on the Diffuse Cavitation of Isolated Grain-Boundary Facets in Creeping Polycrystal,' Acta Met., 1981, 29, 675.
66. DYSON, B.F., VERMA, A.K., and SZKOPIAK, Z.C., 'The Influence of Stress State on Creep Resistance: Experiments and Modeling,' Acta Met., 1981, 29, 1573.
67. CHEN, I-W., private communication.
68. SKELTON, R.P., 'Diffusional Creep Strain due to Void Nucleation and Growth,' Metal Sci., 1975, 9, 192.
69. LONSDALE, D., and FLEWITT, P.E.J., 'The Effect of Hydrostatic Pressure on the Uniaxial Creep Life of a 2-1/4% Cr, 1% Mo Steel,' Proc. Roy. Soc., 1981, A373,
70. BASSANI, J.L., 'Macro and Micro-Mechanical Aspects of Creep Fracture,' in Advances in Aerospace Structures and Materials - I, Eds. Wang, S.S., and Renton, W.J., ASME: New York, 1981, p. 1.
71. MacGREGOR, C.W., and FISHER, J.C., 'A Velocity Modified Temperature for the Plastic Flow of Metals,' J. Appl. Mech., 1946, 13, A11.
72. LARSON, F.R., and MILLER, J., 'A Time-Temperature Relationship for Rupture and Creep Stresses,' Trans. ASME., 1952, 74, 765.
73. COCKS, A.C.F., and ASHBY, M.F., 'Creep Fracture by Void Growth,' in Creep in Structures, Eds. Ponter, A.R.S., and Hayhurst, D.R., Springer: Berlin, 1981, p.
74. DAVIS, E.A., and MANJOINE, M.J., 'Effect of Notch Geometry on Rupture Strength at Elevated Temperatures,' in Strength and Ductility of Metals at Elevated Temperatures, STP No. 128, ASTM: Philadelphia, 1953, p.
75. BRIDGMAN, P.W., Studies in Large Plastic Flow and Fracture, McGraw-Hill: New York.
76. HAYHURST, D.R., and LECKIE, F.A., 'The Effect of Creep Constitutive and Damage Relationships upon the Rupture Time of a Solid Circular Torsion Bar,' J. Mech. Phys. Solids, 1973, 21, 431.
77. LECKIE, F.A., and HAYHURST, D.R., 'Creep Rupture of Structures,: Proc. Roy. Soc., 1974, A340, 323.
78. HAYHURST, D.R., LECKIE, F.A., and MORRISON, C.J., 'Creep Rupture of Notched Bars,' Proc. Roy. Soc., 1978, A360, 243.
79. HAYHURST, D.R., and HENDERSON, J.T., 'Creep Stress Redistribution in Notched Bars,' Int. J. Mech. Sci., 1977, 19, 133.
80. HAYHURST, D.R., LECKIE, F.A., and HENDERSON, J.T., 'Design of Notched Bars for Creep-Rupture Testing Under Tri-axial Stresses,' Int. J. Mech. Sci., 1977, 19, 147.
81. BASSANI, J.L., 'Creep Crack Extension by Grain-Boundary Cavitation,' in Creep and Fracture of Engineering Materials and Structures, Eds. Wilshire, B., and Owen, D.R.J., Pineridge Press: Swansea, U.K., 1981, p. 329.

82. RIEDEL, H., and RICE, J.R., 'Tensile Cracks in Creeping Solids,' in ASTM STP 700, ASTM: Philadelphia, 1980, p.
83. RIEDEL, H., 'Creep Deformation at Crack Tips in Elastic-Viscoplastic Solids,' Brown Univ. Report MRL E-114, Brown University: Providence, R.I., 1979.
84. McCLINTOCK, F.A., and BASSANI, J.L., 'Problems in Environmentally-Affected Creep Crack Growth,' in Three Dimensional Constitutive Relations and Ductile Fracture, Ed. Nemet-Nasser, S., North Holland: Amsterdam, 1981, p. 123.
85. HUI, H., and RIEDEL, H., 'The Asymptotic Stress and Strain Field Near the Tip of a Growing Crack under Creep Conditions,' Int. J. Fracture, 1980, 16, p.
86. RAJ. R., and BAIK, S., 'Creep Crack Propagation by Cavitation near Crack Tips,' Metal Sci., 1980, 14, 385.
87. EVANS, A.G., and HSUEH, C.H., 'Creep Fracture in Ceramic Polycrystals,' in Creep and Fracture of Engineering Materials and Structures, Eds. Wilshire, B., Owen, D.R.J., Pineridge Press: Swansea, U.K., 1981, p. 409.

Chapter 2

THE HIGH TEMPERATURE FAILURE OF CERAMICS*

A.G. Evans

Materials and Molecular Research Division, Lawrence Berkeley Laboratory and Department of Materials Science and Mineral Engineering, University of California, Berkeley, California 94720, U.S.A.

ABSTRACT

This review describes the various cavitation mechanisms that operate in single phase and two phase ceramic polycrystals at elevated temperatures. Analysis of these mechanisms has been used to develop failure models and thereby, to provide failure time expressions suitable for the interpretation of rupture experiments and for the eventual prediction of failure. The failure processes have been distinguished on the basis of crack nucleation control and crack propagation control. The former yields either Monkman-Grant failure expressions or probabilistic Orr-Sherby-Dorn failure relations, depending upon the cavitation homogeneity and the level of the applied loading. The available creep rupture data has been analyzed from an informed perspective provided by the failure models.

1. INTRODUCTION

The mechanical failure of ceramics at elevated temperatures is accompanied by permanent deformation and exhibits a strong dependence on temperature and strain-rate (Fig. 1). The failure usually evolves by the nucleation, growth and coalescence of cavities at preferred microstructural sites (Fig. 1). The deduction of comprehensive engineering expressions for high temperature failure requires that the cavity evolution process be understood at the fundamental

---

*This work was supported by the Director, Office of Energy Research, Office of Basic Energy Sciences, Materials Science Division of the U.S. Department of Energy under Contract No. DE-AC03-76SF00098.

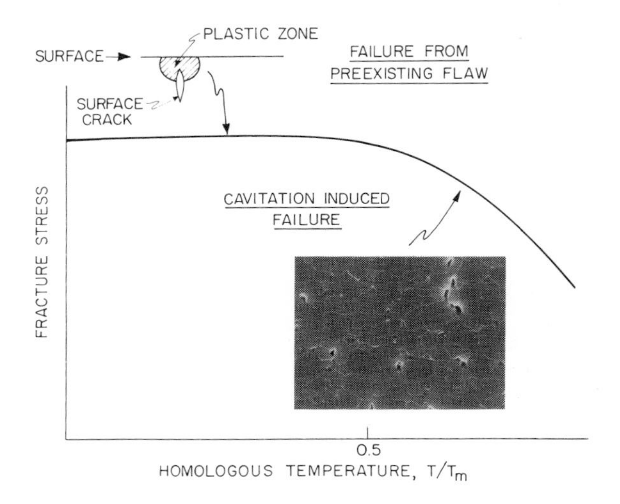

Fig. 1. A schematic of the temperature dependence of the strength of a typical ceramic polycrystal.

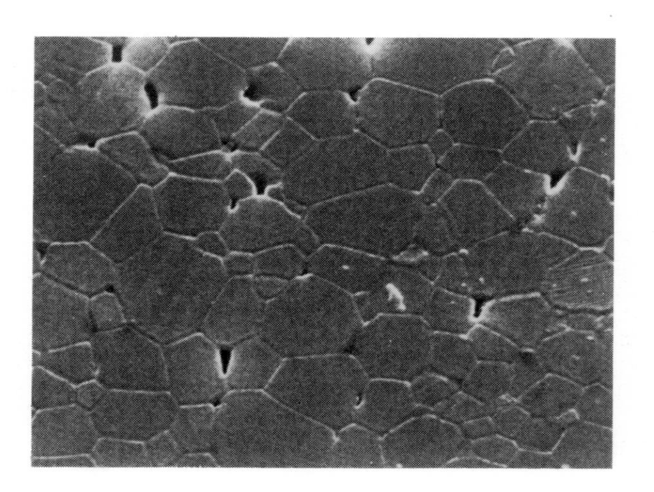

Fig. 2a. Scanning electron micrograph illustrating the processes of crack nucleation, involving nucleation and propagation.

level. This review describes the essential details that underlie the engineering analysis of failure.

High temperature failure typically involves several sequential processes: cavity nucleation, cavity propagation, crack nucleation and crack propagation (Fig. 2). Each process must be comprehensively characterized in order to establish a generalized description of failure. Consequently, the first part of this review is concerned with a characterization of crack nucleation (by a cavity nucleation, propagation and coalescence sequence); while the second part is devoted primarily to crack propagation processes. The implicatons for microstructural design are presented in the final section.

The analysis of high temperature failure in ceramics must be cognizant of the variety of microstructures that may exist and their influence on the specific failure evolution mechanisms. The two most important classes of ceramic microstructure are emphasized in this paper: predominantly single phase polycrystals (albeit, in some instances, with small isolated second phase particles at two grain interfaces) (Fig. 2a), and two phase systems with a continuous second phase (Fig. 3). The former microstructure typifies materials fabricated by solid state sintering, chemical vapor deposition etc., while the latter microstructure is characteristic of liquid phase sintered ceramics. An important theme of this paper will be the vital influence of second phases on the failure process at high temperatures.

The creep deformation of most commercial ceramics occurs by diffusion, viscous flow or solution/reprecipitation and consequently, exhibits a stress exponent, n, in the range $1 \leqslant n < 2$ (dislocation creep is rarely observed) [1,2]. The analysis of creep rupture in ceramics [3] is thus frequently based on a deformation linearity premise (which appears to provide an adequate first order characterization of the observed rupture behavior). The analytic simplicity afforded by linearity will be adopted in the present review.

The paucity of comprehensive creep rupture data for ceramics, especially data accompanied by microstructural observations of failure evolution, limits the present ability to provide a well-balanced view of the creep rupture process. The intent of this review is thus to provide a description of the underlying failure phenomena that establishes the eventual basis for interpreting failure data and predicting failure.

2. CRACK NUCLEATION CONCEPTS

Crack nucleation during high temperature creep generally occurs by the nucleation, growth and coalescence of cavities [3-8]. Each of these processes requires separate consideration,

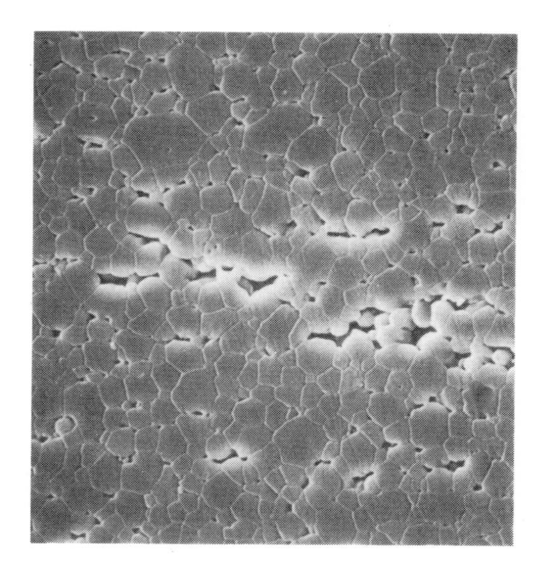

Fig. 2b. Scanning electron micrograph illustrating the process of crack nucleation, involving cavity coalescence.

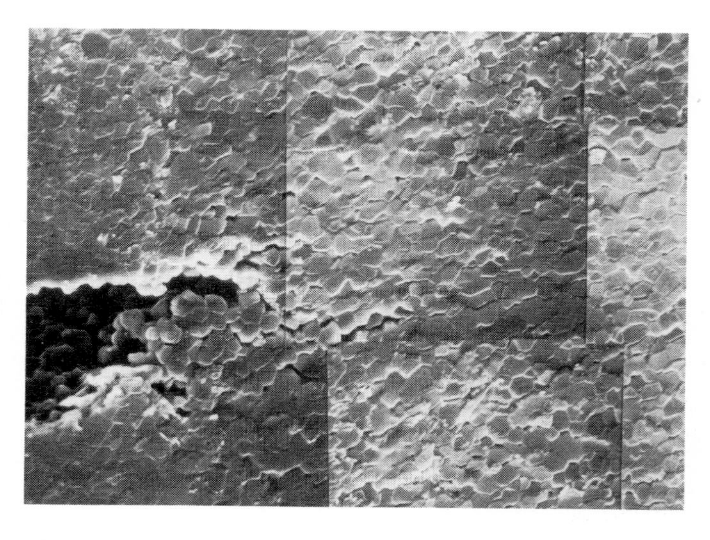

Fig. 2c. Scanning electron micrograph illustrating the process of crack propagation.

as detailed in the following sections. A common theme will be the role of microstructural heterogeneity on the observed cavitation (Fig. 4) and on the resultant crack nucleation characteristics. The inhomogeneity undoubtedly exhibits a direct link with the probabilistic aspects of failure and thus constitutes an essential ingredient in the formulation of engineering failure relations. Additionally, inhomogeneous cavitation can result in appreciable changes in the local stress (constraint) that both modifies the cavity growth rates and exerts a strong influence on the cavity coalescence process. Constraint in the presence of inhomogeneous cavitation is thus afforded separate attention.

2.1 The development of constraint

When cavities form within an isolated microstructural region, the local volume change is constrained by the surrounding material and induces modified local stresses [5,9,10]. The resultant stresses are dictated by the relative rates of cavity volume change and creep relaxation. The stress distributions in a polycrystalline aggregate are complex, and their rigorous determination requires extensive numerical computation. However, an approximate analytic solution pertinent to linear materials permits both the identification of the important creep rupture parameters and elucidates the essential trends. The analysis [5] is based upon a continuum solution for the transformation of a particle in a viscoelastic solid, and requires cavitation to occur within a zone of diameter d (Fig. 5), such that matter deposition on the intervening boundaries proceeds at a rate which differs from the average mass transport rate in the surrounding material (Fig. 5a). The enhanced matter deposition, δ, that occurs in time, Δt, induces rigid body displacements of the juxtaposed grains which, if unconstrained, would produce a shape change in the zone comprising these grains (Fig. 5b). The unconstrained shape change is analgous to a transformation strain, e_{ij}^T, [11] as depicted in Fig. 5c. Maintaining conformance of the 'transformation' zone with the surrounding, 'matrix' grains induces a constraint p_{ij}^I on the transformation zone and corresponding stresses in the matrix (Fig. 5d).

The constraint p_{ij}^I is dictated by the unconstrained transformation strain rate, $\dot{e}_{ij}^T$, and by the effective viscosities η of the transformation zone and matrix. The unconstrained strain-rate is the net cavity volume change that occurs within a specified time increment, Δt. Hence, since cavitation proceeds in response to stresses normal to the cavitating boundary, the appropriate $\dot{e}_{ij}^T$ derives from the cavity volume change in the presence of the <u>resultant</u> normal stress acting during the interval, Δt. The transformation strain is partially accommodated by viscous relaxation of

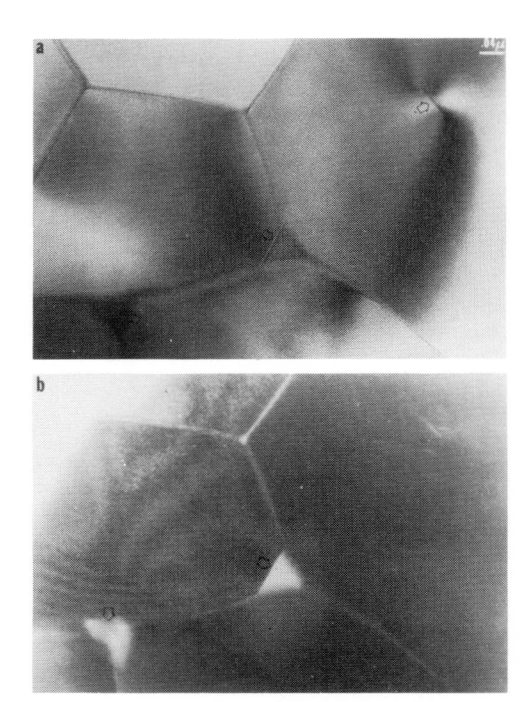

Fig. 3. Transmission electron micrographs of a Si_3N_4 material with a continuous second phase, transmission electronmicrograph.

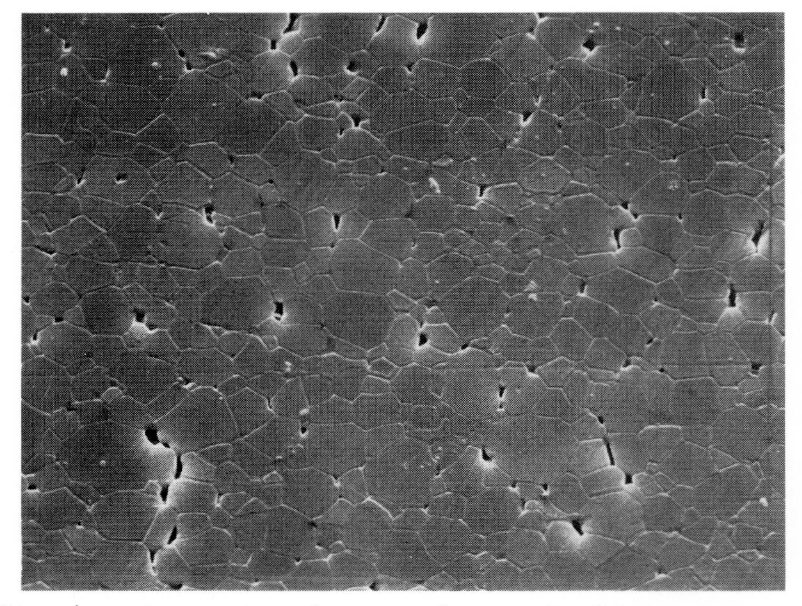

Fig. 4. A scanning electron micrograph of heterogeneous cavitation in polycrystalline Al_2O_3.

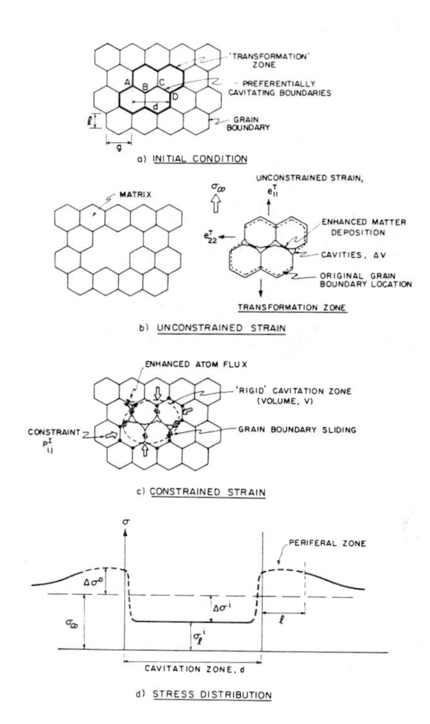

Fig. 5. A schematic of the constraint developed by inhomogeneous cavitation, (a) the initial condition, (b) the unconstrained strain, (c) the constrained strain, (d) the stress distribution.

the shear stresses during Δt (within both the transformation zone and the matrix). The resultant constrained stress determines p_{ij}^{I}.

The viscous deformation involves grain boundary sliding and diffusive flow [12]. The viscosity assigned to this mode of deformation depends upon the number of grains participating in the relaxation process (especially those grains at the periphery of the cavitation zone, Fig. 5c, where the shear stresses are most intense). The viscosity approaches the continuum value for the polycrystalline aggregate, $\eta_{cont.}$, [12] when a sufficiently large number of grains are involved;

$$\eta_{cont.} \equiv \sigma_{\infty}/\dot{\varepsilon}_{\infty} = 3\sqrt{3}\ell^{3}kT/14\Omega(\sqrt{3}\ell D_{\ell} + \pi D_{b}\delta_{b}) \quad , \qquad (1)$$

where $D_b\delta_b$ is the grain boundary diffusion parameter, D_ℓ is the lattice diffusivity, Ω is the atomic volume, ℓ is the grain facet length, σ_∞ is the applied stress and $\dot{\varepsilon}_\infty$ is the steady-state creep rate. It is assumed, for present purposes, that cavitation zones consisting of at least three grain facets (Fig. 5) embrace an adequately large number of peripheral grains (~30 peripheral grains for the three-dimensional zone subject to analysis).

The transformation strain-rate $\dot{e}_{ij}^{T}$ is determined by the distribution of matter deposition within the cavitation zone. It is a function of both the total cavitation volume, ΔV, the distribution of grain boundary orientations within the cavitation zone, and the zone shape. The general solution, which contains both deviatoric and dilational components, is unwiedly. Hence, specific results are presented for the two limits of most significance.

When the cavitation zone diameter is relatively small (such as the three cavitating facets depicted in Fig. 5), the constraint is essentially the same as that expected for a purely dilational transformation. The zone can therefore be considered subject to a dilation dictated exclusively by the cavitation volume, ΔV. Consequently, by equating the cavitation zone volume to that of a spheroidal region of equivalent size (Fig. 5c);

$$V \approx (\pi/3)d^{2}\ell \quad , \qquad (2)$$

the transformation strain rate becomes;

$$\dot{e}^{T} \approx 3\Delta\dot{V}/\pi d^{2}\ell \quad . \qquad (3)$$

This dilational strain results in a shape independent, upper bound constraint, given by;

$$p^I = -4\dot{e}^T\eta/3 \equiv -4\Delta\dot{V}\sigma_\infty/\pi d^2\ell\dot{\varepsilon}_\infty \quad , \tag{4}$$

and the local tension $\sigma_\ell{}^i$ normal to the cavitating boundaries becomes;

$$\sigma_\ell^i = \alpha\sigma_\infty + p^I/3 \quad , \tag{5}$$

where α is determined by the inclination of the boundary to the applied stress axis. Conservation of matter within the zone requires that;

$$\Delta\dot{V} \approx (\pi/3)d^2\dot{\delta} \quad , \tag{6}$$

where $\dot{\delta}$ is determined by the cavity growth mechanism. Hence,

$$p^I = -4\dot{\delta}\sigma_\infty/3\ell\dot{\varepsilon}_\infty \quad . \tag{7}$$

Specifically, for viscosities characterized by equation (1), the constraint becomes;

$$p^I \approx -\left(\frac{6\sqrt{3}}{7}\right)\frac{\ell^2\dot{\delta}kT}{\Omega(\sqrt{3}\ell D_\ell + \pi D_b\delta_b)} \quad , \tag{8}$$

and the local tensile stress normal to the cavitating boundaries depicted in Fig. 5 is given for fine grained materials ($\ell D_\ell << D_b\delta_b$) by

$$\sigma_\ell^i \approx \left(\frac{3}{4}\right)\sigma_\infty - \left(\frac{2\sqrt{3}}{7\pi}\right)\frac{\ell^2\dot{\delta}kT}{\Omega D_b\delta_b} \quad . \tag{9}$$

Other grain orientations yield slightly different results. The constraint reduces to a lower level than given by equation (9) when the zone approaches a free surface or, when an array of such zones, separated by $<d$, interact.

When the cavitation zone enlarges, such that $d > 6\ell$, an appreciable deviatoric stress develops, and the problem resembles that of a crack, diameter d, subject to opening displacements that accommodate the enhanced matter deposition along the intervening grain facets. The crack solution, pertinent to the large zone size limit, provides a constraint along the applied stress axis [5];

$$p_{11}^{I} \equiv -\frac{3\pi\eta\dot{\delta}}{2d} = -\frac{9\sqrt{3}}{28}\frac{\dot{\delta}\ell^{2}kT}{\Omega D_{b}\delta_{b}}\left(\frac{\ell}{d}\right) . \quad (10)$$

The tangential stresses outside the original cavitation zone are enhanced by the constraint on the cavity volume change. The stresses on those boundaries contiguous with the cavitating boundaries are of principal interest, because these stresses dictate the zone expansion and cavity coalescence processes that result in eventual crack nucleation. The stresses relate to the continuum stresses, as redistributed by local grain boundary sliding and diffusion. It may be assumed that the stress redistribution is confined primarily to those boundaries immediately adjacent to the cavitation zone; such that the average stress on the peripheral boundaries is similar to the average continuum stress. Cavity growth in the peripheral zone can then proceed at a rate dictated by this average stress. The upper bound continuum stress for the small cavitation zone, subject to dilation, is;

$$\sigma_{\ell}^{0} \approx (p^{I}/3)(\ell/x)^{3} + (3/4)\sigma_{\infty} , \quad (11a)$$

where x is the distance from the center of the cavitation zone. The average stress on the first peripheral zone is thus;

$$\langle\sigma\rangle_{\ell}^{0} \approx p^{I}/8 + (3/4)\sigma_{\infty} . \quad (11b)$$

The equivalent solutions at the large zone limit are;

$$\sigma_{\ell}^{0} = \sigma_{\infty} + p_{11}^{I}\left\{1 - x\left[x^{2}-(d/2)^{2}\right]^{-1/2}\right\} , \quad (12a)$$

$$\langle\sigma\rangle_{\ell}^{0} = \sigma_{\infty} + p_{11}^{I} - \frac{p_{11}^{I}}{4\ell(\ell+d)}\left[2(d+2\ell)\sqrt{(\ell^{2}+\ell d)} + d^{2}\ln\left(\frac{d+2\ell+2\sqrt{(\ell^{2}+\ell d)}}{d}\right)\right] . \quad (12b)$$

2.2 Cavity nucleation

In the absence of pre-existent porosity, the first step in the high temperature failure process consists of the nucleation of either cavities on grain boundaries or holes within viscous phases. The nucleation process is generally considered to involve a critical nucleus formed by the local accumulation of vacancies in a region subject to tensile stress. Cavity nucleation by vacancy coalescence can be treated using standard nucleation concepts [13] to demonstrate that a critical stress σ_c is needed to induce stable cavities and that this critical stress depends on the location of the nucleation site. The critical stress is determined by firstly identifying the condition, during cavity formation, that dictates the maximum change in thermodynamic potential. The potential contains terms due to the work done against the local stress during a cavity volume increment, ΔV, and terms associated with the change in surface and grain boundary areas, ΔA_s and ΔA_b, respectively, such that

$$\Delta\phi = -\sigma\Delta V + \gamma_s \Delta A_s - \gamma_b \Delta A_b \quad . \tag{13}$$

A critical nucleus exists when $\Delta\phi = 0$, as given by;

$$r^* = 2\gamma_s/\sigma^*$$

$$\sigma^* = \gamma_s (dA_s/dV)^* - \gamma_b (dA_o/dV)^* \quad , \tag{14}$$

where the asterisk indicates that parameters are evaluated at critical size r*. The critical nucleus concept can be used to deduce the critical stress by obtaining the cavity nucleation rate from the product of the number of nuclei at the critical size and the probability that a vacancy will be added to the critical nucleus [13]. The general result for a grain boundary located cavity is given by [13];

$$(\sigma^*)^2 \approx \frac{4\gamma_s^3 F_V(\psi)}{kT} \ \ell n \left[4\pi z \gamma_s D_b \delta_b \eta_o / \sigma^* \Omega^{4/3}\right] \quad , \tag{15}$$

where z is Zeldovich's factor ($\sim 10^{-2}$), ψ is the dihedral angle, $F_V(\psi)$ is a function that depends on the void location and η_o is the number of available nucleation sites per unit area of grain boundary. For a nucleus located on a two grain interface (Fig. 6a) [13];

$$F_V(\psi) = (2\pi/3) \left[2-3 \cos(\psi/2) + \cos^3(\psi/2)\right] \quad .$$

A typical trend in the critical stress for nucleation on two grain interfaces in Al_2O_3 (plotted in Fig. 7) indicates that, for typical dihedral angles ($\psi \gtrsim \pi/2$), the stress is many times larger than the applied stress levels ($\lesssim$ 50 MPa) known to initiate cavities at grain boundaries during creep tests. The resolution of this dilemma probably resides in a combination of two effects. A substantial reduction in the critical nucleation stress can be obtained when nucleation occurs either at three or four grain interfaces, at inclusions or within second phases. Also, stress concentrations can develop in creeping solids in the presence either of grain boundary sliding transients or of microstructural inhomogeneity.

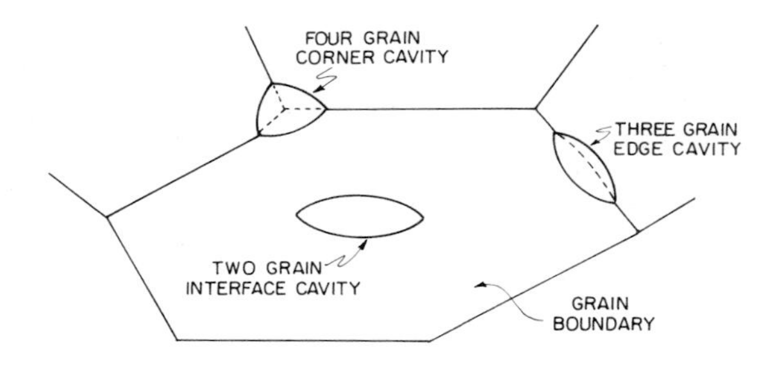

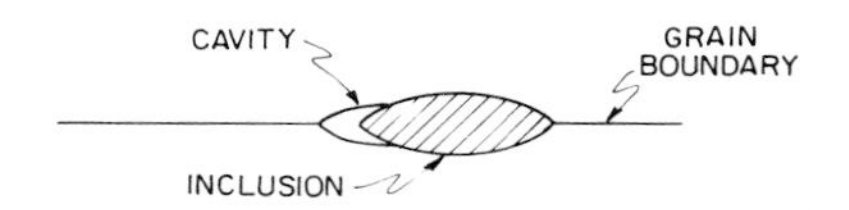

Fig. 6. Schematics of potential cavity nucleation sites in polycrystals, (a) cavities on grain interfaces, (b) a cavity at a grain boundary inclusion.

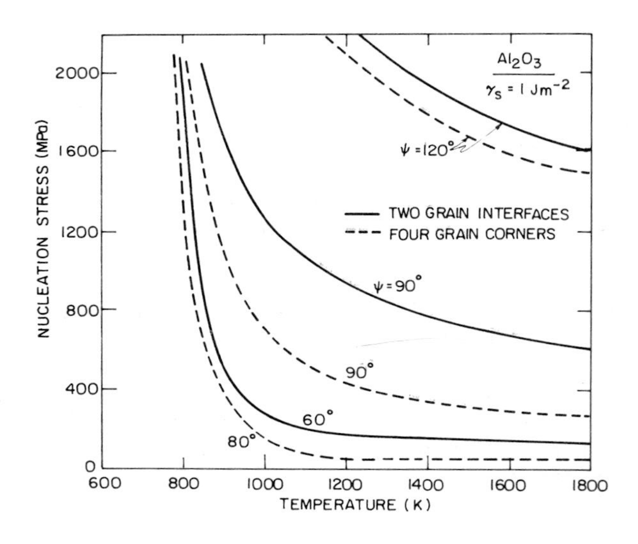

Fig. 7. The critical cavity nucleation stress in Al_2O_3 for cavities on two grain interfaces and at four grain corners, plotted for several dihedral angles.

A reduced critical nucleation stress is generally the direct consequence of a cavity morphology change that increases the relative change in volume to surface area during cavity enlargement (Eq. 2): as achieved by inducing maximum deviations from sphericity. Thus cavitation at three or four grain corners (Fig. 6 a) often occurs more readily than on two grain interfaces. Additionally, the critical stress exhibits a strong dependence on dihedral angle. For example, cavities on four grain corners nucleate spontaneously (at zero stress) when $\psi = 70°$ and quite readily when $\psi < 90°$ (Fig. 7), as expressed by the critical stress parameter [13];

$$F_V(\psi) = 8\left\{\pi/3-\cos^{-1}\left[\left[\sqrt{2}-\cos(\psi/2)\ (3-A^2)^{1/2}\right]/A\ \sin(\psi/2)\right]\right\}$$

$$+ A\cos(\psi/2)\left[(4\sin^2(\psi/2) - A^2)^{1/2} - A^2/\sqrt{2}\right]$$

$$- 4\cos(\psi/2)\left[3-\cos^2(\psi/2)\right]\sin^{-1}\left[A/2\ \sin(\psi/2)\right] , \qquad (16)$$

where,

$$A = (2/3)\ \left[\sqrt{2}(4\sin^2\psi/2-1)^{1/2} - \cos(\psi/2)\right] .$$

Equivalently, inclusions can be a major source of premature cavity nucleation, whenever the inclusion has an associated dihedral angle appreciably smaller than the equivalent matrix angle (Fig. 6b). A typical nucleation stress in the presence of a grain boundary inclusion is given by the nucleation parameter [13];

$$F_V(\psi) \approx (4\pi/3)\ \{2-3[\cos(\psi/2+\beta-\mu)/2] + \cos^3(\psi/2+\beta-\mu)/2\} ,$$

where

$$\beta = \cos^{-1}[(\gamma_{ib}-\gamma_i)/\gamma_s] \quad \text{and} \quad \mu = \cos^{-1}(\gamma_b/2\gamma_{ib}) , \qquad (17)$$

where γ_i is the energy of the inclusion free surface and γ_{ib} is the energy of the inclusion/matrix interface.

A continuous amorphous phase often provides an alternate source of premature cavitation in certain ceramics, by virtue of the relatively smaller surface energy of the amorphous phase ($\sim 0.2\ Jm^{-2}$ for many amorphous phases compared with $\sim 1\ Jm^{-2}$ for crystalline phases). Nucleation occurs by the formation of holes with the amorphous phase. Such materials are a typical consequence of liquid phase sintering and the existence of amorphous enclaves at three grain channels is common [14,15] (Fig. 8).[i] The amorphous pockets can be

---

[i] The enclaves are a consequence of anisotropy in the interface energy, resulting in facetted interfaces.

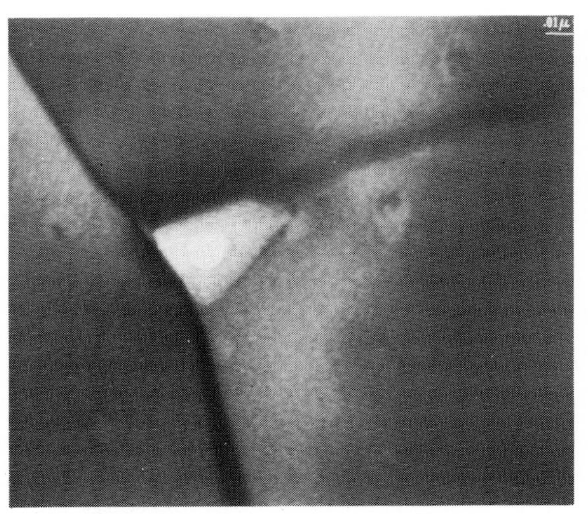

Fig. 8. A transmission electron micrograph of a hole formed within an amorphous pocket in Si_3N_4.

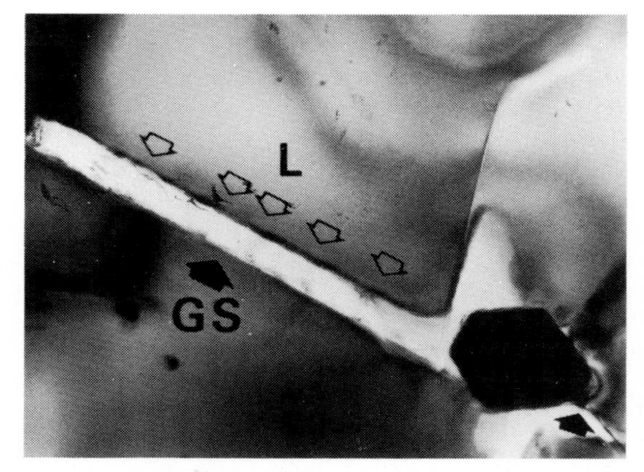

Fig. 9. A transmission electron micrograph of oblate holes within an amorphous phase along a two grain channel in Si_3N_4 (courtesy of D.R. Clarke).

where ℓ is the length of the freely sliding boundary (i.e., either the total length of the sliding boundary or the distance between impeding ledges, or inclusions), G is the shear modulus, and t_ℓ is the duration of the sliding transient. The maximum stress occurs at a distance [18], (Fig. 11)

$$z_\ell \approx 1.3(GD_b\delta_b\Omega t_\ell/(1-\nu)kT)^{1/3} , \tag{21}$$

from the position (ledge or corner) that impedes the sliding. It is tempting to invoke these transient sliding induced stresses as the source of grain boundary cavities [17]. However, caution should be exercised in applying transient related nucleation concepts until adequate attention has been devoted to a comprehension of the mechanism and duration of grain boundary sliding transients and to the duration of the stress concentration, $\hat{\sigma}$ (vis-a-vis, the time needed to nucleate a stable cavity). Neither of these topics has yet been examined in convincing detail.

Although a reasonable comprehension of the important issues involved in cavity nucleation in ceramic polycrystals has recently been developed, a satisfactory quantitative interpretation does not exist (primarily by virtue of an incomplete knowledge of grain boundary sliding transients). Nevertheless, the strong effects of dihedral angle that result in relatively easy nucleation in the range $70° < \psi < 110°$ (Eq. (16)), have lead to suggestions that cavities nucleate readily in polycrystalline ceramics [4,5], on that proportion of grain boundaries that exhibit low dihedral angles. However, direct evidence for this hypothesis does not exist, and nucleation should be regarded as an unresolved issue until further study has been completed.

The nucleation requirement can, of course, be negated if the as-sintered material contains appreciable remnant porosity at grain boundaries. In this context, it is interesting to note that the final removal of porosity during sintering becomes increasingly difficult when the dihedral angle approaches 70°. The tendency for porosity retention on low dihedral angle boundaries is consistent with the relative ease of cavity nucleation (and growth) along these same boundaries. The spectrum of dihedral angles in polycrystalline ceramics (and its dependence on the range of grain boundary and surface energies) thus emerges as a central feature of high temperature failure in ceramics.

2.3 Cavity growth mechanisms

The cavity growth rates in ceramics depend upon the spatial density and location of cavity nucleation sites and upon the mechanism of cavity growth. The cavity growth mechanisms are sufficiently distinct for materials that include a continuous, amorphous phase that these materials are examined separately. All other materials are regarded as 'single' phase with regard to their cavity growth behavior (inclusions, or precipitates, often dictate cavity nucleation propensities, but are presumed to exert a minimal influence on cavity propagation).

Before embarking upon specific analyses of cavity growth, some general considerations are presented. Problems of cavity growth are invariably analyzed by performing a series of inter-related calculations. For diffusive cavitation, the diffusion equations pertinent to the atom flux over the cavity surface (dictated by the curvature gradient) and along the grain boundary (determined by the normal stress gradient) are firstly solved, yielding the relations [5,20,21];

$$\Delta\dot{V} = \dot{a}F_1\left[D_s\delta_s\Omega\gamma_s/kT,\psi,a,\ell\right] , \tag{22a}$$

$$\dot{\delta} = \left[\sigma_\ell^i-\sigma_o h(f)\right]F_2\left[\Omega D_b\delta_b/kT,a,\ell\right] , \tag{22b}$$

where $D_s\delta_s$ is the surface diffusivity, σ_o is the sintering stress, h is a function of the relative cavity size and F_n are functions that depend upon the cavity morphology and spacing. A similar pair of equations pertain to hole growth by viscous flow (with the diffusivity terms replaced by a viscosity). Conservation of matter is then invoked, by allowing the atom flux leaving the cavity tip to equal the flux entering the grain boundary, to obtain [5,21],

$$\Delta\dot{V} = \dot{\delta}\, F_3(a,\ell) . \tag{22c}$$

Combining Eqs. (22) permits the cavity velocity to be expressed as;

$$\dot{a} = (F_3F_2/F_1)\,(\sigma_\ell^i-\sigma_o h) . \tag{23}$$

Finally, the local stress is obtained (Eqs. 5 and 7) as;

$$\sigma_\ell^i \equiv \alpha\,\sigma_\infty - p^I/3 = \sigma_\infty\left[\alpha-\Delta\dot{V}/\dot{\varepsilon}_\infty F_4(d\underline{V})\right] , \tag{24}$$

where $\underline{V}$ is the volume of the cavitation zone.

Combining Eqs. (23) and (24) the cavity growth rate becomes,

$$\dot{a}\left[1+\left(\frac{\sigma_\infty}{\dot{\varepsilon}_\infty \ell}\right)\frac{F_3F_2}{F_4}\right] = \frac{F_2F_3}{F_1}\,(\alpha\sigma_\infty - h\sigma_o) . \tag{25}$$

The second term in the parentheses on the left hand side is simply the modification to the velocity that derives from the constraint (i.e., $\dot{a} \to \dot{a}^u$, the unconstrained velocity, as $F_4 \to \infty$ and $p^I \to 0$). Hence, the cavity growth rate can invariably be expressed in the form;

$$\dot{a} = \dot{a}^u[1 + (\sigma_\infty/\dot{\varepsilon}_\infty)(F_2F_3/F_4)]^{-1} \quad . \tag{25a}$$

Consequently, the cavity propagation time, t_p, obtained by applying the integral,

$$t_p = \int_{a_o}^{\ell} (1/\dot{a})da \quad , \tag{26}$$

where a_o is the initial cavity size, always separates into two components [5,10];

$$t_p = \int_{a_o}^{\ell} (1/\dot{a}_u)da + (1/\dot{\varepsilon}_\infty)\int_{a_o}^{\ell} (F_1/F_4)\,[\alpha-\sigma_o h/\sigma_\infty]^{-1}\,da$$

$$\equiv t_p^u + t_p^c \quad , \tag{27}$$

where t_p^u is the propagation time in the absence of constraint and t_p^c is the additional contribution to the propagation time provided by the constraint. Note that t_p^c has the form;

$$t_p^c\dot{\varepsilon}_\infty = T(\sigma_\infty, \psi, D_s\delta_s\Omega\gamma_s/kT, \ell, \sigma_o) \quad . \tag{28}$$

The product $t_p^c\dot{\varepsilon}_\infty$ emerges because relaxation of the volume strain within the cavitation zone is dictated by the creep rate (viscosity) of the surrounding material. Highly constrained cavitation thus anticipates Monkman-Grant behavior ($t_p\dot{\varepsilon}_\infty$ = constant) in a natural way, irrespective of the specific mechanisms of creep and cavitation.

Comparison of t_p^u with t_p^c frequently indicates that the latter is large and dominates the failure process. The circumstances appropriate to such a comparison are afforded explicit attention in the following section. In the present section, t_p^u and the 'fully-constrained' t_p^c are presented for each cavitation mechanism. It is noted, however, that the F_2 independence of $t_p^c\dot{\varepsilon}_\infty$ indicates that fully constrained cavity growth is dictated by cavity geometry and volume conservation requirements (invariant with the atom flow rate between cavities, which determines F_2). Several different atom transport mechanisms can thus be expected to yield similar fully-constrained cavitation times.

2.3.1 'Single' phase materials

Cavities in single phase ceramic polycrystals invariably form at grain boundaries, in accord with one of three morphological types: equilibrium, crack-like or finger-like. Equilibrium cavities prevail at low stresses and large cavity spacings, crack-like cavities develop at high stresses or small cavity spacings, while finger-like cavities become important at high cavity growth rates (especially in coarse-grained polycrystals). The specific rates of growth of cavities exhibiting these general morphological characteristics depend upon the cavity nucleation sites (at two, three or four grain interfaces). Observations performed on polycrystal ceramics suggest that cavity nucleation on three/four grain interfaces (Fig. 2a) prevails in fine-grained materials [4], whereas nucleation on two grain interfaces (Fig. 12) becomes increasingly important in more coarse-grained materials [22].

(a) Fine-grained materials

Cavity nucleation in fine grained materials presumably initiates at those grain corners with low dihedral angles (section 2.2), especially when subject to the transient sliding of juxtaposed grain boundaries. The cavities initially extend from their nucleation sites to occupy a three grain channel (Fig. 13a). This process occurs relatively quickly. The resultant equilibrium-shaped cavities then expand (Fig. 13b) and retain their equilibrium morphology, while the cavity is small. Retention of the equilibrium shape requires that the surface flux be sufficiently large (by virtue of a large surface curvature) that attempted deviations from curvature uniformity (as motivated by the atom flux into the boundary from the cavity tip) are instantly removed [20]. Continued expansion of the cavity reduces the surface curvature and eventually, attempted deviations from curvature uniformity are retained. Thereupon, a transition to a crack-like cavity morphology ensues, (Fig. 13c) and the resultant cavity extends preferentially along boundaries approximately normal to the applied tension [5,20]. The crack-like cavity continues to extend along the grain interface, to form a full-facet cavity (Fig. 13d). The full-facet cavities are generally resistant to extension along the contiguous boundaries and consequently increase their volume by a thickening process [4]. The length stability of the full-facet cavities is presumably a consequence of the grain boundary sliding that occurs in response to the atom flux along the cavitating boundary. This reduces the normal tension on the sliding boundaries (section 2.2) and thus reduces the boundary flux at the grain corner to a level that can be accommodated by a surface flux acting over the full length of the cavity.

Fig. 12. Scanning electron micrograph of cavities on two grain interfaces in coarse grained (15 μm) Al_2O_3.

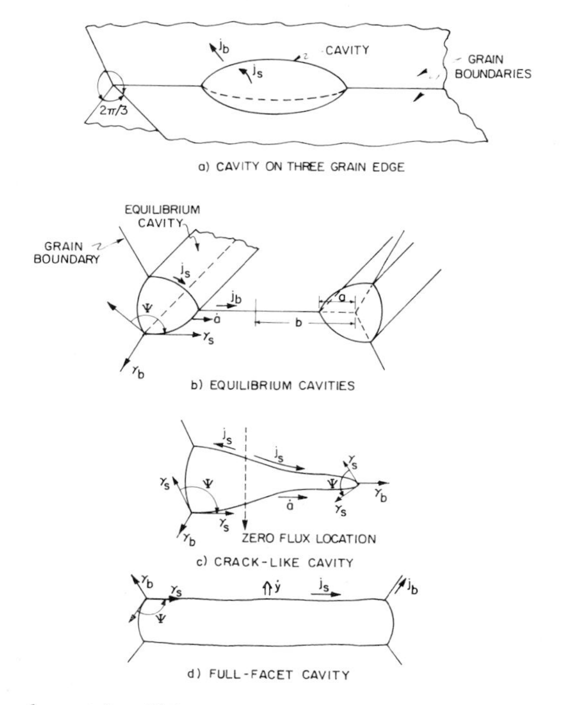

Fig. 13. A schematic illustrating the progression of a cavity from (a) nucleus on a three or four grain site to, (b) a cylindrical equilibrium cavity, (c) a crack-like cavity, (d) a full facet cavity.

The growth rate of the equilibrium cavities can be deduced by firstly evoking matter conservation which, for unit width of the cavitation zone, requires that [5,21],

$$\dot{V}_{eqm} \approx \dot{\delta} \ell \quad , \tag{29}$$

where $\dot{V}$ is the rate of volume change of an individual cavity. The volume of an equilibrium-shaped, cylindrical, triple junction cavity is (for unit width) [5]

$$V_{eqm} = 3\sqrt{3}\, a^2\, F(\psi)/4 \quad , \tag{30}$$

where a is the distance of the cavity tip from the original site of the triple junction (Fig. 13b) and

$$F(\psi) = 1 + \frac{\sqrt{3}[\psi - \pi/3 - \sin(\psi - \pi/3)]}{2\sin^2(\psi/2 - \pi/6)} \quad . \tag{31}$$

The rate of volume change is thus;

$$\dot{V}_{eqm} = 3\sqrt{3}\, a\, \dot{a}_{eqm} (F(\psi)/2) \quad . \tag{32}$$

The cavity velocity is related to the additional matter deposition, from Eqs. (29) and (32), by;

$$\dot{\delta} = (3\sqrt{3}/4) \dot{a}_{eqm}\, f\, F(\psi) \quad , \tag{33}$$

where $f = 2a/\ell$. The matter deposition is also related to the level of the local stress over the intervening boundaries [5];

$$\dot{\delta} = \frac{12 \Omega D_b \delta_b}{kT\ell^2} \frac{[\sigma_\ell^i - (1-f)\sigma_o]}{(1-f)^3} \quad , \tag{34}$$

where σ_o, the sintering stress, is given by;

$$\sigma_o = 2\gamma_s h(\psi)/\sqrt{3}\, a$$

$$h(\psi) = \sin[\psi/2 - \pi/6] \quad . \tag{35}$$

The cavity velocity is thus,

$$\dot{a}_{eqm} = \frac{16 \Omega D_b \delta_b}{\sqrt{3}\, kT\ell^2} \frac{[\sigma_\ell^i - \sigma_o(1-f)]}{F(\psi)\,(1-f)^3} \quad . \tag{36}$$

The magnitude of the local stress pertinent to Eq. (36) is deduced by noting that the matter deposition given by Eq. (34) must be compatible with the development of the local stress induced by the constraint of the surrounding material. For example, using the upper bound constraint (Eq. 9) pertinent to a small cavitation zone (e.g., Fig. 5), the local stress becomes;

$$\sigma_\ell = [(3/4)(1-f^3)\sigma_\infty + (24\sqrt{3}/7\pi)(1-f)\sigma_o]$$

$$[(1-f)^3 + (24\sqrt{3}/7\pi)]^{-1} \quad . \tag{37}$$

Combining Eqs. (36) and (37) the final relation for the highly constrained cavity velocity, expressed in dimensionless form, becomes;

$$\dot{a}^c_{eqm}\left(\frac{kT\ell^3}{\Omega D_b\delta_b\gamma_s}\right) = \frac{(16/\sqrt{3})[(3/4)(\sigma_\infty\ell/\gamma_s)f - (4/\sqrt{3})h(\psi)(1-f)]}{F(\psi)f^2[(1-f)^3 + (24\sqrt{3}/7\pi)]} , \tag{38a}$$

which reduces for $\sigma_\infty \gg \sigma_o$ to;

$$\dot{a}^c_{eqm}\left(\frac{kT\ell^3}{\Omega D_b\delta_s}\right) \approx \left(\frac{7\pi}{6}\right)\frac{\sigma_\infty\ell}{F(\psi)} \quad . \tag{38b}$$

For unconstrained conditions, the equivalent result ($\sigma_\infty \gg \sigma_o$) is;

$$\dot{a}^u_{eqm}\left(\frac{kT\ell^3}{\Omega D_b\delta_b}\right) \approx \left(\frac{12}{\sqrt{3}}\right)\frac{\sigma_\infty\ell}{F(\psi)f(1-f)^3} \quad . \tag{38c}$$

The variations of the cavity velocity with cavity length and with the dominant variables ($\sigma_\infty\ell/\gamma_s,\psi$) are exemplified in Figs. 14a and 15. The corresponding change in the local stress is plotted in Fig. 14(b).

The equivalent analysis of crack-like cavity growth can be conducted by noting that both the cavity profile and the atom flux at the tip of well developed crack-like cavities depend on the instantaneous cavity velocity; viz. the prior, equilibrium morphology of the cavity is of minor significance [23]. The growth process can thus be adequately treated by focusing on the tip region, and neglecting complex morphological changes that may be occurring in the vicinity of the cavity center. Commencing with the expression for the surface flux at the tip of a crack-like cavity [20]

$$\Omega J_s = 2\sin(\psi/4)\dot{a}^{2/3}_{crack}(D_s\delta_s\Omega\gamma_s / kT)^{1/3} , \tag{39}$$

and noting that the surface flux is related to the volume rate of matter removal, up to the zero flux position (Fig. 13c), by;

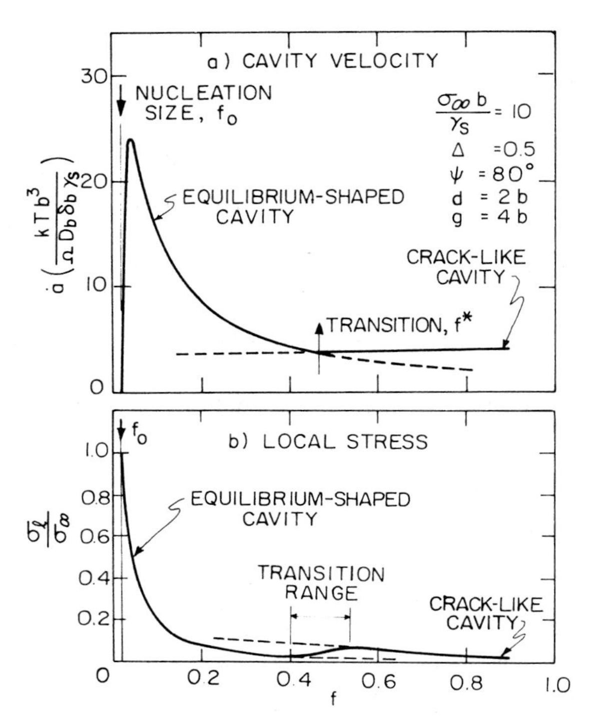

Fig. 14. (a) The variation in cavity velocity with cavity length, indicating the crack like transition, (b) the change in the local stress with cavity length.

$$J_s = \dot{V}_o/2\Omega \quad , \tag{40}$$

and that the matter removed from the cavity tip must be deposited on the grain boundary, in order to satisfy matter conservation,

$$\dot{V}_o \approx \dot{\delta}\ell/2 \quad , \tag{41}$$

the boundary 'thickening' rate becomes;

$$\dot{\delta} = 8 \sin(\psi/4)\dot{a}_{crack}^{2/3}(D_s\delta_s\Omega\gamma_s/kT\ell^3)^{1/3} \quad . \tag{42}$$

Combining Eq. (42) with the relations for the boundary transport problem (Eq. (34)) and for the upper bound local constraint (Eq. (9)) permits the cavity velocity to be derived. The velocity is given by [5];

$$\nu^{2/3}\xi^{1/3}[(2/3)(1-f)^3 + (16\sqrt{3}/7\pi)] + 2\nu^{1/3}(1-f)\xi^{-1/3}$$

$$= (3/4)\sigma_\infty\ell/\gamma_s \sin(\psi/4) \quad , \tag{43}$$

where

$$\nu = \dot{a}_{crack}(kT\ell^3/D_b\delta_b\Omega\gamma_s) \quad \text{and} \quad \xi = D_s\delta_s/D_b\delta_b \quad .$$

For situations of practical interest,

$$\sigma_\infty\ell/\gamma_s \sin(\psi/4) \geq 1 \quad .$$

Hence, Eq. (43) reduces, for this highly constrained situation, to,

$$\nu_c^{2/3} \approx 0.15[\sigma_\infty\ell/\gamma_s \sin(\psi/4)]\ \xi^{-1/3} \quad . \tag{44a}$$

An almost constant velocity is thus anticipated in the crack-like region during the highly constrained initial state of cavitation. However, when the constraint is reduced in the later stages of cavitation, cavity acceleration is anticipated, and the cavity velocity attains an unconstrained level given by;

$$\nu_u^{2/3} \approx (3/2)\ [\sigma_\infty\ell/\gamma_s \sin(\psi/4)]\ (1-f)^{-3}\ \xi^{-1/3} \quad . \tag{44b}$$

At very low stress levels ($\sigma_\infty\ell/\gamma_s \sin(\psi/4) << 1$), the unconstrained cavity velocity, given by,

$$\nu_u^{1/3} \approx (3/8)\ [\sigma_\infty\ell/\gamma_s \sin(\psi/4)]\ (1-f)^{-1}\ \xi^{-1/3} \quad . \tag{44c}$$

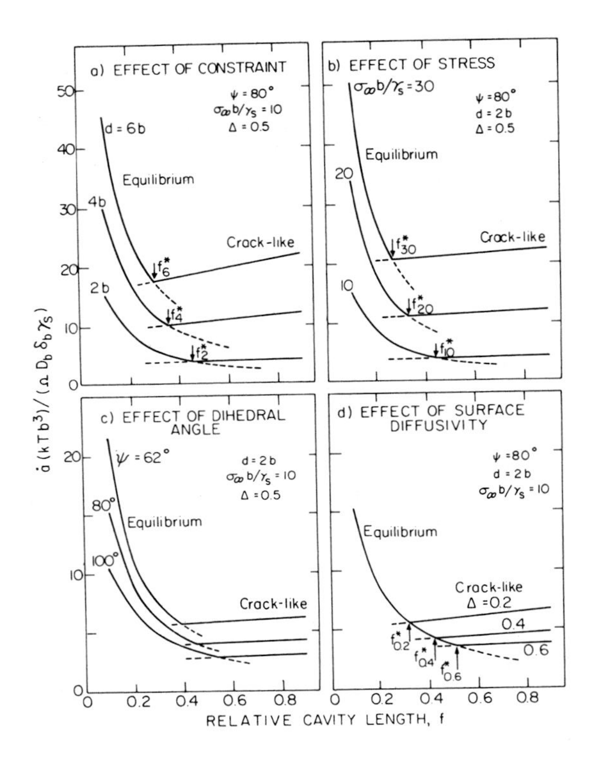

Fig. 15. Trends in the cavity velocity with cavity length.

exhibits a stronger stress dependence ($\nu_u \propto \sigma_\infty$). But, this regime is not likely to be encountered in most practical situations. The trends in the cavity velocity predicted by the above analyses are plotted in Figs. 14a and 15, and the local stress is plotted in Fig. 14b. The transition between the equilibrium and crack-like morphologies is considered favorable when the crack-like velocity exceeds that for equilibrium cavities, and the local stress is assumed to adjust to the crack-like value, over the transition range (Fig. 14b).

The important trends in cavity velocity are illustrated in Fig. 15. Firstly, the strong influence of the constraint upon initial cavitation is noted (Fig. 15a). The effect is manifest at the very earliest stages of cavity growth and continues to be amplified as the extension proceeds. The magnitude of the applied stress (Fig. 15b) also has a substantial effect on the cavity velocity, over the entire range.

The material parameters with the dominant influence upon inhomogeneous cavitation are the local values of the dihedral angle, ψ (Fig. 15c), and the ratio ξ of the surface to boundary diffusivity (Fig. 15d). Smaller values of these quantities encourage cavitation. This may account for the observation that crack-like cavities exhibit relatively small dihedral angles (Fig. 16).

The time t_p taken for cavities to extend across grain interfaces can be deduced from the cavity velocities using,

$$t_p/\ell = \int_{f_o}^{f^*} \frac{df}{\dot{a}_{eqm}} + \int_{f^*}^{f} \frac{df}{\dot{a}_{crack}} \quad . \tag{45}$$

Some typical propagation times for highly constrained cavitation are plotted in Fig. 17. When the dihedral angle or the local surface diffusivity are small and/or the stress is relatively large, most of the time required to develop a full facet length cavity is dominated by the growth in the crack-like mode (as might be anticipated from the velocity diagrams). The initial cavitation that occurs in local regions of a creeping polycrystal (due to small local values of ψ or D_s) can thus be approximately characterized for highly constrained, localized cavitation by a constant cavity velocity, whereupon the propagation time (except at impractically low stresses, $\sigma_\infty \ell/\gamma_s \sin(\psi/4) < 1$) becomes;

$$t_p^c \dot{\varepsilon}_\infty \approx 50\,\pi(\gamma_s/\sigma_\infty \ell)^{1/2} \sin(\psi/4)^{3/2} \xi^{1/2} \quad . \tag{46}$$

The equivalent propagation time, for uniform unconstrained cavitation is dominated (for most practical stress levels) by equilibrium cavity growth (Eq. (38d)), and given ($\sigma_\infty >> \sigma_o$) by;

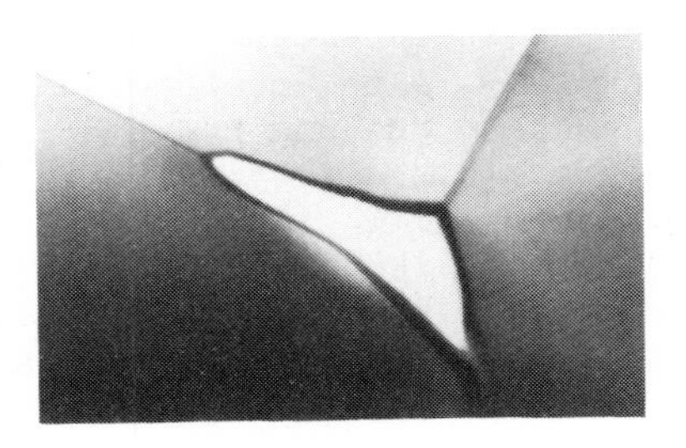

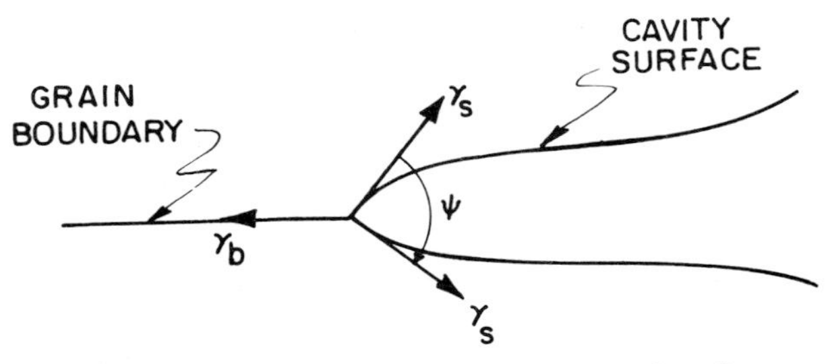

Fig. 16. A transmission electron micrograph of a crack-like cavity in Al_2O_3.

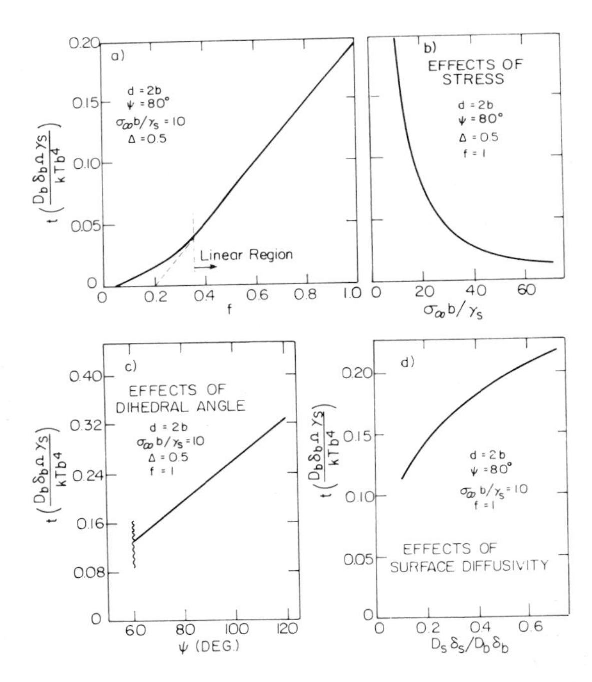

Fig. 17. Cavity propagation times plotted as a function of (a) cavity length, (b) stress, (c) dihedral angle and, (d) diffusivity ratio.

$$t_p^u \left(\frac{\sigma_\infty \Omega D_b \delta_b}{kT\ell^3} \right) \approx \frac{F(\psi)}{80\sqrt{3}} \left[1 - 10 f_o^2\right] , \qquad (47)$$

where f_o is the initial cavity size (a_o/ℓ). The full-facet cavities, once formed, exhibit a thickening rate, $\dot{y}$, given by [4];

$$\dot{y} = \frac{3D_b \delta_b \Omega}{kTb^2 \lambda[1-\lambda+\lambda^2/\xi']} \left\{ \frac{\sigma_\infty}{(1-\lambda)} - \frac{\gamma_s \tan(\psi/2-\pi/3)}{b\lambda} \right\} , \qquad (48)$$

where 2b is the spacing between neighboring full facet cavities and $\lambda = \ell/b$.

(b) Coarse-grained materials

In coarse-grained ceramics, cavity nucleation on two grain interfaces becomes more prevalent [22] (Fig. 12). The source of these cavities has not been studied. Analogy with metallic systems [17] would suggest nucleation at grain boundary precipitates, especially on sliding boundaries. However, in sintered coarse-grained materials, the remnant porosity during final stage sintering occurs primarily along two grain interfaces (Fig. 18). The cavities in ceramics could thus be equally plausibly associated with fine residual pores. Cavity nucleation at grain boundary ledges that impede sliding (Fig. 19) is another possibility [24]. This issue requires resolution, because the spacing between cavity sites, b, has a profound influence on the cavity growth rates.

The cavities on two grain interfaces also exhibit equilibrium, crack-like and full-facet growth morphologies, but with different geometric characteristics [21] than the cavities at three-grain edges. However, the method of analysis essentially duplicates that described for cavities on three grain corners and hence, only the pertinent cavity growth expressions are presented. Of primary interest are the times required to coalesce adjacent cavities (spacing b) and thereby, to create a full-facet cavity. The principal difference between the full-facet cavity formation times for cavities on two and three grain interfaces resides in the interchange between the cavity spacing b and the grain facet length, ℓ (the effective cavity spacing for three grain edge cavities). The geometric effects also result in important differences for equilibrium cavities at small dihedral angles ($\psi \sim \pi/3$). Otherwise the results generally deviate by less than an order of magnitude. For stress levels at which equilibrium cavity growth dominates, $\sigma_\ell a/\gamma_s < 1.9(1+\xi b/a)$, the coalescence times (subject to the constraint derived from Eq. (10), with $a = \ell$) are given (for $\sigma_\infty \gg \sigma_o$) by [10,21];

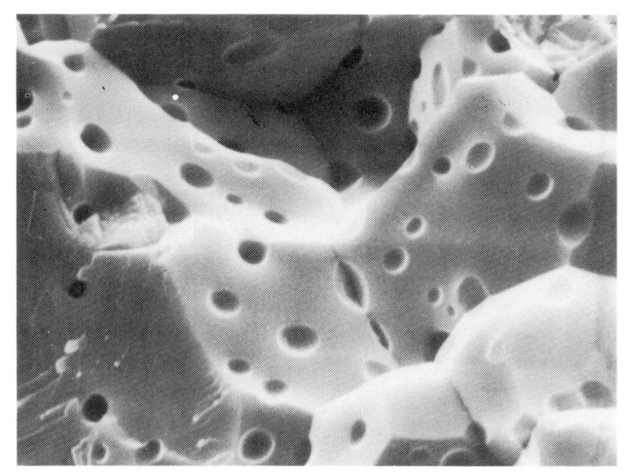

Fig. 18. A scanning electron micrograph of remnant porosity on two grain interfaces in MgO.

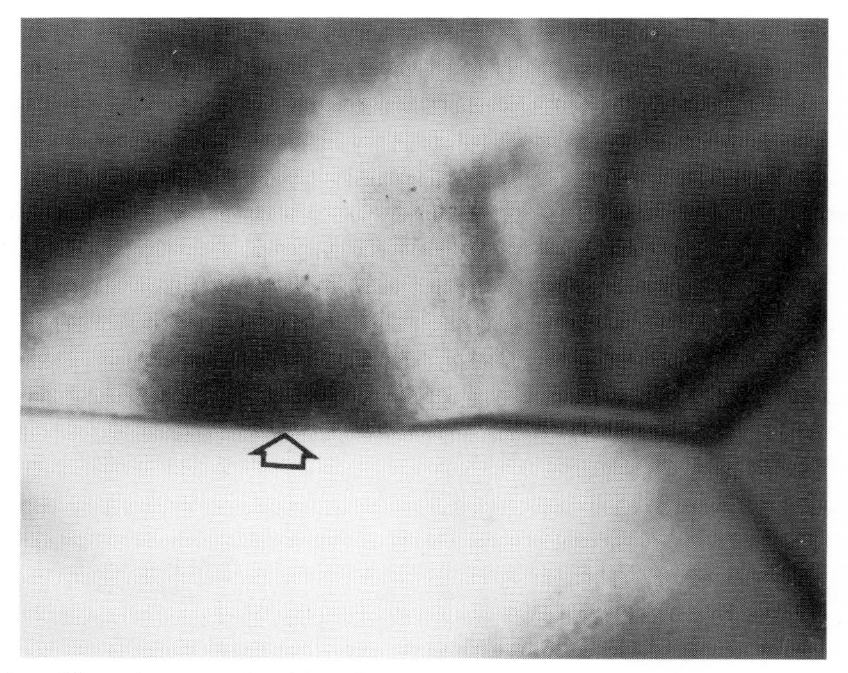

Fig. 19. A transmission electron micrograph of strain contours in the vicinity of a grain boundary sliding impediment in Al_2O_3 (courtesy of J. Porter).

$$t_p^c \dot{\varepsilon}_\infty \approx (2\pi/3) \left[1-(a_o/b)^3\right] g(\psi) (b/\ell) \quad ,$$

$$\frac{t_p^u \sigma_\infty \Omega D_b \delta_b}{kTb^3} \approx 0.07\left[1-(a_o/b)^2\right]^3 g(\psi) \quad , \qquad (49)$$

where, $g(\psi) = \text{cosec}(\psi/2)[(1+\cos\psi/2)^{-1}-(1/2)\cos\psi/2]$. Crack-like cavity growth has not yet been analyzed under constrained conditions. However, in the absence of constraint, at stress levels of most practical concern ($\sigma_\infty b\xi/\gamma_s \sin(\psi/4) > 10$) the coalescence time due to crack-like growth is given by [21];

$$\frac{t_p^u \sigma_\infty^{3/2} D_b \delta_b \Omega}{kT\gamma_s^{1/2} b^{5/2}} \approx 0.33\, \xi^{1/2} \sin^{3/2}(\psi/4)\, H(a_o/b) \quad , \qquad (50a)$$

where,

$$H(a_o/b) = [1-a_o/b] \left\{1-(a_o/b)(1+a_o/b)\ [19-16(a_o/b)-5(a_o/b^2)]/16\right\}$$

while at very low stress levels ($\sigma_\infty b\xi/\gamma_s \sin\psi/4 < 1$), cavity coalescence in the crack-like mode occurs within a time [21];

$$\frac{t_p^u \sigma_\infty^3 D_b \delta_b \Omega}{kT\gamma_s^2 b} \approx 3.6 \sin^3(\psi/4) H(a_o/b)\xi^{-1} \quad . \qquad (50b)$$

Several remarks concerning the latter results are appropriate at this juncture. Similar expressions describe unconstrained crack-like cavity growth from three grain edges in fine grained materials (as derived from Eq. (44b) and (44c) respectively); but these expressions were not presented in the preceding discussion because, in that case, they do not appear to describe a behavioral realm of practical significance. The practical utility of Eq. (50) has yet to be adequately assessed. Nevertheless, the stress dependences provided by these growth processes are of interest. The non-linearity, which arises from a stress dependent cavity width (and extends into descriptions of constrained crack-like cavity growth, as evidenced by Eq. (46)) has been invoked [25] as a source of non-linear rupture behavior observed in the presence of (intrinsically linear) diffusive cavity growth mechanisms. However, some caution must be exercised, by recognizing the limited local stress conditions that must obtain when these modes of cavity growth operate. This issue will be more extensively addressed in a subsequent section.

(c) Finger-like growth

Cavities propagating in the crack-like mode are capable of attaining velocity levels at which the cavity tip becomes unstable in the presence of small perturbations [26]. Subsequent cavity growth then proceeds at an accelerated rate by the growth of finger-like entities from the cavity front (Fig. 20). Analysis of this instability at the tip of a crack-like cavity provides information pertinent to the wavelength, λ_c, of the fastest growth disturbance [26];

$$\lambda_c \approx 2\pi \left(\frac{\gamma_s \Omega D_b \delta_b}{kT\dot{a}}\right)^{1/3} \xi^{-1/6} \quad . \tag{51}$$

Presumably, therefore, a crack-like cavity is capable of developing instabilities when λ_c is appreciably smaller than the total length, z, of the cavity front, e.g., when $\lambda_c \tilde{<} z/10$. Hence, for cavities emanating from three grain corners, with a cavity front length $z \approx \ell$, insertion of the highly constrained cavity velocity from Eq. (44) into Eq. (51) suggests that finger-like cavities develop when the grain facet length exceeds a critical value, ℓ_c, given by;

$$\ell_c \sigma_\infty / \gamma_s \ \sin(\psi/4) \ \tilde{>} \ 5.10^3. \tag{52}$$

Inserting some typical values for creep loading of ceramic materials ($\sigma_\infty \approx 10^8$ Pa, $\gamma_s \approx 1$ Jm^{-2}) indicates that finger-like growth under highly constrained conditions is unlikely in fine grained materials ($\ell < 5$ μm) but probable in more coarse grained polycrystals.

Less stringent limitations on finger formation obtain for unconstrained conditions, because the cavity velocity becomes unbounded as $f \rightarrow 1$. For this case, the transition to finger-like growth occurs when;

$$\ell_c \sigma_\infty / \gamma_s \ \sin(\psi/4) \ \tilde{>} \ 5.10^2 (1-f)^3 \quad . \tag{53}$$

Hence, finger-like growth should be observed, under unconstrained conditions, even in very fine grained materials ($\ell \sim 1$ μm), when the cavity size $f > 1/2$. The observation of finger-like entities should thus provide an indication of locally unconstrained cavitation.

2.3.2 Materials with a continuous amorphous phase

The critical stress levels for hole formation within an amorphous phase at three and two grain channels (Eqs. (18) and (19)) provide the basis for interpreting and predicting high temperature failure in liquid phase sintered ceramics [15]. At stress levels below the stress needed to nucleate holes at three grain channels, no known failure mechanism exists and the material should deform continuously, without failure, by a

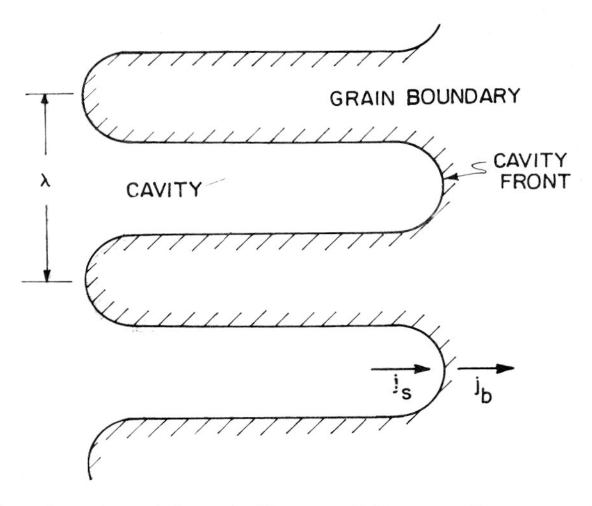

Fig. 20. A schematic of finger-like cavity growth.

solution/reprecipitation mechanism [27]. The lower critical stress thus constitutes a failure threshold. The practical utility of this threshold is dictated by the magnitude of the tensile stress concentrations that develop at three grain channels, in response to grain boundary sliding and grain rotation. Any tendency toward local dilation by a combination of sliding and rotation substantially enhances the tension within the amorphous pockets and inevitably nucleates holes [28]. However, the detailed analyses (of sliding, viscous flow and solution/reprecipitation that accompany deformation in these materials) needed to compute the magnitude and duration of the stress concentration have yet to be performed. The existence of a practical threshold will thus be regarded an ambiguous issue until further theoretical and experimental work has been conducted. Nevertheless, it is interesting to note that abrupt grain boundary sliding events introduce highly transient tensile stress concentrations (Fig. 11) that are relieved by viscous flow within the pocket and that the net hydrostatic stress within symmetric pockets (Fig. 21a) is zero [18]. Net dilational stresses only occur in certain assymetric pockets (Fig. 21b). The resultant stress duration is dictated by the rate of flow of amorphous material from the neighboring two grain channels. The morphology of the amorphous pockets and the thickness of the two grain channels thus exert an important influence on the failure threshold.

At stress levels above that needed to nucleate holes within two grain channels (Fig. 10, Eq. (19)), facet-sized cavities can develop by the growth and coalescence of the oblate holes within the channel (Fig. 22a). The specific hole growth rate can be computed from standard expressions for the pressure distribution that develops within a parallel sided channel containing a flowing viscous fluid [29];

$$\nabla^2 \sigma = 12\eta\dot{\delta}/\delta^3 \quad , \qquad (54)$$

where δ is the thickness of the fluid channel. When the fluid perfectly wets the solid (as required for liquid phase sintering) and contains an array of holes with spacing 2b and radius a, the boundary conditions needed to solve Eq. (54) can be specified as follows: symmetry at the mid position between holes requires that $d\sigma/dr = 0$ at $r = b$ (r is the distance from the center of the hole) and chemical potential continuity at the hole surface demands that $\sigma = \gamma_s(1/a + 2/\delta)$ at $r = a$. Then, imposing volume conservation;

$$V_\ell = \delta(2\sqrt{3}\, b^2 - \pi a^2) \quad , \qquad (55)$$

where V_ℓ is the fluid volume in the region between prospective cavities, the solution to Eq. (54) can be expressed in terms of the hole growth rate as [15,29];

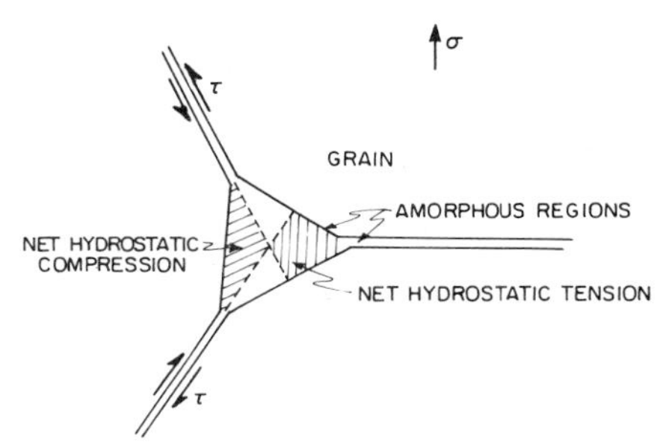

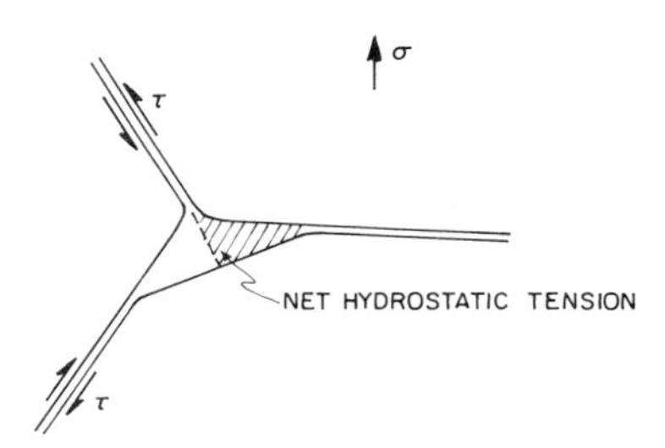

Fig. 21. Schematics illustrating the stressed zones in amorphous pockets induced by grain boundary sliding, (a) a symmetric pocket, (b) an asymmetric pocket.

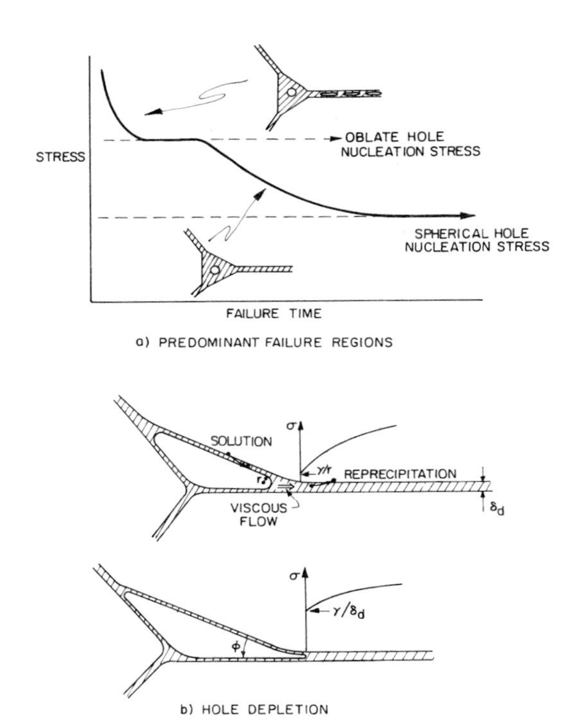

Fig. 22. A schematic of the potential failure processes that can occur in a liquid phase sintered material containing a continuous second phase, (a) general effects of stress, (b) depletion of amorphous pocket.

$$\dot{\alpha} = \left(\frac{\delta_o^2}{5b^2\eta}\right) \frac{(1-0.9\alpha_o^2)\ [\sigma_\ell^i-\gamma_s(1/\alpha b+2/\delta)(1-0.9\alpha^2)]}{\alpha\ (1.1-\alpha^2)\ [0.96\alpha^2-\ell n\alpha-0.72-0.23\alpha^4]} \quad , \tag{56}$$

where $\alpha = a/b$ (α_o is the initial value at hole nucleation) and δ_o is the initial channel thickness. The local stress can be deduced by firstly differentiating Eq. (55),

$$\dot{\delta} = \frac{2\pi V_\ell \alpha\dot{\alpha}}{(2\sqrt{3}-\pi\alpha^2)^2 b^2} \equiv \frac{4\sqrt{3}\pi\delta_o\alpha\dot{\alpha}}{(2\sqrt{3}-\pi\alpha^2)^2} \quad , \tag{57}$$

and combining with Eq. (7) to obtain (for a boundary normal to the applied stress);

$$\sigma_\ell^i/\sigma_\infty = 1 - 16\sqrt{3}\pi\ \delta_o\alpha\dot{\alpha}/9\ell(2\sqrt{3}-\pi\alpha^2)^2\dot{\varepsilon}_\infty \quad . \tag{58}$$

Inserting the local stress into Eq. (56) and integrating between α_o and 1 gives the fully-constrained time;

$$t_p^c\ \dot{\varepsilon}_\infty = (8\sqrt{3}/9\pi)\ (\delta_o/\ell)\ T_1^c\ (a_o/b, \sigma\delta_o/\gamma_s) \quad , \tag{59a}$$

where, for the important case $\delta < a_o$,

$$T_1^c = [4.6\gamma_s/\sigma_\infty\delta_o-1]^{-1} \left\{(1.1-\alpha_o^2)^{-1} - 10 + [(2.3-\sigma_\infty\delta_o/2\gamma_s]^{-1} \ell n\left[\frac{(22-5\sigma_\infty\delta_o/\gamma_s)(1.1-\alpha_o^2)}{1.2+\alpha_o^2-0.5\sigma_\infty\delta_o/\gamma_s}\right]\right\} ,$$

which reduces for $\sigma_\infty\delta_o/\gamma_s >> 1$ to

$$T_1^c = 10 - [1.1 - \alpha_o^2]^{-1}$$

or

$$t_p^c\dot{\varepsilon}_\infty \approx \left(\frac{80\sqrt{3}}{9\pi}\right)\left(\frac{\delta_o}{\ell}\right)\left\{1 - 0.1\ [1.1-(a_o/b)^2]^{-1}\right\} .$$

Note that the hole spacing b has a relatively minor influence on t_p^c (at least when $a_o << b$) and that $t_p^c\dot{\varepsilon}_\infty$ increases as δ_o increases. (However, t_p^c may not increase as δ_o increases because $\dot{\varepsilon}_\infty$ may exhibit an inverse dependence on δ_o, e.g., for diffusion limited solution/reprecipitation creep [27]). The equivalent result for unconstrained hole growth is [15];

$$t_p^u \sigma_\infty / \eta = T_1^u(b/\delta_o, a_o/b) \quad , \tag{59b}$$

where T_1^u is the function plotted in Fig. (23).

At stress levels below that required to nucleate holes on two grain interfaces, but above that needed to nucleate holes within amorphous pockets, the first event to initiate during creep rupture is the viscous expansion of the hole within the pocket. This leads to the essential depletion of the pocket (Fig. 22b) by viscous flow of the amorphous phase into the surrounding two grain channels (resulting in an increased film thickness along the two grain interfaces). The process is controlled either by the stress at the hole surface or by the flow rate into the channel. Specific results can be obtained by considering a triangular channel(ii) and rigid displacements of the grain in the direction of the applied stress. Viscous flow along the boundary channel is characterized by the two-dimensional equivalent of Eq. (54);

$$\frac{d^2\sigma(x)}{dx^2} = \frac{-12\eta\dot{\delta}}{\delta^3} \quad . \tag{60}$$

Symmetry requires that $d\sigma/dx = 0$ at the grain facet center ($x=\ell/2$); while the stress at the mouth of the channel ($x=d$) is governed by the stress at the surface of the hole,

$$\sigma(a) \approx \beta\gamma_\ell / r \quad , \tag{61}$$

where r is the radius of curvature of the hole and β ranges between 2 (when the hole is spherical) and 1 (when the hole becomes cylindrical); this stress prevails at the channel mouth provided that viscous flow within the pocket occurs at a sufficiently rapid rate that the stresses in the region between the hole and the channel opening are equilibrated. Integration of Eq. (60) subject to these boundary conditions gives the stress distribution within the channel as;

$$\sigma(x) = \frac{6\eta\dot{\delta}}{\delta^3}\left[(d^2-x^2) + \ell(x-d)\right] + \frac{2\gamma_\ell}{r} \quad . \tag{62}$$

and the separation rate becomes,

$$\dot{\delta} = \frac{\delta^3\left[\sigma_\ell^i - \beta\gamma_\ell(1-2f)/r\right]}{\eta\ell^2\left[1-6f+9f^2-4f^3\right]} \quad , \tag{63}$$

(ii) The channel slope in the presence of complete wetting between the solid and liquid phases must be dictated by anisotropy of the solid/liquid interface energy.

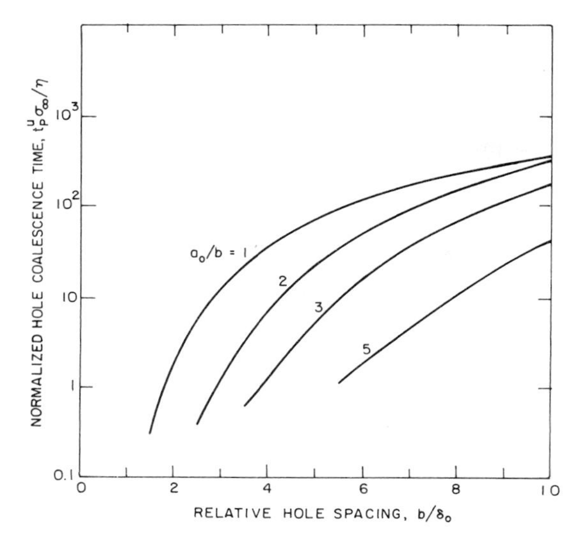

Fig. 23. A plot of the unconstrained hole coalescence time as a function of the initial film thickness for several choices of the hole nucleation size.

Fig. 24. A transmission electron micrograph of a thin amorphous film over a cavity surface observed in Si_3N_4.

where $f = d/\ell$. The surface tension term, γ_ℓ/r, is of significant magnitude (relative to the local stress) during the two extremes of the depletion process: both immediately after nucleation of the hole and as complete hole depletion is approached. In the presence of stress concentrations (section 2.2) the initial expansion of the cavity occurs rapidly. Hence, it is surmised that the surface tension term does not significantly affect the depletion rate until depletion of the channel is nearly complete (when the surface tension stress increases continuously as the second phase is drawn into the mouth of the two grain interface, Fig. 22b). This process exhibits a relaxation time [15], $\tau_d \approx 2\eta\ell(\tan\phi - \phi)/\sigma_\ell$. Hence, when $t \ll \tau_d$ and the retardation effects that occur in the final stage can be neglected, the time t_d for 'depletion' of the pocket can be expressed directly in terms of the displacements, from Eq. (7) and (63), as ($\gamma_s/r \approx 0$) [15];

$$t_d^c \dot{\varepsilon}_\infty = (4/9)\,(\delta_d - \delta_o)/\ell \quad ,$$

$$t_d^u \sigma_\infty/\eta = \ell^2\,[1 - 6f + 9f^2 - 4f^3]\,[1/\delta_o^2 - 1/\delta_d^2] \quad , \qquad (64)$$

where δ_d is the thickness of the two grain channel at depletion;

$$\delta_d \approx \delta_o + d^2 \tan\phi/2\ell(1-f) \quad ,$$

and ϕ is the angle contained by the triangular pocket (Fig. 22b). The 'depletion' time, t_d, is the time that expires while sufficient material is removed from the pocket that the increase in curvature of the liquid surfaces causes the surface tension stress to increase to a significant fraction of the local stress. It should thus be recognized that a small quantity of residual liquid may remain within the pocket, for a period $t \gg t_d \approx \tau_d$. In fact, the presence of residual liquid, as influenced by the level of the local stress, has a direct association with the next step in the failure sequence: which may occur either by viscous hole extension into the two grain channel or by solution/reprecipitation.

When the local stress after hole depletion exceeds γ_ℓ/δ_d, the liquid meniscus can be drawn into the channel mouth (Fig. 22b), by allowing a positive pressure gradient to be retained within the liquid. Under these conditions, the thickness of the liquid film increases, as the meniscus enters the channel, causing both the stress at the meniscus to decrease and allowing nucleation of holes within the two grain channel (Eq. (19)). A combination of finger-like hole growth from the channel mouth (see Eq. (75)) and expansion of oblate holes (Eq. (59)) then permits full-facet cavitation to proceed at a rate presumably in excess of that attained by solution/reprecipitation.

However, when the local stress is not high enough to induce full-facet cavitation by visous flow mechanisms, the depleted holes extend (more slowly), by a solution/ reprecipitation process. Solution/reprecipitation is motivated by a stress difference between the liquid film at the cavity surface(iii) and the liquid within the channel, and entails the transport of the solid phase, through the liquid, from the cavity surface onto the grain surfaces within the channel. Hence, further redistribution of the liquid phase is not necessarily involved. The process can be either diffusion controlled or interface controlled; both possibilities will be examined. Before proceeding, it is noted that solution/reprecipitation can occur concurrently with pocket depletion (by viscous flow) and hence, that a rigorous analysis would examine this concurrence. But, for simplicity, solution/reprecipitation is considered to initiate when hole depletion is essentially complete because, for most practical systems, the solution/reprecipitation process is the most time-consuming constituent in full-facet cavitation (as substantiated in the subsequent analysis).

Diffusion controlled solution/reprecipitation requires that a chemical potential gradient exists along the channel to motivate migration of the solid phase. This can only be achieved in the presence of a pressure gradient within the liquid. A suitable pressure gradient exists if the liquid phase continues to flow within the channel, throughout the process; suggesting that viscous flow is a necessary accompaniment to solution/reprecipitation. However, it is also noted that the viscous flow prerequisite may be obviated in the presence of grain boundary ledges that substantially impede viscous flow through the channel. In the absence of an appreciable spatial density of ledge-like impediments, the stress distribution within the channel (Eq. (62)) dictates the mass transport rate along the channel. The specific rate of diffusion is given by the differential equation

$$\frac{d^2\sigma(x)}{dx^2} = -\frac{kT\dot{\Delta}}{\Omega D_\ell C_\ell \delta} , \tag{65}$$

(iii) The complete wetting requirement for liquid phase sintering suggests that a liquid film is likely to exist over the cavity surfaces, as generally observed (Fig. 24). However, it is emphasized that the complete wetting between solid and liquid required for sintering only specifies a zero dihedral angle pertinent to the liquid along a grain boundary within the solid and does not relate to solid, liquid, vapor equilibrium.

where $\dot{\Delta}$ is the rate of separation of the grain centers induced by matter deposition on the grain surfaces, D_ℓ is the diffusivity of the solid phase in the liquid and C_ℓ is the solubility of the solid in the liquid. Integration of Eq. (65), subject to the same boundary conditions pertinent to viscous flow in the channel, gives;

$$\sigma(x) = \frac{kT\dot{\Delta}}{2\Omega D_\ell C_\ell \delta}\left[(d^2-x^2) + \ell(x-d)\right] + \frac{2\gamma_\ell}{r} . \tag{66}$$

Comparison of Eqs. (62) and (66) indicates that;

$$\frac{\dot{\Delta}}{\dot{\delta}} = \frac{12\Omega D_\ell C_\ell \eta}{kT\delta^2} . \tag{67}$$

This result illustrates that cavity growth by diffusion controlled solution/reprecipitation is likely to proceed most rapidly when the liquid film thickness is small, contrary to intuitive expectations (a tendency that will not, of course, apply when appreciable ledge-like impediments exist along the grain interfaces). Additionally, solution/reprecipitation cavity growth is encouraged by a high diffusivity (as expected) and a high liquid viscosity. Combining Eqs. (7), (63), and (67) and recognizing that the local stress (Eq. (7)) is dictated by the sum, $\dot{\delta} + \dot{\Delta}$, the differential equation describing the grain separation rate due to matter deposition can be derived as,

$$\dot{\Delta}(t)\left\{\frac{\ell^2 kT\left[1-6f(t)+9f^2(t)-4f^3(t)\right]}{12\Omega D_\ell C_\ell \delta(t)} + \frac{4\sigma_\infty}{9\dot{\varepsilon}_\infty}\left[1 + \frac{kT\delta^2(t)}{12\Omega D_\ell C_\ell \eta}\right]\right\}$$

$$= \sigma_\infty - \beta\gamma_\ell(1-2f)/r . \tag{68}$$

Cavity growth is controlled in this instance by the rate of transport of the solid phase through the liquid that exists within the two grain channel and along the cavity surface. Specifically, domination of the cavity growth rate by diffusion through the liquid over the cavity surface, vis-a-vis diffusion along the channel, is presumed to be dictated by the relative film thickness, δ_s/δ_b, at the two locations (the role of δ_s/δ_b being analagous to that of the diffusivity ratio $D_s\delta_s/D_b\delta_b$ in the previously discussed cavity growth problem for single phase systems). However, for the present problem, the requirements on the contained angle, ϕ, are not readily specified and an exact solution for the cavity growth rate awaits further consideration of this issue. Several simplifying assumptions (concerning the trends in ϕ with cavity expansion) can, of course, be made in order to obtain preliminary solutions. In this context, it is tempting to propose that diffusion through the fluid film over the cavity surface is rapid enougth to maintain a constant contained angle, ϕ, during cavity growth. However, this assumption yields an inadmissible solution for unconstrained cavity growth; as well as being at variance with available observations (Fig. 25), which suggest that ϕ decreases as the cavity expands. Instead, the observations are used to suggest the assumption that the cavity only extends along the boundary normal to the applied stress and does not progress along the inclined boundaries. Then, volume conservation of the solid phase (the liquid phase volume is necessarily constant) permits Δ to be expressed as;

$$\dot{\Delta} = \ell f_o \dot{f} \quad . \tag{69}$$

A solution for t_p can, in principle, be obtained by combining Eqs. (63), (68) and (69). However, the solutions are unwieldy. Hence the only results presented herein pertain to the case, $\dot{\Delta} >> \dot{\delta}$ (as dictated by Eq. (67)) and for $\sigma_\infty >> \gamma_s/r$; the implied loss of driving force (that accompanies an increase in δ) and neglect of the stress at the channel mouth results in lower bound cavitation times;

$$t_p^c \dot{\varepsilon}_\infty > (2/9)(d_o/\ell)\left[1-2d_o/\ell\right] \quad , \tag{71}$$

$$t_p^u \sigma_\infty \left(\frac{C_\ell D_\ell \delta_\ell \Omega}{kT\ell^3}\right) > (1/192)(d_o/\ell)$$

$$\left\{1-16(d_o/\ell)+48(d_o/\ell)^2-48(d_o/\ell)^3+16(d_o/\ell)^4\right\} \quad .$$

The constrained time t_p^c (being insensitive to the atom transport rate) can be expected to exceed the lower bound by up to $\sim$2, (as obtained when $\dot{\delta} \approx \dot{\Delta}$); while the unconstrained time could be substantially in excess of the lower bound.

Cavity growth by solution/reprecipitation could be interface limited [27]. Then, the stress within the liquid layer at the grain interface can equilibrate at the level of the locally applied stress; except near the cavity tip where the stress must decrease to $\sim$0 as the cavity surface is approached. Assuming that the stress gradient with the liquid can be supported by fluid flow into the channel mouth and over the cavity surface, the matter deposition rate along the grain interface is simply;

$$\Delta \dot{V}_{int} = \sigma_\ell \Omega k_1 \ell(\ell-d)/kT \quad , \tag{72}$$

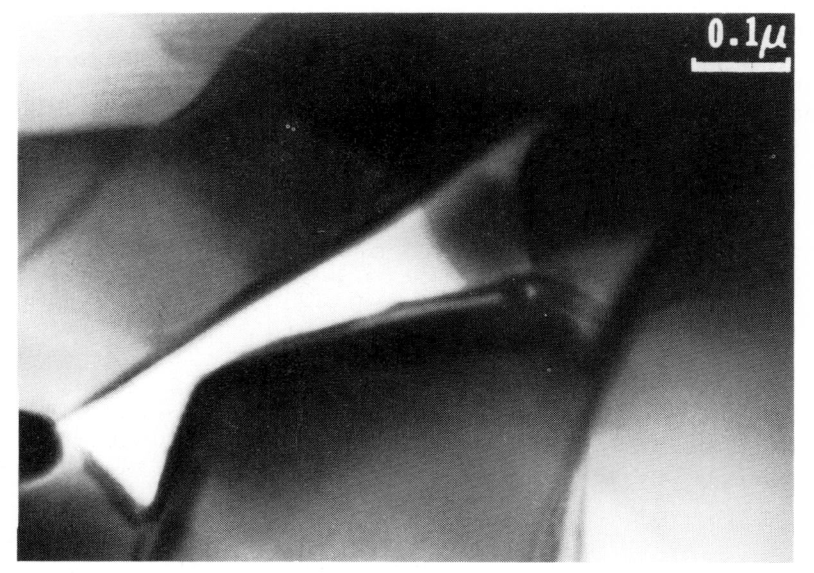

Fig. 25. A transmission electron micrograph of a cavity in Si_3N_4 extended by solution/reprecipitation.

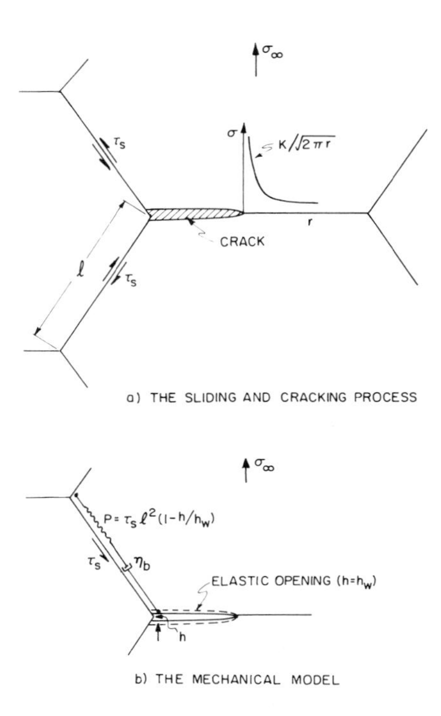

Fig. 26. A schematic of the grain boundary sliding and brittle cracking process, (a) the stress field, (b) the mechanical model.

where k_1 is the rate of condensation of atoms from the liquid film onto the grain surface: a parameter that can be determined from crystallization or creep studies [27]. Then, subject to volume conservation (Eq. (69)), the cavitation time for equilibrium shaped cavities, when $\dot{\Delta} >> \dot{\delta}$, becomes,

$$t_p^c \dot{\varepsilon}_\infty = (2/9)(d_o/\ell)[1-2d_o/\ell] \quad ,$$

$$t_p^u \sigma_\infty k_1 \Omega / kT = \ell[1-(d_o/\ell)] \quad . \qquad (73)$$

Note that the constrained result is necessarily the same as that for the diffusion controlled process.

Comparison of the cavitation times for each of the prospective processes suggests that, under fully-constrained conditions (when only the cavity morphology and matter conservation requirements are important) the cavitation times are similar for all processes; with distinctions between mechanisms depending on specific hole spacings, pocket shapes and sizes, film thicknesses etc. Some typical choices for liquid phase sintered Si_3N_4 (Table I), indicate that the

Table I

Constrained Processes for Si_3N_4

$\delta_o \approx 20Å$, $\ell \sim 1$ μm, $d_o \sim 0.1$ μm

Process	$t^c \dot{\varepsilon}_\infty$
Oblate Hole Coalescence	$\sim 10^{-2}$
Pocket Depletion	$\sim 5 \times 10^{-3}$
Solution/Reprecipitation	$\sim 2 \times 10^{-2}$

fastest constrained process for this material is pocket depletion, while solution/reprecipitation is the slowest. This is consistent with the frequent observation of both hole depletion (Fig. 24) and of cavity expansion by solution/reprecipitation (Fig. 25), and the infrequent detection of holes along two grain interfaces (Fig. 9). When the constraint relaxes, during subsequent cavity growth, appreciably different cavitation rates can be attributed to the various growth mechanisms, depending primarily on the viscosity of the amorphous phase and the diffusivity of the solid in the liquid.

Finger-like hole growth can also initiate within the amorphous layer at high hole growth rates. The critical wavelength for this process is given by;

$$\lambda_c = \pi \left(\frac{\gamma \delta^2}{3\eta \dot{a}} \right)^{1/2} . \tag{74}$$

For unconstrained growth of a hole, by viscous flow, from a three grain corner, finger-like growth will occur when;

$$\frac{\sigma_\infty \ell}{\gamma (1-f)^4} \mathrel{\tilde{>}} \pi^2/4 . \tag{75}$$

2.3.3 The role of grain boundary sliding

Brittle cracking or cavity growth along grain boundaries, motivated by the sliding of neighboring grain boundaries, may occur during creep (Fig. 26a). However, this process is confined to the limited set of situations for which appreciable mass transport by diffusion or viscous flow is prohibited [30]; because grain boundary and surface diffusion modify both the surface profile [20] and the stress state ahead of the crack and thereby create a diffusive cavity that propagates in accord with the mechanisms described in the preceding sections (with grain boundary sliding as an incidental, rather than a motivating, process, c.f. diffusive creep [12]). When significant diffusion or viscous flow is not admitted, and the only permissable viscous motion occurs by grain boundary sliding, brittle cracks are tenable. The occurrence of this condition must be limited (since grain boundary sliding itself usually involves diffusive processes [12], because of the presence of ledges and of nonplanarity). Nevertheless, important situations can be conceived wherein the proposed process might be encountered. For example, ceramics prepared by liquid phase sintering (e.g., Si_3N_4) often have planar boundaries, and contain a second phase at the boundaries that is too narrow (a few lattice spacings, Fig. 8) to admit significant viscous modifications of the crack tip[iv], but wide enough to facilitate boundary sliding.

If diffusion or viscous flow are excluded, grain boundary sliding will result in elastic stress concentrations at grain triple points. The stress concentration (in the absence of dislocation plasticity) has the square root singularity

---

[iv] The brittle cracking in this instance refers to fracture of the thin amorphous phase in a manner analagous to the fracture of liquids.

dislocation plasticity) has the square root singularity typical of shear cracks [30]. A crack will develop from the triple point either if the singularity attains the critical level required for grain boundary (or second phase) fracture, $K_c^{g \cdot b}$, or if a defect of sufficient size pre-exists at the grain boundary. The onset of cracking will thus depend on the local conditions at individual triple points.

The presence of the crack will relax the elastic stress concentration at the triple point and permit the adjacent boundaries to slide; thereby producing an opening displacement at one end of the crack. As the sliding progresses, the stress intensity factor at the micro-crack tip (Fig. 26a) increases, causing additional crack extension and further sliding. Also, the stress concentration at the neighboring triple points becomes enhanced, leading to an increased probability of microcrack initiation at these locations. Once the crack reaches the opposing triple point, the singularity at its tip will begin to relax, by sliding of the intersecting boundaries, and further crack extension will be suppressed. The ultimate formation of open facet-sized cracks is thus to be anticipated. Failure will presumably occur when sufficient contiguous boundaries have developed cracks (forming a macrocrack of critical size) [31]. The interaction of propagating cracks with performed cracks is thus undoubtedly involved in the failure process.

The extension of a wedge crack the emanates at a triple point can be analyzed using a cracked dislocation solution, [30,32]

$$K \approx \sigma_n\sqrt{\pi(a/2)} + hE/\sqrt{2\pi a}(1-\nu^2) \quad , \qquad (76)$$

where 2h is the wedge opening, a is the crack length, K is the stress intensity factor, E is Young's modulus and σ_n is the component of the applied stress normal to the crack. The first term is due to the normal opening of the crack and the second derives from the wedge opening produced by sliding.

A crack will propagate whenever the stress intensity factor K exceeds the critical value $K_c^{g \cdot b}$. The motion of a crack on a susceptible boundary can thus be directly obtained from Eq. (76) once the time dependence of the opening h has been established. The resultant crack propagation exhibits the general characteristics depicted in Fig. (27). The initial growth is constrained by the grain boundary sliding rate. However, when the crack reaches a critical length a*, at which it can continue to propagate at a fixed wedge opening h* (due to the action of the normal stress, σ_n), the constraint imposed by the boundary sliding rate becomes relaxed and the crack will extend catastrophically up to the stable length ℓ. The propagation time t_p is determined by the time taken for the crack to attain the critical length a*. Hence, t_p can be ascertained if both a* and the driving force for sliding in the range $0 < a < a^*$ can be deduced.

The critical length a* is given by the coupled requirement that K increases with crack length at a fixed wedge opening h, i.e., the crack is driven by the normal stress, and that K at this condition, K*, is equal to $K_c^{g\cdot b}$. The first condition is established by determining when K, for a fixed h, is a minimum; because, upon exceeding the minimum, K will increase monotonically with crack length and the crack will become unstable. Hence, differentiating Eq. (76) with respect to a at fixed h and setting to zero, gives

$$a^* = h^* E/\sigma_n \pi(1-\nu^2) \quad . \tag{77}$$

Substituting h* from Eq. (77) into Eq. (76), and requiring that the resultant $K = K_c^{g\cdot b}$, then gives;

$$a^* = \frac{1}{2\pi}\left(\frac{K_c^{g\cdot b}}{\sigma_n}\right)^2 \quad . \tag{78}$$

The rate of wedge opening $\dot{h}$ is determined by a conventional spring, dashpot approach; wherein the opening is motivated by the elasticity of the material and resisted by the viscosity η_b of the sliding boundary (Fig. 26b). The wedge opening permitted by the elasticity of the material is governed exclusively by grain boundary sliding and therefore relates to the resolved shear stress τ_s. The normal stress is not involved because it generates an opening at the crack center but not at the wedge. The problem is most conveniently posed using the illustration shown in Fig. 26b. The wedge crack releases the constraint of the surrounding grains on the triple point and the grain on one side of the sliding boundary exerts an elastic force on the triple point. As the crack extends and the wedge opening increases the force decreases, and must reduce to zero at elastic equilibrium, i.e., at the stable value of the elastic wedge opening, h_w (Fig. 26b). The reduction in force occurs in accord with the linear elastic properties of the grain, such that the average effective shear stress τ_B acting on the boundary at any instant is

$$\tau_B = \tau_s(1 - h/h_w) \quad . \tag{79}$$

But, the average sliding rate $\langle\dot{u}\rangle$ can be related to the average effective stress τ_B by:

$$\langle\dot{u}\rangle = \tau_B \delta_b/\eta_b \equiv (\tau_s \delta_b/\eta_b)(1 - h/h_w) \quad . \tag{80}$$

Hence, noting that;

$$\langle u\rangle = \frac{\pi}{4} h \operatorname{cosec}\phi \quad , \tag{81}$$

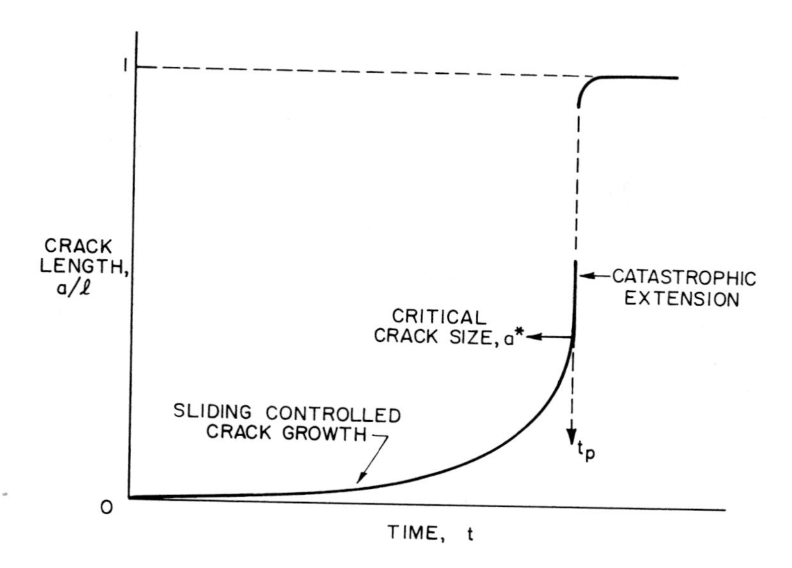

Fig. 27. The growth characteristic of a brittle crack restricted by the sliding rate of the neighboring grain boundary.

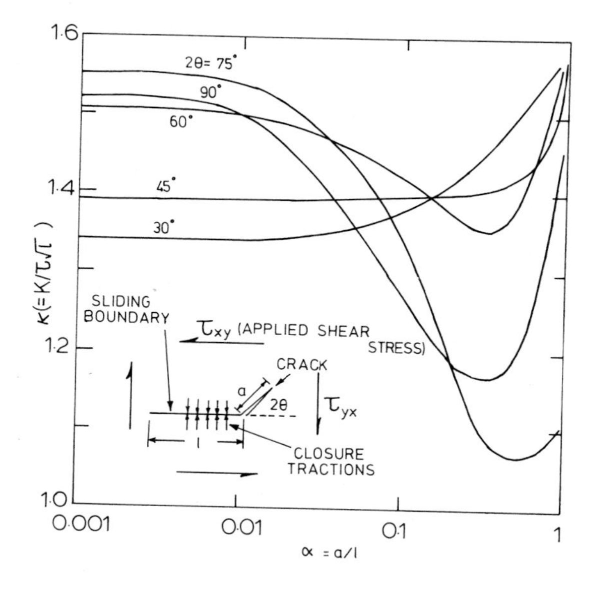

Fig. 28. A plot of the normalized stress intensity factor pertinent to grain boundary sliding and brittle cracking.

the wedge opening becomes;

$$\dot{h} = \frac{4 \sin \phi \tau_s \delta_b}{\pi \eta_b} [1 - h/h_w] \quad . \tag{82}$$

The equilibrium elastic opening h_w (at which the elastic driving force is zero) is dictated exclusively by the grain boundary sliding and represents the opening at which sliding would cease. The equilibrium opening will depend on both the instantaneous crack length and the in-plane shear stress [30];

$$h_w = \frac{\sqrt{2\pi}(1-\nu^2)a\tau_s}{E} \left[\frac{\kappa^\tau(a/\ell,\phi)}{\sqrt{a/\ell}} - \sqrt{\pi/2}\, \sin 2\phi \right] , \tag{83}$$

where κ^τ is plotted in Fig. 28. The instantaneous wedge opening h prior to catastrophic extension is determined by applying the crack extension requirement ($K = K_c^{g \cdot b}$) to Eq. (76), giving

$$h = \frac{\sqrt{2\pi}(1-\nu^2)a}{E} \left[(K_c^{g \cdot b}/\sqrt{a}) - (\sqrt{\pi/2}\, \sigma_n) \right] \quad . \tag{84}$$

Differentiation then yields the opening rate $\dot{h}$, as

$$\dot{h} = \frac{\sqrt{2\pi}(1-\nu^2)\dot{a}}{E} \left[(K_c^{g \cdot b}/\sqrt{a}) - (\sqrt{\pi/2}\, \sigma_n) \right] \quad . \tag{85}$$

Substituting the above results for h, h_w and $\dot{h}$ into Eq. (82) and rearranging gives

$$\dot{a}(1-\nu^2)\left(\frac{\eta_b}{E\delta_b}\right) = \frac{4 \sin\phi\sqrt{a}}{\pi^2} \left\{ \frac{\left[\sqrt{a}-2\sqrt{a^*} + (\tau_s/\sigma_n)(\kappa\sqrt{2\ell/\pi}-\sqrt{a}\, \sin 2\phi)\right]}{(\kappa\sqrt{2\ell/\pi}-\sqrt{a}\, \sin 2\phi)(\sqrt{a^*}-\sqrt{a})} \right\} . \tag{86}$$

The significant features of Eq. (86) to note are: (a) there is a threshold for crack development (obtained by setting the numerator to zero) given by the condition

$$\kappa\tau\sqrt{\ell}\Big|_{a=0} > K_c^{g.b} ,$$

(b) the velocity becomes unbounded as $a \to a^*$, (c) the velocity above the threshold condition is zero at a=0. The general trends are complex. However, some conditions wherein simplified crack velocity relations pertain can be explored by re-expressing Eq. (86) in terms of a threshold stress, through the term

$$a^*_{th} \equiv \ell(\tau_s/\sigma_n)^2 \kappa^2/2\pi = (\pi/2)(K_c^{g \cdot b}/\sigma_n)^2_{th} , \tag{87}$$

to obtain:

$$\dot{a}(1-\nu^2)\left(\frac{\eta_b}{E\delta_b}\right) = \frac{4\beta \sin\phi (a/a^*)^{1/2}}{\pi^2}$$

$$\frac{[(a/a^*)^{1/2}(1-\beta \sin 2\phi)+2(a^*_{th}/a^*)^{1/2}-2]}{[2(a^*_{th}/a^*)^{1/2}-\beta(a/a^*)^{1/2}\sin 2\phi][1-(a/a^*)^{1/2}]} , \tag{88}$$

where $\beta = \tau_s/\sigma_n$. For stresses considerably in excess of the threshold, such that $a^* \ll a^*_{th}$ and for β values typical of crack-susceptible boundaries ($\beta \sim 1$), Eq. (88) reduces to:

$$\dot{a}(1-\nu^2)(\eta_b/E\delta_b) \approx 4 \sin\phi\beta(a/a^*)^{1/2}/\pi^2[1-(a/a^*)^{1/2}] . \tag{89}$$

Note that the facet length ℓ does not enter this limit solution.

The time of propagation t_p of a crack across the boundary facet can be obtained by direct integration of the velocity relation. For the limit solution, Eq. (89), the following simple result obtains:

$$t_p = \frac{\pi(1-\nu^2)}{8 \sin\phi} \frac{(K_c^{g\cdot b})^2}{\sigma_s \tau_s} \left(\frac{\eta_b}{E\delta_b}\right) . \tag{90}$$

This result provides useful insights into the relative roles of the important microstructural variables, η, δ_b, $K_c^{g\cdot b}$, ϕ, and of the relative stress conditions σ_n, τ_s. For a uniaxial tension σ_∞ inclined at an angle ψ to the sliding boundary,

$$\tau_s = \sigma_\infty \sin\psi \cos\psi ,$$

$$\sigma_n = \sigma \sin^2(\psi + \phi) ,$$

Equation (90) becomes;

$$t_p = \frac{\pi(1-\nu^2)}{8\sigma_\infty^2} \left[\frac{(K_c^{g\cdot b})^2 \eta_b}{E\delta_b}\right] [\sin\phi \sin\psi \cos\psi \sin^2(\psi+\phi)]^{-1} . \tag{91}$$

The non-linear stress dependence is significant: a result that essentially derives from the condition that the crack length at critically a* is proportional to the inverse square of the normal stress. Differentiation of Eq. (91) with respect to ϕ and ψ suggests that the most fracture susceptible boundaries (those with minimum t_p) pertain for the condition $\psi = 35°$, $\phi = 55°$, yielding

$$(t_p)_{min} \approx \left(\frac{K_c^{g \cdot b}}{\sigma_\infty \sqrt{\delta_b}}\right)^2 \frac{\eta_b}{E} \quad . \tag{92}$$

An extension of this analysis to incorporate limited diffusion or viscous flow can be contemplated. In the presence of limited atom transport rates along the cavitating boundary, the neighboring grains can deform in a predominantly elastic manner and a behavior analagous to brittle cracking results. This mode of cavitation is tenable whenever the atom transport rate is small enough to prohibit the rigid separation of grains bordering the cavitating interface. Consideration of relaxation times for grain boundary diffusion over the cavitating facet[(v)], in the presence of elastic deformation, reveals that elastic cavity growth prevails when[21]

$$(\langle\sigma\rangle/G) \gg a/(\ell-a) \quad ,$$

where $\langle\sigma\rangle$ is the average stress between the cavity tip and the adjacent three grain corner. Such behavior only occurs, therefore, at high stress levels (e.g., in the vicinity of a macrocrack tip). When these conditions are satisfied, solutions for the propagation time can be deduced for specific cavity growth mechanisms. This is achieved by invoking a relation between $\dot{a}$ and K characteristic of the operative mechanism (for example, Eq. (112), pertinent to elastically driven cavity growth by grain boundary diffusion) and inserting into an expression for K deduced from Eqs. (76), (82), and (85);

$$K = \sigma_\eta \sqrt{\pi a/2} + \frac{E h_w \left[1-\exp(-t/\tau_b)\right]}{\sqrt{2\pi a}(1-\nu^2)} \quad , \tag{93}$$

where $\tau_b = \pi\eta_b h_w / 4 \sin\phi \tau_s \delta_b$. Integration of the resultant differential equations yields t_p. However, specific solutions remain to be determined.

(v) A diffusion process confined to the cavitating facet is probably appropriate in the presence of grain boundary sliding.

A simplified procedure for estimating t_p, suggested by Tsai and Raj [28], is based on a constant cavity velocity assumption and a cavity growth rate $\dot{a} \propto K^2$. Applied to viscous hole growth motivated by grain boundary sliding, this procedure yields;

$$t_p \approx 10^{-4}(\ell/\delta_o)^3(E\eta/\sigma_\infty)^2 \quad . \tag{94}$$

2.4 Cavity Coalescence

The ultimate failure of polycrystalline ceramics occurs when facet-sized cavities coalesce to form an identifiable crack, which then extends to a critical dimension. Cavity coalescence is thus an important phase in the high temperature failure process. The coalescence process is sensitively dependent upon the existence of constraint. Two extremes are amenable to analysis and provide a useful basis for the interpretation of experimental results. When the cavities are very narrow, as pertinent either to crack-like cavities or to cavities formed by hole growth (within a thin viscous grain boundary phase) at high local stress levels, the small cavity volume may be elastically accomodated, by the sliding of the contiguous boundaries [31]. Then, the cavities may be regarded as independent statistical entities subject to a stress at the level of the applied stress (Fig. 29). Similarly, when many cavities develop simultaneously along planar zones within the microstructure (as apparently occurs in certain ceramics), the matrix constraint is small (large d in Eq. (10)) and cavity growth independence can again be regarded as an approximately valid basis for analysis. The independent formation of facet-sized cavities, and their coalescence to form a crack, can be treated using standard probabilistic procedures to obtain expressions relating the crack nucleation time to stress and microstructure. The resultant failure time expressions are necessarily specimen-size dependent.

When inhomogeneous cavitation occurs within small microstructural regions (of the order of several grain dimensions) at low stress levels, such that the cavity widths are large enough to exclude elastic volume accomodation, constraint exerts an important influence on the cavity coalescence mechanism [5]. In this instance, the tensile stress concentrations that develop outside the initial cavitation zone induce a zone spreading process along planes normal to the applied stress (Fig. 30). Analysis of the zone spreading conditions provides the basis for predicting specimen-size independent crack nucleation times.

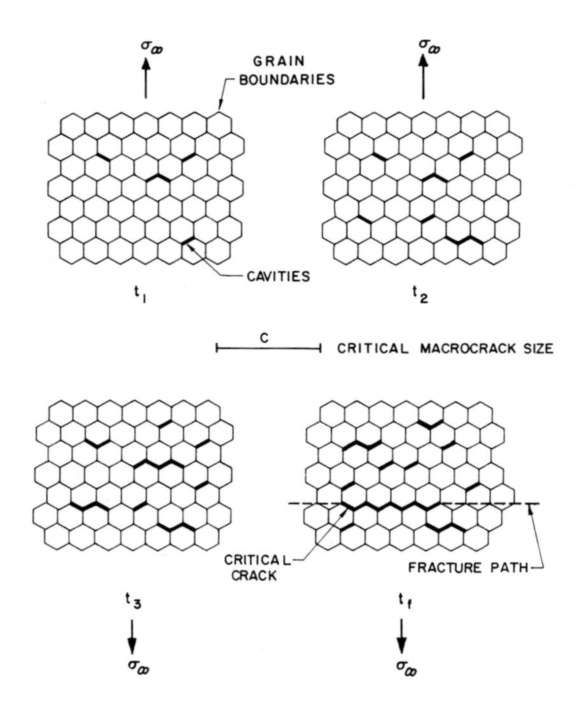

Fig. 29. A schematic of cavity coalescence by a statistical accumulation of contiguous cavities.

2.4.1 Probabilistic cavity coalescence

The development of a probabilistic model for the failure time requires that a critical size, a_c, be defined at which a coalesced array of cavities constitutes a macrocrack. This is, in general, a rather nebulous concept, because the critical size depends on the microstructure, and on the size of the crack tip 'process zone' (i.e., the enhanced cavitation zone created by the crack). The zone size depends, in turn, on the level of the applied stress or, more likely, on the stress intensity factor K. The simplest approach capable of yielding useful results is to assume that a macrocrack has developed when K attains a specified level; namely, when K reaches a certain fraction K_f of the critical stress intensity factor K_c. The critical size for a penny-shaped crack can then be written as;

$$a_c = \pi K_f^2/4\sigma_\infty^2 \quad . \tag{95}$$

The formation of a macrocrack of size a_c will be determined by the distribution of times, t_p, for the development of facet-length cavities (section 2.3). This time is statistically distributed, because of the variability of grain boundaries with regard to second phase content and composition, boundary energy, diffusivity, etc. It is required for present purposes that a distribution function be assigned to this propagation time; namely, a function that can be expected to accurately describe the grain boundary variability typical of ceramics. This decision is facilitated by noting that only a small fraction of the boundaries usually cavitate prior to failure: these being the boundaries with the greatest cavitation susceptibility. When such conditions pertain, a distribution that characterizes the large extreme of cavitation susceptibilities should apply. There are only three extreme value distributions [31] and therefore the choice of functions can be considerably restricted by applying the extreme value condition. Initially, a distribution based on the second type of extreme value function is selected, because this distribution has previously been found to describe extreme value characteristics of ceramic microstructures [31]. An additional rationale for the chosen distribution is provided later. The assumed cumulative probability, $p(t_p)$ of observing facet sized cavities at time t_p is,

$$p(t_p) = 1 - \left[\exp - (t_p/t_o)^m\right] \quad , \tag{96}$$

and for $t_p < t_o$,

$$p(t_p) = (t_p/t_o)^m \quad ,$$

where m is the shape parameter and t_o is the scale parameter. The scale parameter t_o will depend on the specific mechanism of cavity growth, and will be a function of the stress level, viscosity, (diffusivity), boundary energy, etc. All cavity growth mechanisms yield times that are stress dependent and proportional to an Arrhenius factor (through a diffusivity or viscosity). Hence the scale parameter can be expressed by the general relation;

$$t_o = \zeta\sigma_\infty^{-n} \exp(Q/kT) \quad , \tag{97}$$

where Q is the activation energy for the cavity growth process, n is the stress exponent ($1 < n < 3$) and ζ is the parameter that contains the remaining cavity growth variables.

Analysis of the macrocrack formation process can now proceed by assuming that t_p is not appreciably influenced by the prior existence of cavities on adjacent facets. Then, the probability P of forming contiguous facet-sized cavities of sufficient extent to produce a macrocrack of length a (Fig. 30) can be obtained from McClintock's result [33],

$$P = \left(\frac{A_T \sqrt{p}\ \ell np}{\ell^2(0.5\ell np-1)}\right) \exp[(a/\ell)\ell np] \quad , \tag{98}$$

where p is the probability that a given facet has cavitated at time t and A_T is the total grain boundary area subject to the stress σ_∞. For small p (the case of present interest) combining Eqs. (96) and (98) gives the probability $P(t_i)$ of macrocrack formation at time t_i as;

$$P(t_i) = \left(\frac{2A_T}{\ell^2}\right)(t_i/t_o)^{m[0.5+\pi K_f^2/4\sigma_\infty^2\ell]} \quad . \tag{99}$$

At a specific probability level, e.g., the median level ($P = 0.5$), the macrocrack incubation time becomes;

$$\ell n(t_i/t_o) = \frac{\ell n(\ell^2/4A_T)}{m(0.5+\pi K_f^2/4\sigma_\infty^2\ell)} \quad . \tag{100}$$

For most conditions of interest in ceramics $K_c^2 \gg \sigma^2\ell$, whereupon Eq. (100) reduces to;

$$\ell n t_i = \ell n t_o - \left(\frac{4\sigma_\infty^2\ell}{\pi m K_f^2}\right) \ell n(4A_T/\ell^2) \quad . \tag{101}$$

Invoking the general requirement that t_o be proportional to an Arrhenius and a stress term (Eq. (97)), Eq. (101) can be written as;

$$\ell n[t_i \exp(-Q/kT)] = \ell n\zeta - n\ell n\sigma_\infty - \sigma_\infty^2(4\ell/\pi m K_f^2)\ell n(4A_T/\ell^2) \quad , \tag{102a}$$

or

$$\ell n\theta_r = B - n\ell n\sigma_\infty - C\sigma_\infty^2 \quad , \tag{102b}$$

where θ_r is the Orr-Sherby-Dorn rupture parameter, C is a parameter that depends on the grain size and toughness and on the sample size (through A_T/ℓ^2), and B and n are constants that depend on the details of the cavity propagation process.

The appearance of Orr-Shelby-Dorn behavior is satisfying because it implies that all of the temperature effects, correlated by other investigators through this parameter, will also apply to the present model. The general shape of the failure time, stress curve predicted by Eq. (102b) is plotted in Fig. 31a. The general utility of the predicted failure time relation can be explored by plotting the logarithm of the rupture parameter θ_r as a function of σ_∞^2, and treating A, n and C as adjustable parameters (with n being confined to the range, $1 \leq n \leq 3$). The results of Walles [34] on Al_2O_3 and SiC fibers (the only comprehensive data presently available) taken from the correlation developed by Charles [35] are plotted in Fig. 31b,c. The correlation is very good. Such a correlation does not, of course, substantiate the validity of the model, because alternate models can provide correlations of nearly equal quality. It does, however, permit the model to be regarded as a serious candidate. Further experimental comparisons are presented in a subsequent section.

2.4.2 Zone spreading

The incidence of zone spreading is contingent upon the presence of appreciable constraint and the resultant development of enhanced tensions around the periphery of the initial cavitation zone. If cavitation firstly occurs along several contiguous boundaries, for which one of the parameters that dominate the cavitation rate (ψ, D_s) deviates from the average value, the local stress outside the cavitation zone, on the contiguous boundaries, then exceeds the applied stress (Fig. 30). The cavitation rates in this peripheral zone are presumably non-uniform and hence a complete solution of peripheral cavity growth constitutes a formidable problem. Nevertheless, the essence of the process can be established by adopting a simplified, intermittent spreading procedure. Cavity growth in each peripheral zone is assumed to occur

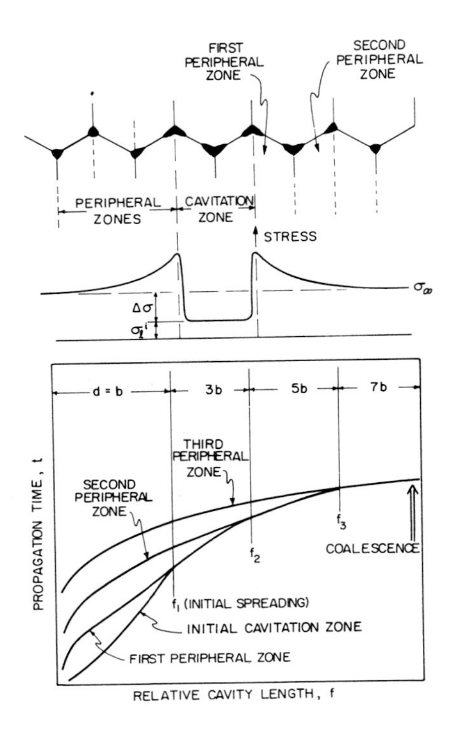

Fig. 30. A schematic of the zone spreading process.

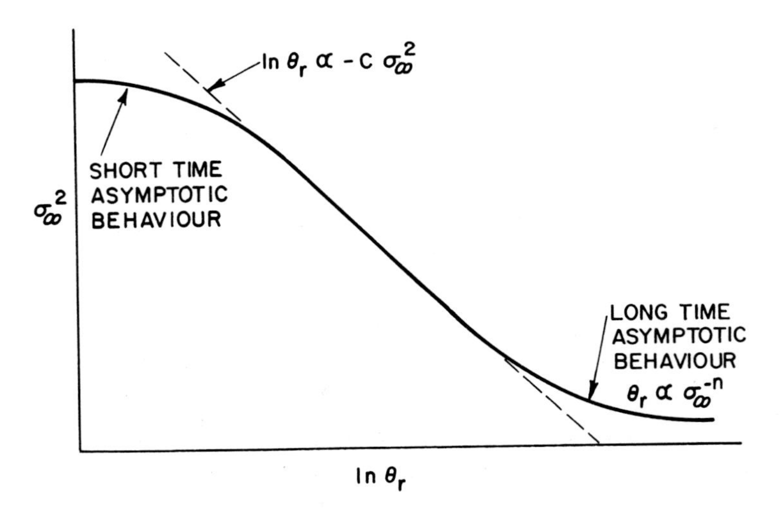

Fig. 31a. Failure times predicted by the statistical accumulation model: schematic.

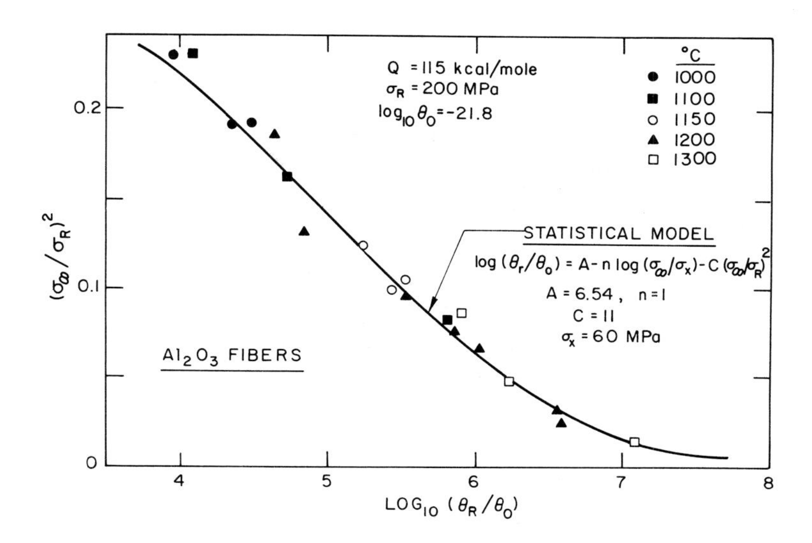

Fig. 31b. Failure times predicted by the statistical accumulation model: comparison with results for Al_2O_3 fibers.

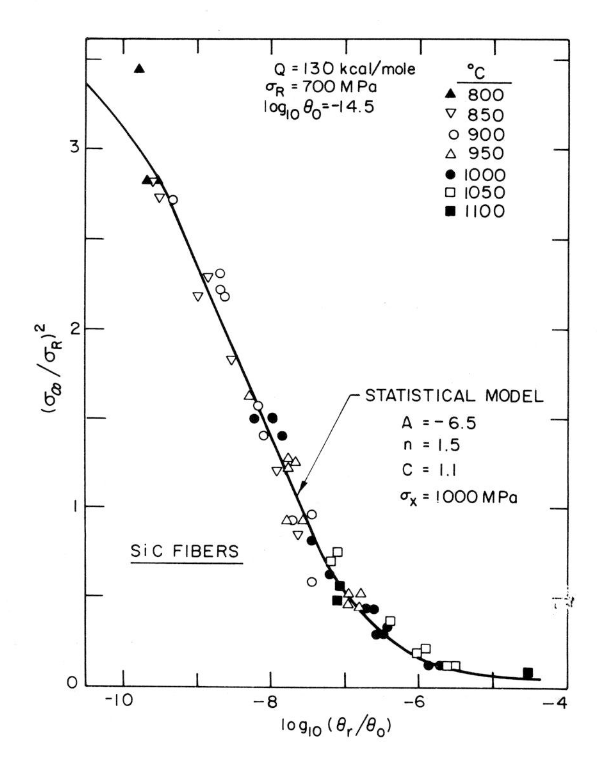

Fig. 31c. Failure times predicted by the statistical accumulation model: comparison with results for SiC fibers.

uniformly (i.e., two uniformly approaching cavities on each peripheral boundary, Fig. 30) at a stress equal to the average stress over that boundary (Eq. (11b)), while cavitation on the original boundary continues at the initially deduced local stress. Then, at a time when the cavity lengths in the cavitation zone and in the peripheral zone become equal, the cavitation zone is considered to advance to the boundary of the first peripheral zone. The process is continued by considering the growth in the next peripheral zone, with a new value of the local stress assigned to the cavitation zone (based on the increase in the zone size, d). Proceeding in this way the time t_i needed to form a discrete macrocrack can be deduced [5].

The zone spreading problem is illustrated for the case of a single phase polycrystal, for which preferred cavitation is based on local deviations in ψ or D_s (Fig. 32). The zone spreading process can be conveniently separated into three regimes. Firstly, large local deviations in the dihedral angle and in the diffusivity appear to be relatively innocuous, because cavity extension along the grain facet can proceed without the generation of appreciable stress in the peripheral zone. Hence, the cavity extends fully along the grain facet before inducing significant cavitation on the contiguous boundary (Fig. 32a). This cavitation behavior is likely to pertain in isolated regions during the early stages of failure, and explains the observation of premature full-facet sized cavities [4].

Secondly, when cavity propagation occurs in regions containing several contiguous boundaries with significant (but not large) deviations in dihedral angle, the trends in constraint (Fig. 32b) suggest continuous zone spreading. The failure time is then dictated by the spreading process and occurs relatively rapidly. Such regions consequently exert the primary influence on high temperature failure. In this circumstance, a large proportion of the failure time is consumed while cavitation is confined to a small number of contiguous grain facets. A strong interdependence of the failure time on the steady state creep rate (Monkman-Grant behavior) is thereby, anticipated, and the distribution in failure times is related primarily to the creep rate variability of the surrounding material.

Finally, it is noted that in regions of relative uniformity, cavitation develops homogeneously, by virtue of a rapid zone spreading process (Fig. 32c). The stress thus remains at a level essentially similar to the applied stress, and the unconstrained failure time relations pertain. However, failure does not evolve quickly. This behavior is not well understood; but presumably, the long failure times obtain either because cavity nucleation is inhibited in these

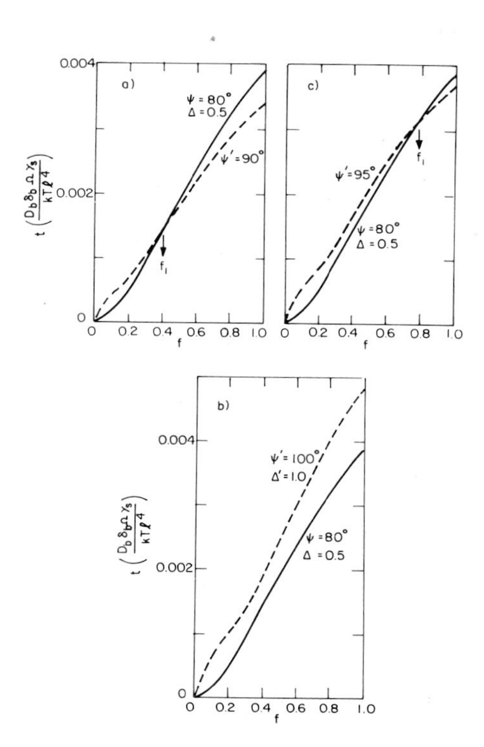

Fig. 32. Cavity propagation times in the cavitation zone and the first peripheral zone predicted for cavities growing from three grain corners.

regions (by the larger ψ) or because the larger cavitation times associated with the larger ψ or D_s are not counteracted by the loss of constraint (Fig. 32c).

3. CRACK PROPAGATION

3.1 Morphological observations

Observations of the crack tip region at relatively high applied loadings in materials subject to crack growth under creep conditions indicate the concurrent existence of a damage zone and of crack tip blunting [3,36] (Fig. 33). The damage zone, which consists of individual and coalesced cavities, is undoubtedly a consequence of enhanced cavitation rates in the crack tip stress field. The crack advance under this circumstance appears to be incremental [3]. Specifically, the crack tip remains stationary and blunts until the damage level attains a sufficient intensity that the adjacent cavities merge with the crack. This constitutes a crack advance. The process then repeats, and a quasi-steady-state velocity results. The most intense damage is generally not coplanar with the crack (Fig. 33) and consequently, the crack path is typically quite irregular (relative to the more planar crack surfaces created during brittle fracture).

At lower applied loadings the damage rate ahead of the crack decreases relative to the crack tip blunting rate. A condition is then reached wherein the crack continues to blunt without perceptible crack advance (Fig. 33), resulting in an apparent creep crack growth threshold [3,37]. The existence of the threshold is an important concept in creep crack growth. Finally, at very low load levels, crack healing may occur [36,38] (Fig. 33), by a diffusion mechanism, involving neck growth within segments of the crack surface in mutual contact.

The crack opening and blunting processes are accompanied by surface displacements that form a impression ahead of the crack (Fig. 34) and a ridge over the crack surface (Fig. 34) [3]. These displacements are related to the stress fields around the crack and thus, provide a means of characterizing the crack tip field under creep conditions.

3.2 Crack tip fields

The characterization of crack extension-rates is typically determined by the parameter that dictates the amplitude of the singular field near the crack tip. For example, stress corrosion cracking rates in elastic materials are adequately characterized by the stress intensity factor, K [39]. The situation is more complex under creep conditions [40]. The important singularity depends upon the manner in which the crack growth proceeds. For present purposes, it is

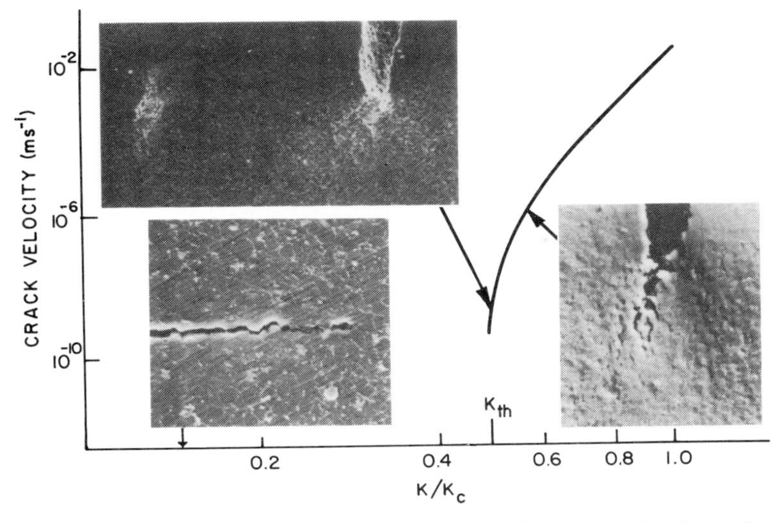

Fig. 33. A schematic illustrating the characteristic behavior of pre-existing cracks under creep conditions

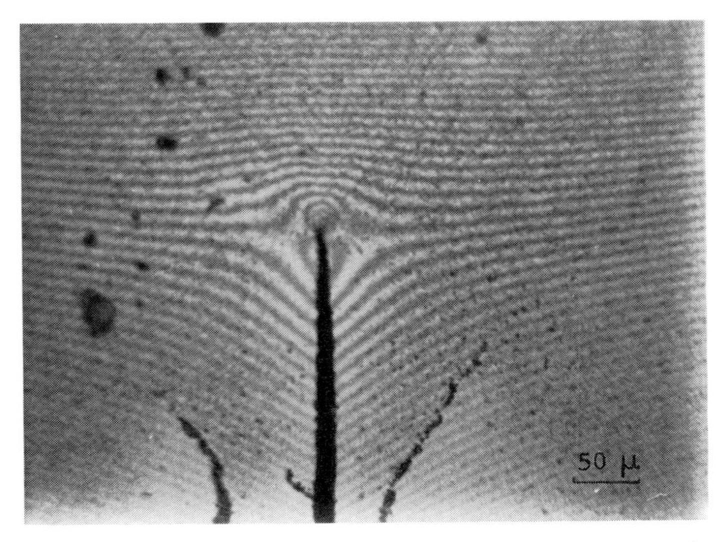

Fig. 34. Interference fringe displacements of the surface of a crept, pre-cracked Al_2O_3 illustrating the surface displacements that accompany crack blunting.

required that the crack advance incremently. Hence, immediately following a crack advance, the crack tip zone is subject to primary creep, characterized by

$$\dot{\varepsilon} = \dot{\varepsilon}_{op}(\sigma/\sigma_o)^{n_p}(\varepsilon/\varepsilon_{op})^{-m} , \qquad (103)$$

where n_p and m are primary creep exponents and ε_{op}, σ_o and ε_{op} are primary creep coefficients. The crack tip field under primary creep conditions is given by;

$$\sigma_{ij}/\sigma_o \propto [C_p(t)/r]^{(m+1)/(m+n_p+1)} , \qquad (104)$$

where r is the distance from the crack tip and $C_p(t)$ is the stress field amplitude. If the primary creep field is embedded in an elastic field (a likely situation following a crack increment), then;

$$C_p(t) = \frac{K^2(1-\nu^2)}{E}\left[\frac{m+1}{(m+n_p+1)t}\right]^{1/(m+1)} , \qquad (105)$$

where E is Young's modulus and ν is Poisson's ratio. The amplitude is thus expressible in terms of a time modified K(vi). Hence, if crack advance occurs while primary creep prevails at the crack tip, and before the creep front advances to the specimen boundaries, the crack velocity should be adequately characterized by K.

For larger intervals following crack advance, the primary creep region will extend to the specimen boundaries and secondary creep will prevail at the crack tip,

$$\dot{\varepsilon} = \dot{\varepsilon}_{os}(\sigma/\sigma_o)^{n_s} , \qquad (106)$$

where n_s and ε_{os} are the secondary creep parameters. The crack tip field is then characterized by

$$\sigma_{ij}/\sigma_o \propto [C_s(t)/r]^{1/(1+n_s)} , \qquad (107)$$

where

$$C_s(t) = \left[\frac{(n_s+m+1)C_p(t)}{(m+s)(n_s+1)}\right] t^{-m/(m+1)} .$$

(vi)J would be more appropriate if the far field were subject to plastic deformation.

However, the far field is dictated by a primary creep region rather than an elastic region and K is thus an inadequate loading parameter. The crack growth behavior is best approximated by the asymptotic value of C_p;

$$C_p^* = \lambda a \, \sigma_\infty^{(m+n_p+1)/(m+1)} , \qquad (108)$$

where σ_∞ is the applied stress, a is the crack length and λ is a proportionality constant that depends on the primary creep parameters and the far field loading.

Ultimately, for long crack advance waiting periods, as pertinent to low crack velocities, steady-state creep prevails throughout the specimen. The crack tip field is still characterized by Eq. (107), but now C_s can be related to the applied loading by the time independent parameter;

$$C_s^* = F_s \sigma_s \dot{\varepsilon}_{os} a (\sigma_\infty/\sigma_o)^{1+n_s} , \qquad (109)$$

where F_s is a parameter that depends on the specimen geometry and loading.(vii) Note that, for $n_s = 1$,

$$C_s^* \propto a\sigma_\infty^2 \equiv K^2/\pi , \qquad (110)$$

and the stress amplitude at the crack tip is uniquely determined by K. For typical practical ceramics, $1 < n_s < 2$ [1,24]; hence K should be a reasonable correlating parameter for most crack growth data. However, some non-uniqueness should be expected at low crack velocities, where C_s^* provides a more appropriate association between the crack tip field and the applied loading.

3.3 Crack growth data

Most high temperature crack growth data for ceramics have been evaluated using K as the appropriate loading parameter. The uniqueness of K has been confirmed at high crack velocities [41], but its utility at low velocities has yet to be fully explored. However, crack opening and surface displacement measurements performed on polycrystalline Al_2O_3 [36] suggest the approximate validity of K (and of the linear stress field amplitude) at relatively low applied loadings; as demonstrated by good correlations with the crack tip displacement field expected for a linear material (Fig. 35); despite the non-linearity of the creep rate ($n \approx 1.8$) measured at low strain rates.

(vii) The equivalent parameter for elastic loading is, $F_E = K/\sigma\sqrt{\pi a}$.

Several interesting features emerge from the existent crack growth data. The critical stress intensity factor for single phase materials, K_{IC}, decreases with increase in temperature, but can increase in materials that contain a continuous amorphous second phase at the grain boundaries [42]. The crack growth susceptibility increases as the temperature increases or as the viscosity of amorphous second phases decreases [43]. Consequently, the exponent n_v that characterizes the crack velocity, v,

$$v/v_o = (K/K_{IC})^{n_v} , \tag{111}$$

can exhibit a wide range of values (typically $6 < n_v < 10^3$), dependent upon temperature and composition [37]. Adequate crack growth models must account for this range of possibilities. Finally, an apparent crack growth threshold is observed [37], and probably relates, as noted above, to dominance of the crack tip blunting rate (relative to the damage rate ahead of the crack).

3.4 Crack growth models

Explicit crack growth models exist for cracks extending along the boundary between two grains by a process involving surface and grain boundary diffusion [44] (Fig. 36). The analysis predicts that [44]

$$K/K_G = 0.85 \left[(v/v_{min})^{1/12}+(v/v_{min})^{-1/12}\right] , \tag{112}$$

where

$$K_G^2 = E(2\gamma_s-\gamma_b)(1-\nu^2) ,$$

$$v_{min} = 8(D_s\delta_s)^4 \, \Omega[E/(1-\nu^2)D_b\delta_b]^3/kT\gamma_s^2 .$$

This relation anticipates a threshold K (Fig. 36), as well as conforming with selected data. However, the mechanism is not representative of the crack growth behavior in polycrystalline aggregates; a process which involves incremental crack advance into a damage zone. An alternate, damage-zone, model is thus required.

A comprehensive damage zone model should incorporate the following features. The crack tip field in the absence of damage should be expressible in terms of K, C_p^* or C_s^*, depending upon the waiting period for crack advance. The damage should reduce the stress in vicinity of the crack tip by virtue of constraint on the local volume expansion by the surrounding material. The stress at the crack tip should be consistent with chemical potential continuity where grain

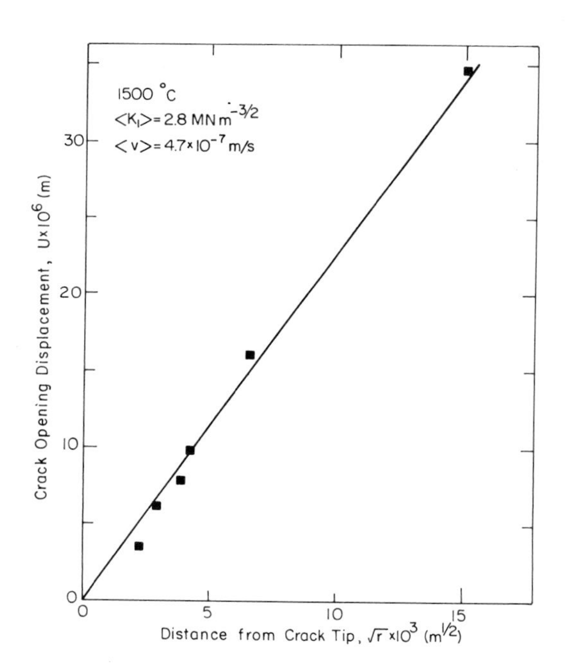

Fig. 35. Creep crack opening displacement measured on an Al_2O_3 specimen.

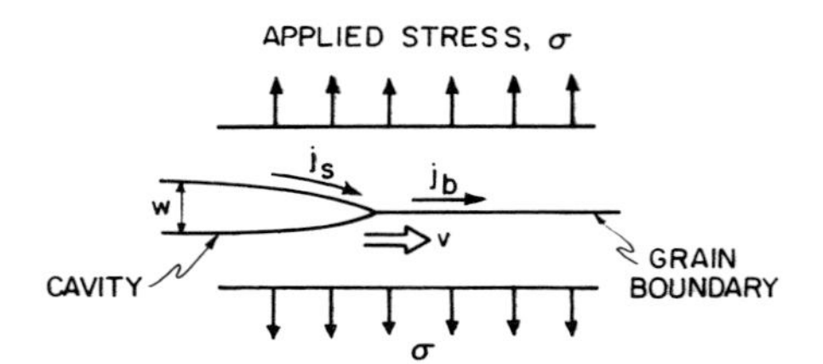

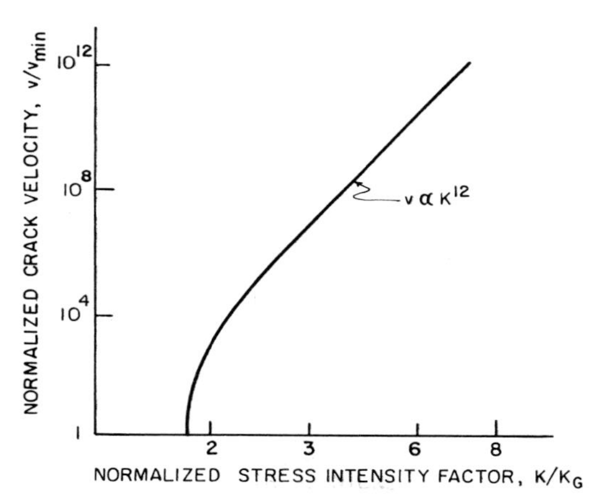

Fig. 36. Crack velocity behavior predicted for a single crack in a bicrystal.

boundaries emerge at the crack surface. The damage should be in the form of grain boundary cavities activated by the normal stress; a requirement which would induce non-coplanar cavitation,(viii) in accord with observation. Opening of the crack and coalescence with the cavities (to constitute a crack advance) should incorporate grain boundary sliding. Such a comprehensive model has not been developed. However, certain of the important requirements have been invoked in several recent attempts. Bassani [40] has examined the growth of an individual coplanar cavity within the various important singular fields. However, constraint effects have not yet been incorporated. Raj and Baik [45] have developed a bi-crystal model with coplanar damage (Fig. 37). The stress field amplitude is considered to be dictated by K and the growth of the damage is allowed to relax the stress near the crack tip. The crack is assumed to advance when the cavities coalesce with the crack tip. A threshold is also invoked, based on the threshold stress for cavity nucleation.

Finally, Tsai and Raj [28] have developed a generalized damage zone concept. The elements of this model undoubtedly provide the closest available representation of creep crack growth by a damage mechanism. The model invokes a damage zone size, z_d, that exists in quasi-steady state when the crack is propagating at a velocity v. The crack growth rate can consequently be expressed in terms of the time, t_g, taken to form full-facet cavities of the zone periphery, because steady state requires that the crack tip advance by one facet length, ℓ, when the time t_g has expired ($v = \ell/t_g$). The time t_g may be computed by determining the stress at the damage zone boundary and inserting this stress into the specific relation that characterizes the active cavity growth mechanism (section 2.3). This procedure requires recognition of the effect on the stress at z_d of both the principal crack and the intervening damage. For a linear material, the stress is given by;

$$\sigma_\ell \approx [K/\sqrt{2\pi z_d}]\, g\, (e_{ij}^T) \quad , \qquad (113)$$

where g is a function of the cavitation strain e_{ij}^T within the damage zone.

(viii) The maximum tension ahead of a crack occurs at an orientation $\theta \approx \pi/3$.

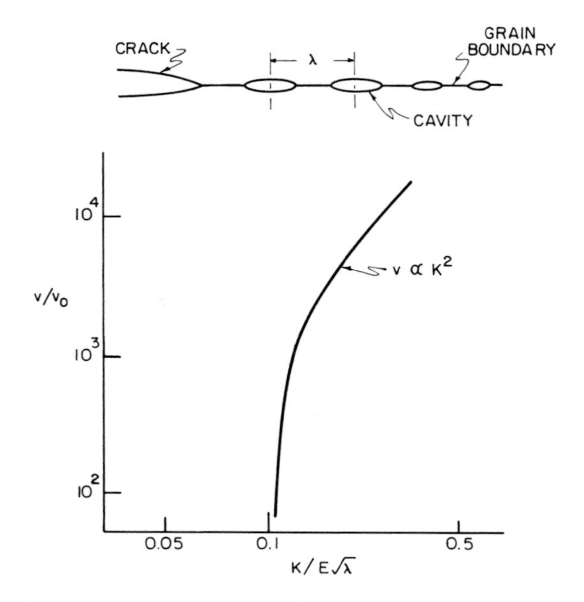

Fig. 37. Crack velocity behavior predicted for a cavity coalescence process of crack advance within a bicrystal.

Precise determination of σ_ℓ are mechanism dependent. For example, mechanisms involving the growth of very narrow crack like-cavities (section 2.3) allow the intervening material to be analyzed as an elastic body containing an array of microcracks. The stress field may then be computed using one of several established techniques that allow for the interaction between the principal crack and the microcrack damage [46,47].

Damage zone size determination is also mechanism dependent. The most plausible determinant of the zone size is the level of the local stress (or strain) vis-a-vis the critical cavity nucleation stress (section 2.2). An approximate solution for z_d, based on the assumption of a Dugdale zone [48] within a linear material, subject to a initial stress σ_c, yields a stress independent zone size,

$$z_d \approx (\pi/8)\,(K^*_{th}/\sigma_c)^2 \quad , \tag{114}$$

where K_{th}^* represents the threshold stress intensity at which damage begins to form ahead of the crack tip. Both K_{th}^* and σ_c are mechanism dependent. More precise formulations based on a variable stress within the damage zone would result in a stress dependent damage zone size.

Combining the zone size relation (Eq. 114) with the local stress (Eq. 113) gives;

$$(\sigma_\ell/\sigma_c) \approx (2/\pi)\,(K/K^*_{th})g(e^T_{ij}) \quad . \tag{115}$$

Inserting Eq. (115) into the appropriate cavity growth relation (section 2.3) yields t_g and hence, v. The presence of an extended damage zone minimizes the constraint at the zone boundary. Hence, expressions for unconstrained cavity propagation can be applied. For example, if cavitation is dominated by the equilibrium growth of the cavity from a three grain corner, Eq. (47), the crack propagation rate becomes;

$$v = \frac{160\sqrt{3}}{\pi}\left(\frac{\Omega D_b \delta_b \sigma_c g(e^T_{ij})}{kT\ell^2 F(\psi) K^*_{th}}\right) K \quad . \tag{116}$$

This result illustrates that the assumption of a Dugdale zone yields a stress intensity factor exponent, n_v, similar(ix) to the stress exponent for the underlying cavity growth mechanism. It is thus unlikely to account for the large range in n_v that obtains for crack propagation in ceramic polycrystals. A more comprehensive treatment of the stress within the damage zone would thus appear to be a prerequisite for the adequate modelling of creep crack growth in ceramics. (Unless the range of n_v is associated with asymptotic approach to the threshold).

3.5 Crack propagation times

Experimentally determined relations between the crack growth rate and the stress intensity factor (Eq. (111)) can be used to predict that component of the failure time attributed to macrocrack propagation [42]. This is achieved by noting that the stress intensity is related to the applied loading by;

$$K = \sigma_\infty Y \sqrt{a} \quad , \tag{117}$$

where Y is a geometric parameter ($2/\sqrt{\pi}$ for a penny-shaped flaw). Differentiating to obtain;

$$dt = \frac{2}{\sigma_\infty^2 Y^2} \left(\frac{K}{v}\right) dK \quad , \tag{118}$$

and integrating between K_f (the stress intensity that characterizes the occurence of a discrete macrocrack) and the critical stress intensity factor, K_c, then gives the crack propagation time;

$$t_c = \frac{2}{Y^2} \int_{K_f}^{K_c} \left(\frac{K}{\sigma v}\right) dK \quad . \tag{119}$$

Inserting the crack growth rate relation given by Eq. (111), the crack propagation time at constant stress becomes,

$$t_c = \frac{2K_c^{n_v}}{\sigma_\infty^2 Y^2 v_o (n-2)} \left[\frac{1}{K_f^{n_v - 2}} - \frac{1}{K_c^{n_v - 2}} \right] , \tag{120}$$

which, for large n_v and $K_c >> K$ becomes;

$$t_c \approx \frac{2}{v_o (n_v - 2)} \left(\frac{K_c}{\sigma_\infty Y}\right)^2 \left(\frac{K_c}{K_f}\right)^{n_v - 2} \quad . \tag{121}$$

(ix) e_{ij}^T exhibits some dependence on K and may thus cause the K exponent in Eq. (116) to deviate from unity.

An accurate definition of K_f is thus of crucial importance to the prediction of crack propagation times. Specifically, if the nucleation process yields a discrete crack when K_f is a specified fraction of K_c (as proposed in section 2.4.1), the the stress dependence of the propagation time must be uniquely characterized by the simple proportionality;

$$t_c \propto \sigma_\infty^{-2} \quad . \tag{122}$$

Alternatively, in the presence of pre-existent cracks with an initial stress intensity, $K_i > K_{th}$, the crack propagation time becomes;

$$t_c \approx \frac{2a_i^{-(n_v-2)}}{v_o(n_v-2)} \left(\frac{K_c}{\sigma_\infty Y}\right)^{n_v} \quad . \tag{123}$$

A much larger stress exponent, $t_c \propto \sigma_\infty^{-n}$ thus obtains and the failure time depends upon the initial magnitude, a_i, of pre-existing cracks (as characterized by proof testing or NDE).

4. PREMATURE FAILURE FROM LARGE SCALE INHOMOGENEITIES

There are several important microstructural sources of premature high temperature failure in ceramics; especially zones of exceptional grain size and isolated amorphous regions in otherwise single phase material. A large grained region in a fine-grained solid subject to creep deformation has a higher viscosity than the matrix, because of the strong grain size dependence of the creep rate (either Herring-Nabarro or Coble creep). This region, and the surrounding fine-grained matrix, must therefore experience stresses in excess of the applied stress, by up to $\sim 2\sigma_\infty$ [22] (Fig. 38). This enhanced tension can accelerate the cavity propagation process and thus prematurely initiate a crack. The magnitude of the effect can be discerned by incorporating the stress concentration factor into the cavity propagation times derived in section 2. Generally, the maximum reduction in crack nucleation times is in the range, 2-4, depending upon the stress dependence of the dominant cavitation mechanism [22]. Fine grained, or amorphous, zones can also induce stress concentrations of similar magnitude within the surrounding material and reduce the crack nucleation time to a comparable extent (2-4). Additionally, however, the potential for a reduced cavity spacing in fine-grained or amorphous zones (vis-a-vis the surrounding matrix) can cause rapid internal failure of these zones, despite the reduced local stress level. For example, cavitation occurring from three grain edges within a fine grained zone can cause internal failure of the zone at times up to an order of magnitude less than the time needed to induce a crack in the coarse grained matrix. The maximum reduction in local rupture time occurs when the grain size ratio is ~0.2 and the stress within the fine grained region is 0.1 σ_∞. However it is emphasized that internal rupture of this region does not necessarily result in premature failure because the rupture must be capable of extending into the matrix. This topic has not yet been addressed. But, presumably, the stress intensity associated with the local rupture should exceed K_{th} in order to induce eventual failure.

Large amorphous regions may exert an additional detrimental influence on the high temperature failure resistance of ceramics: especially when the amorphous phase exhibits good wetting characteristics and is capable of rapid viscous flow. Then, the amorphous material can flow into cavities created within the solid phase and accelerate their growth. This may be achieved by reducing the dihedral angle (based on good wetting characteristics) and enhancing the matter transport rate from the cavity surface to the cavity tip (i.e., an effective increase in $D_s\delta_s$). For example, noting that the crack-like cavity propagation times at high stress levels (Eqs. (46) and (50a)) are characterized by, $t_p \propto \sin^{3/2} (\psi/4) (D_s\delta_s/D_b\delta_b)^{1/2}$, a perfectly wetting amorphous phase ($\psi \sim 0$) is predicted to reduce the propagation time to ~0 by allowing the formation of crack-like cavities of negligible width. The propagation times are then limited by the flow rate of the viscous material into the cavity, in accord with the principles discussed in section 2.3.2. This process would allow the amorphous zone to spread along a plane normal to the applied tension and thereby induce a substantial reduction in the failure time. The details of this process have not yet been evaluated.

5. INTERPRETATION OF EXPERIMENTAL RESULTS

Plots of the available creep rupture data for ceramics, using logarithmic scales, (Fig. 39) indicates that the temperature dependence of the failure time can be adequately incorporated into an Arrhenius parameter, as anticipated by both the Monkman-Grant and Orr-Sherby-Dorn parameter. The stress exponents, n, are mostly in the range 3-6, except for sintered SiC [49] which has an exponent of ~30. The latter is undoubtedly a consequence of crack propagation controlled failure [50]; while the smaller n values are probably associated with crack nucleation controlled failure. However, explicit correlations with the failure models pertinent to nucleation controlled rupture are limited by the paucity of ancilliary data, such as creep rates and diffusivities. Before embarking upon a closer scrutiny of specific rupture results, it is appropriate to recognize the data correlation

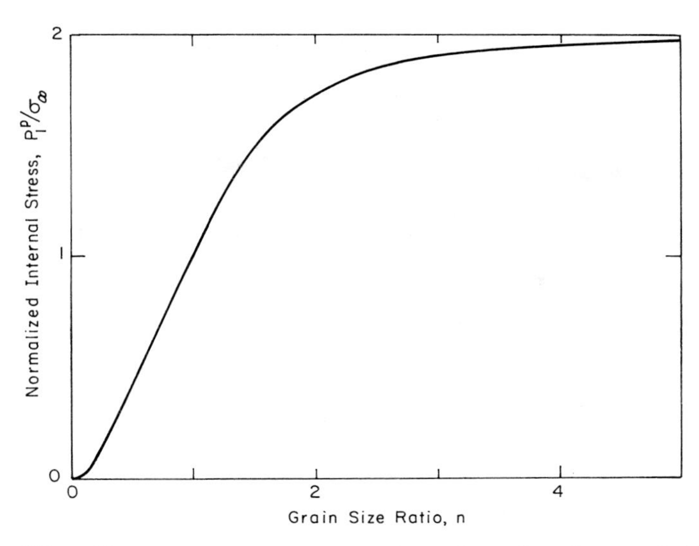

Fig. 38. The stress concentration that develops within a large grained region.

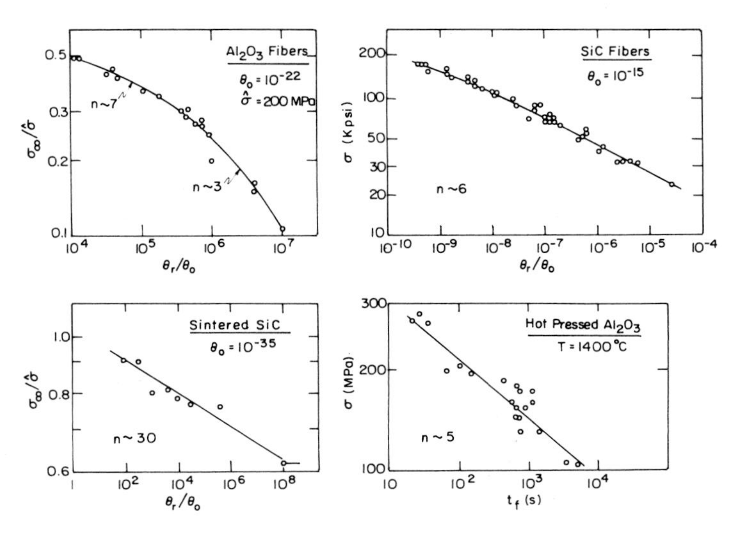

Fig. 39. A summary of creep rupture data for polycrystalline ceramics.

scheme devised by Charles [35]. Based upon an assumption of crack propagation controlled failure and a hypothesized crack growth rate relation,

$$\dot{a} = (D/a) \exp[\beta(2\sigma_\infty\sqrt{a/\rho} - \gamma_s/\rho)] \quad , \qquad (124)$$

where D is a diffusivity, β is a coefficient and ρ is the crack tip radius, the rupture time t_r was derived as;

$$t_r e^{-Q/kT} = \theta_o (N^3/R + 3N^2/R^2 + 6N^3 + 6/R^4)\ (N^3 + 3N^2 + 6N + 6)^{-1} \quad , \qquad (125)$$

where θ_o and N are fitting parameters unique to a specific material and R is the ratio, $\hat{\sigma}/\sigma_\infty$, where $\hat{\sigma}$ is the intrinsic (zero time) strength of the material. All available creep rupture data for ceramics can be fitted by Eq. (125), which thus presents a useful basis for comparing the creep rupture performance of different materials. However, more specific inferences, concerning data extrapolation and underlying mechanisms, should be treated with caution.

A mechanistic interpretation of the rupture data ultimately requires subsidiary microstructural information. Interpretation attempted in the absence of such information should be regarded as speculative. It is tempting to account for the observed stress rupture exponent by invoking the unconstrained, low stress, crack-like cavity growth process ($n = 3$). However, the experimental results have been obtained in a stress range ($\sigma_s \ell/\gamma_s \sin(\psi/4) > 10$) that appreciably exceeds the stresses at which this mechanism ostensibly operates. An interpretation based on crack-like cavity growth should thus be regarded cautiously, in the absence of discrete information concerning cavity shapes. An alternative interpretation, based on the Monkman-Grant relation, is also without basis, because creep rate information has not generally been obtained on the same materials used to determine rupture characteristics. In the one instance (fine grained Al_2O_3) where comparative creep rate and creep rupture data has been acquired the creep exponent ($n_s = 1.8$) is not large enough to account for the stress dependence of the rupture time ($n \approx 5$) and a simple Monkman-Grant relation (e.g., based on highly constrained equilibrium cavity growth) does not appear to rationalize the data.

The compatibility of the available failure data with the probabilistic model of unconstrained cavity growth and coalescence (Fig. 31), is not a sufficient basis for acceptance of a probabilistic interpretation; the parameters of the model must also be consistent with the basic mechanisms of cavity growth. Also, the predicted existence of specimen size effects requires substantiation. The necessary mechanistic information is exemplified by attempting to interpret the rupture data obtained for the Al_2O_3 fibers. Assuming that the Al_2O_3 contains a continuous thin amorphous phase (typical of liquid phase sintered Al_2O_3) and that failure occurs by hole growth within the second phase, the probabilistic model predicts [31]

$$\ell n\theta_r = \ell n[0.3\eta_o(\ell/\delta_o)^2] - \ell n\sigma_\infty - \sigma_\infty^2(4\ell/\pi m K_f^2)\ell n(4A_T/\ell^2) \quad , \tag{126}$$

where η is the viscosity coefficient ($\eta=\eta_o e^{-Q_\eta/kT}$). Correlation of Eq. (126) with the test data for Al_2O_3 fibers (Fig. 31b) is achieved by firstly evaluating the mechanistic parameters from the test data;

$$(4\ell/\pi\eta K_f^2)\ \ell n(4A_T/\ell^2) = 3 \times 10^{-16}\ N^2 m^{-1} \quad , \tag{127a}$$

$$\log\ [0.3\eta_o(\ell/\delta_o)^2] = -5.5 \quad . \tag{127b}$$

Then, by inserting the pertinent dimensional information (area tested, $\sim 1.2 \times 10^{-5}\ m^2$, grain facet length, $5\mu m$) the consistency of the remaining microstructral parameters can be assessed. For example, assuming that $m \approx 1$ (as determined from creep tests on liquid phase sintered Si_3N_4) [31], Eq. (127a) gives $K_f \approx 0.5\ MPa\sqrt{m}$; which compares with $K_c \approx 2\ MPa\sqrt{m}$ for typical polycrystalline aluminas at comparable test temperatures. Additionally, by assuming that $\delta_o \approx 5$ nm (as, again, observed for Si_3N_4) Eq. (127b) yields a viscosity coefficient $\eta_o \approx 10^{-1}$ poise, which in conjunction with the experimentally determined activation energy ($Q_\eta = 115$ kcal/mole), gives a resultant viscosity [31] within an order of magnitude of the viscosity of SiO_2 (a predominant second phase constituent in liquid phase sintered Al_2O_3). This correlation is, of course, based on too many assumed parameters (although, all of the parameters exhibit reasonable values); but it illustrates the detailed microstructural information needed to substantiate failure models and consequently, to develop a prediction capability.

6. IMPLICATIONS AND CONCLUSIONS

The observations and analysis of high temperature cavitation summarized in the present review indicate the inhomogeneous nature of high temperature failure in ceramics. A possible consequence of the inhomogeneity (and the resultant development of constraint) is the inverse dependence of the failure time on the steady-state creep rate of the material (Monkman-Grant behavior). Under crack nucleation controlled conditions within this regime, any microstructural modification that reduces the creep rate should thus produce a proportional increase in the failure time. This correlation provides an invaluable basis for the design of failure resistant microstructures.

Monkman-Grant behavior may be violated under certain conditions; notably especially in the presence of a high proportion of cavitation susceptible boundaries. Constraint effects are then minimal and crack nucleation controlled failure is based on the statistical accumulation of contiguous cavities. A probabilistic analysis of this process indicates that failure in this instance is governed by an Orr-Sherby-Dorn parameter, such that the activation energy term in the parameter is related to that for the dominant cavitation process.

The failure times in both the Monkman-Grant and Orr-Sherby-Grant regimes are predicted to depend on several microstructural characteristics. In single phase materials, low values of the dihedral angle and of surface diffusivity are found to be deleterious. Low dihedral angles (high grain boundary energies) may be inevitable in ceramics (by virtue of covalent or ionic bonding characteristics). However, there may be important influences (both beneficial and deleterious) of solutes, which merit further study. A low surface diffusivity may also be inevitable for typical ceramics, as required for the initial stage sintering. But again, explorations of the temperature dependence of the diffusivity and of solute effects may indicate situations which retard cavitation without detracting from the sinterability.

In two phase materials with a continuous second phase, the predominant material variables are the thickness of the second phase, its viscosity, and the diffusivity of the major second phase constituent. Large values of the second phase thickness and diffusivity, or low viscosities, encourage rupture, as might be intuitively expected. Chemical control is thus a central concern for the creep rupture of these materials.

It also has been demonstrated that several important sources of premature crack nucleation can exist in typical ceramics:[x] in particular, zones of amorphous material in otherwise single phase materials and atypically grained zones. Premature failure results from the development of stress concentrations and/or region of high local cavitation

---

[x] It is notable that these heterogeneities differ in character from those that typically dictate the brittle fracture process at lower temperatures.

susceptibility. The elimination of large scale heterogeneities is thus an essential requirement for the prevention of premature failure.

Crack nucleation controlled creep rupture is expected to pertain to long lifetime behavior, particularly at elevated temperatures. However, the conditions that cause failure to be dominated by crack nucleation, rather than crack propagation, are still rather nebulous; although, observations of crack propagation thresholds begin to suggest crack tip blunting effects which distinguish nucleation control from propagation control.

When crack growth controls failure (as might be expected, for example, in the presence of surface cracks subject to stress intensity levels in excess of the threshold), a characterization of the crack growth rates in term of C_s^* (or K) provides a basis for predicting failure. However, effects of microstructure on the observed crack growth rates have yet to be adequately modelled.

The paucity of comprehensive creep rupture data, and of concomittant microstructural information, obtained on ceramic polycrystals has limited the present ability to distinguish the conditions of stress, temperature and microstructure that dictate the dominant operative realms of the various cavitation mechanisms. Future study should focus upon the acquisition of reliable creep rupture data and the concurrent determination of the important microstructural parameters.

REFERENCES

1. CANNON, R.M., RHODES, W.H. and HEUER, A.H. - Jnl. Amer. Ceram. Soc., 1980, 63, 46.

2. HEUER, A.H., TIGHE, N.J. and CANNON, R.M. - Jnl. Amer. Ceram. Soc., 1980, 63, 53.

3. EVANS, A.G. and BLUMENTHAL, W. - Fracture Mechanics of Ceramics (Eds. R.C. Bradt, D.P.H. Hasselman, F.F. Lange, and A.G. Evans), Plenum, N.Y., 1982, vol. 5.

4. PORTER, J.R., BLUMENTHAL, W. and EVANS, A.G. - Acta Met., 1981, 29, 1899.

5. HSEUH, C.H. and EVANS, A.G. - Acta Met., 1981, 29, 1907.

6. SOLOMAN, A.A. and HSU, F. - Jnl. Amer. Ceram. Soc., 1980, 63, 467.

7. DJEMEL, A., CADOZ, J. and PHILIBERT, J. - 'Creep and Fracture of Engineering Materials and Structures' (Eds. B. Wilshire and D.R.J. Owen), Pineridge, U.K., 1981, p. 381.

8. FOLWEILER, R.C. - Jn. Appl. Phys., 1961, 32, 773.

9. DYSON, B.F. - Can. Met. Quart., 1979, 18, 31.

10. RICE, J.R. - Acta Met. 1981, 29, 675.

11. ESHELBY, J.D. - Proc. Roy. Soc., 1957, A241, 376.

12. RAJ, R. and ASHBY, M.F. - Met. Trans., 1971, 2, 1113.

13. RAJ, R. and ASHBY, M.F. - Acta Met., 1975, 23, 653.

14. CLARKE, D.R. - Ultramicroscopy, 1979, 4, 33.

15. MARION, J., EVANS, A.G., CLARKE, D.R. - to be published.

16. FISHER, J.C. - Jnl. Appl. Phys. 1958, 19, 1062.

17. CHEN, I.W. and ARGON, A.S. - Creep and Fracture of Engineering Materials and Structures, ibid., p. 289.

18. EVANS, A.G., RICE, J.R. and HIRTH, J.P. - Jnl. Amer. Ceram. Soc., 1980, 63, 368.

19. BURTON, B. - 'Diffusional Creep of Polycrystalline Materials,' Trans. Tech. Publ., 1977.

20. CHUANG, T.J. and RICE, J.R. - Acta Met., 1973, 21, 1625.

21. CHUANG, T.J., KAGAWA, K., RICE, J.R. and SILLS, L. - Acta Met., 1979, 27, 265.

22. JOHNSON, S.M., BLUMENTHAL, W.R. and EVANS, A.G., - to be published.

23. PHARR, G.M. and NIX, W.D. - Acta Met., 1979, 27, 1605.

24. LANGE, F.F., CLARKE, D.R., DAVIS, B.I. - Jnl. Mater. Sci., 1980, 15, 601.

25. NIX, W.D. and GOODS, S.H. - 'Fracture', (ed. D. Taplin), University of Waterloo Press, 1977, vol. 2, p. 613.

26. FIELDS, R.J. and ASHBY, M.F. - Phil. Mag., 1976, 33, 33.

27. RAJ, R. and CHYUNG, C.K. - Acta Met., 1981, 29, 159.

28. TSAI, S. and RAJ, R. - to be published.

29. RAJ, R. and DANG, C.H. - Phil. Mag., 1975, 22, 909.

30. EVANS, A.G. - Acta Met., 1980, 28, 1155.

31. EVANS, A.G. and RANA, R. - Acta Met., 1980, 28, 129.

32. STROH, A.N. - Adv. Phys., 1957, 6, 418.

33. McCLINTOCK, F.A. - 'Fracture Mechanics of Ceramics', ibid., 1974, vol. 1, p. 93.

34. WALLES, K.F.A. - Proc. Brit. Ceram. Soc., 1972, 15, 157.

35. CHARLES, R.J. - 'Fracture Mechanics of Ceramics', ibid., 1978, vol. 4, p. 623.

36. BLUMENTHAL, W. - MS Thesis, University of California, Berkeley, 1981, LBL report.

37. LEWIS, M.H., KARUNARATNE, B.S.B., MEREDITH, J. and PICKERING, C. - 'Creep and Fracture of Engineering Materials and Structures', ibid., 1981, p. 365.

38. EVANS, A.G. and CHARLES, E.A. - Acta Met., 1977, 25, 919.

39. WIEDERHORN, S.M. - 'Fracture Mechanics of Ceramics', ibid., 1976, vol. 2, p. 613.

40. BASSANI, J.L. - 'Creep and Fracture of Engineering Materials and Structures', ibid., p. 329.

41. EVANS, A.G., RUSSELL, L.R. and RICHERSON, D.W. - Met. Trans., 1975, 6A, 707.

42. EVANS, A.G. and LANGDON, T.G. - Prog. Mater. Sci., 1976, 21, 171.

43. EVANS, A.G. and WIEDERHORN, S.M. - Jnl. Mater. Sci., 1974, 9, 270.

44. CHUANG, T.J. - Jnl. Amer. Ceram. Soc., 1982, 65, 93.

45. RAJ, R. and BAIK, S. - Metal. Science, 1980, 385.

46. McCLINTOCK, F.A. and MAYSON, H.J. - ASME Applied Mech. Conf., June 1976.

47. HOAGLAND, R.G. and EMBURY, J.D. - Jnl. Amer. Ceram. Soc., 1980, 63, 404.

48. DUGDALE, D.S. - Jnl. Mech. Phys. Solids, 1960, 8, 100.

49. TRANTINA, G.G. and JOHNSON, C.A. - Jnl. Amer. Ceram. Soc., 1975, 58, 344.

Chapter 3

AN EXTRAPOLATION PROCEDURE FOR LONG-TERM CREEP STRAIN AND CREEP LIFE PREDICTION WITH SPECIAL REFERENCE TO $\frac{1}{2}Cr\frac{1}{2}Mo\frac{1}{4}V$ FERRITIC STEELS

R.W. Evans*, J.D. Parker[+] and B. Wilshire*

*Department of Metallurgy and Materials Technology, University College, Singleton Park, Swansea SA2 8PP
[+]Marchwood Engineering Laboratories, CEGB, Marchwood, Southampton. now at Ontario Hydro, 8000 Kipling Avenue, Toronto, Canada.

ABSTRACT

The variations in creep strain, ε, with time, t, for high-precision constant-stress creep curves recorded over a range of stresses from 306 to 125 MNm^{-2} at 838K for $\frac{1}{2}Cr\frac{1}{2}Mo\frac{1}{4}V$ ferritic steel have been shown to be described accurately as

$$\varepsilon = \theta_1(1 - e^{-\theta_2 t}) + \theta_3(e^{\theta_4 t} - 1)$$

The numerical methods, the full computer programme listings and the operating instructions necessary to derive the **θ** functions are detailed in the Appendix. The well-behaved nature of these **θ** functions with respect to stress has enabled predictions to be made of the entire creep curves expected for the low stress levels relevant to service in electricity generating plant (∿40 MNm^{-2}). The serious overestimation of long-term performance obtained by conventional extrapolation of short term stress rupture data is avoided by the new **θ** projection concept developed. Confidence in these new procedures is derived from the fact that the predicted (stress)/(minimum creep rate) relationships coincide exactly with the experimental values available from published low-stress test programmes. Of the many benefits which would follow validation of the proposed forecasting method by laboratories concerned with low stress data accumulation, the most important would be that modern design methods could be exploited through the provision of a full characterization of the long term uniaxial creep and fracture properties of materials rather than just the stress rupture estimates usually forthcoming. Although the maximum test duration of the constant-stress data used in the present analysis was only about three months, the complete behaviour patterns for $\frac{1}{2}Cr\frac{1}{2}Mo\frac{1}{4}V$ ferritic steel with creep lives of up to 30 years and more are offered without reservation for assessment of the **θ** projection concept.

1. INTRODUCTION

Pressure containing components for electricity generating plant must operate for considerable periods under stress at elevated temperatures. In the relevant British Standards, BS 806 and 1113, the design methods for pipework and tubular components serving at temperatures where creep can occur assume operation at constant known pressure and temperature and that the creep life of the component is related to uniaxial creep rupture data through the mean diameter hoop stress σ_{mdh} as

$$\sigma_{mdh} = P\frac{(d - \Delta d)}{2\Delta d} \qquad \text{........(1)}$$

where P is the operating pressure, d is the outside diameter and Δd is the wall thickness. The necessary wall thickness can then be calculated for the operating stress by applying a safety factor (usually $\sim$1.3 to 1.6) to the uniaxial stress resulting in rupture in the stipulated design life [1]. Whilst designs are normally based on 100,000 hour rupture data, considerable economic benefits are derived if even longer lives can be safely achieved. Indeed, components in high-merit power stations may be required to operate for up to 250,000 hours (almost 30 years). It is then necessary to devise reliable means of extrapolating short-term stress rupture data to the design lives involved because of
(a) the major costs incurred by long-term test programmes and
(b) the fact that periods of even ten to twenty years cannot be allowed between the development of new or improved materials and their plant application.

While short term data up to several thousand hours indicate that a linear relationship exists between log (stress) and log (time to rupture), long term studies [2] have demonstrated that extrapolation of short duration tests can lead to serious overestimation of rupture life (t_f) since the stress rupture curves deviate from linearity at longer times (Fig. 1). Even the use of parametric extrapolation techniques such as the Larson-Miller [3] and Manson-Haferd methods [4] overestimate long term life so that only when results are available to, say, 30,000 hours are the 100,000 hour predictions reasonably accurate [2].

Many metallurgical factors may affect the accuracy with which long term uniaxial behaviour can be predicted from the short term creep and stress rupture properties of engineering alloys. The scatter encountered in short term testing is dependent not only on the reliability of the creep equipment but also on small differences in composition within the material specification range and on minor variations in the normal heat treatment procedures employed. Even with high precision testing facilities and precise control of the material composition and microstructure, problems may be encountered because the mechanisms controlling the deformation and fracture behaviour may differ in different stress and temperature regimes. Extrapolation of data from one mechanism zone to another could then overestimate the life at lower stresses and temperatures [5,6]. Moreover, commercial creep resistant alloys generally derive their high temperature strength from the presence of precipitate dispersions which are often unstable during long term creep exposure. The progressive loss of creep and fracture resistance associated with processes such as overaging can then result in the long term properties being inferior to those expected on the basis of short term tests even in the absence of any changes in the mechanisms controlling deformation and failure. The difficulties of predicting long term uniaxial data may therefore be dependent both on the material and on the stress and temperature ranges over which the extrapolation procedures must be employed.

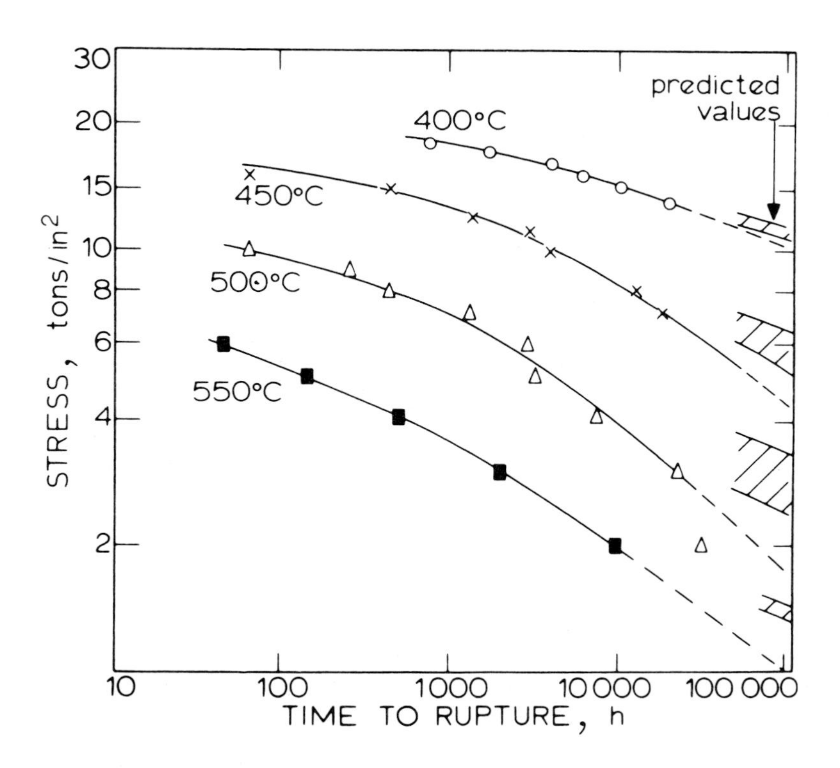

Figure 1. Comparison of the predicted time to fracture with actual data [11].

In order to establish internationally agreed property levels for inclusion in the appropriate materials standards, information provided by approved testing laboratories in Europe, America, Japan and Australia has been collated by the International Organisation for Standardization (ISO). The creep lives recorded at different stresses and temperatures for several casts of the same material are then extrapolated by assuming that a time-temperature parameter can be specified which allows all of the stress rupture data to be presented as a single master curve [7]. Essentially, the ISO has made use of a generalized relationship between absolute temperature and log (time), with the optional inclusion of applied stress, as

$$P(\sigma) = \frac{(\sigma^{-q}\log t) - \log t_a}{(T - T_a)^r} \quad ....(2)$$

where $P(\sigma)$ is the parameter value, σ is the applied stress, t is the testing time, T is the absolute test temperature and q, t_a, T_a and r are selected constants [8]. This generalized equation can then be reduced to specific forms. For example, when $q = 0$ and $r = 1$, equation 2 leads to the Manson-Haferd equation [4],

$$P(\sigma) = \frac{\log t - \log t_a}{T - T_a} \quad ....(3)$$

Similarly, when $q = 0$, $r = -1$ and $T_a = 0$,

$$P(\sigma) = T(\log t - \log t_a) \quad ....(4)$$

which is the Larson-Miller relationship [3].

As a result of the uncertainties associated with procedures involving the use of arbitary constants, extrapolation is generally limited to three times the longest test duration available. Furthermore, the scatter in the multi-laboratory data is quoted as ± 20% of the stress which is equivalent to about an order of magnitude scatter in rupture lives at any specified stress level [7]. Since the lower limit of the scatter band is usually used for design purposes, the procedures adopted may be unnecessarily conservative. An additional problem is then introduced by the need to extend the materials data base in order to take full advantage of modern design procedures [9]. In particular, as improved stress anaylsis techniques become available, reliable long-term creep strain data becomes vital. Unfortunately, relatively little information is available in this area. Consequently, attention is now being focussed on the correlation of creep and rupture strengths which allow the stress to produce a given strain in a specified time to be estimated. Yet, these procedures result in a scatter band which is slightly greater even than the available stress rupture data [9].

In order that all available stress analysis methods can be employed to design electricity generating plant with a degree of conservatism sufficient only to ensure safety, it is therefore necessary that procedures are devised which will provide reliable long-term creep curves for the relevant materials over the stress and temperature ranges encountered in service. The present investigation introduces a new extrapolation technique whereby long-term creep and fracture data may be estimated from short-term test results. Several definitive advantages may then follow if this new procedure can be verified.

(a) The present procedure offers a means of predicting the entire strain/time curve under specified test conditions. The time taken to attain any required creep strain under defined stress and temperature conditions may then be obtained rather than just the rupture life.

(b) The detailed creep curves generated and the rupture lives forecast are for 'constant-stress' conditions. In contrast, most current stress rupture values including those from ISO have been obtained using 'constant-load' tests in which the stress increases steadily as deformation continues throughout the creep life.

(c) The parameters employed in present extrapolation techniques are not arbitary constants but can be interpreted in terms of the detailed mechanisms determining creep and fracture behaviour.

In order that the properties forecast can be assessed in relation to currently available data, the present work was undertaken using results available for $\frac{1}{2}$Cr$\frac{1}{2}$Mo$\frac{1}{4}$V ferrritic steel [10] which is a material widely used in British electricity generating plant.

2. FACTORS AFFECTING THE SHAPE OF CREEP CURVES

The $\frac{1}{2}$Cr$\frac{1}{2}$Mo$\frac{1}{4}$V alloy chosen for the present work was compositionally typical of that used in high temperature components of fossil fuelled power stations. A full analysis is given in Table 1. Prior to testing, the material was normalized for 1 hour at 1238 $\pm$ 15K and then tempered for 3 hours at 973 $\pm$ 10K. These procedures result in complex mixed microstructures of mainly ferritic character with bainitic regions usually forming less than 10% of the volume.

Table 1
Analysis of the $\frac{1}{2}$Cr$\frac{1}{2}$Mo$\frac{1}{4}$V ferritic steel (wt %)

Chromium	0.38	Sulphur	0.035
Molybdenum	0.52	Arsenic	0.02
Vanadium	0.27	Antimony	0.006
Carbon	0.11	Tin	0.02
Manganese	0.47	Aluminium	0.002
Silicon	0.18	Titanium	<0.005
Phosphorus	0.028	Nickel	0.11

Creep tests were carried out mainly at 838K which is the design temperature for components made from $\frac{1}{2}Cr\frac{1}{2}Mo\frac{1}{4}V$ ferritic steel used in electricity generating plant. These tests were performed in tension using high precision, constant stress machines located in a constant temperature room maintained at 303 $\pm$ $\frac{1}{2}$K. Testpieces were prepared having 40 mm gauge length and 8 mm diameter. Changes in gauge length could be resolved to 10 nm using extensometers incorporating a pair of differential capacitance transducers linked to a print out system which enabled a digital record of specimen strain with time to be obtained at any preset time interval of 1 s or more [10].

Over a range of stresses from 125 to 306MNm^{-2}, normal creep curves were invariably recorded. Following the initial specimen strain on loading, the creep rate decreased continuously during the primary stage until a minimum rate was attained before the creep rate eventually accelerated during the tertiary stage which preceded fracture. In all cases, fracture occurred in an intergranular manner. However, the appearance of the true strain/time curves changed considerably over the stress range examined (Fig. 2). The distinct primary stages evident at stress of $\sim$300MNm^{-2} became less pronounced as the total primary creep strain decreased with decreasing applied stress. In contrast, the tertiary stage began progressively to dominate the form of the creep curve with decreasing stress such that, for stresses of $\sim$200 MNm^{-2} and below, the primary stage could be resolved only by detailed examination of the precise strain/time record obtained early in the creep life (Fig. 2).

In view of the dominance of the tertiary stages as the creep lives increased even towards 100 hours, particular attention was directed at clarification of the factors determining tertiary creep behaviour. With constant stress equipment, the acceleration in creep rate at the commencement of the tertiary stage cannot be a result of the gradual increase in stress which occurs with constant load tests when the uniform cross sectional area decreases as creep continues. Instead, this acceleration in rate may be a consequence of necking of the specimen, the development of intergranular cavities and cracks or microstructural instability leading to a progressive loss of creep resistance. The relative importance of these possible causes of tertiary creep in $\frac{1}{2}Cr\frac{1}{2}Mo\frac{1}{4}V$ ferritic steel and their relevance under service conditions can be established by consideration of the results available from microstructural studies carried out not only for the tests of comparatively short durations up to 10,000 hours [10] but also for long term creep tests having rupture lives of over 100,000 hours and for samples taken from components after $\sim$ 100,000 hour service in electricity generating plant [11].

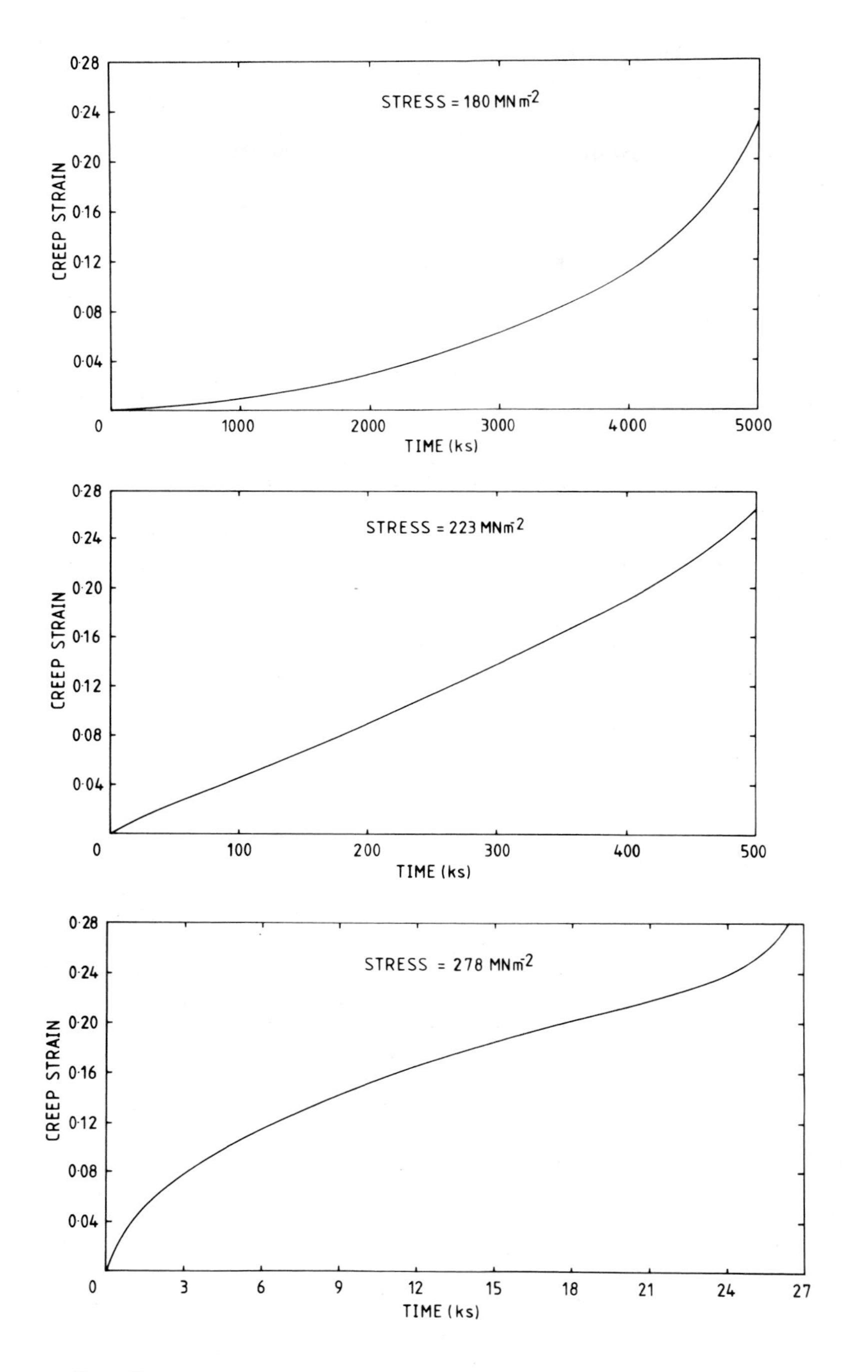

Figure 2. Creep curves recorded in constant stress tests for $\frac{1}{2}Cr\frac{1}{2}Mo\frac{1}{4}V$ steel carried out at applied stresses of 278, 223 and 180 MNm^{-2} at 838K.

2.1 Microstructure of $\frac{1}{2}Cr\frac{1}{2}Mo\frac{1}{4}V$ ferritic steel in the normalized and tempered condition

Several programmes have been aimed at determining the microstructures developed in $\frac{1}{2}Cr\frac{1}{2}Mo\frac{1}{4}V$ steels which had been transformed by continuous cooling from the austenitizing temperature followed by tempering [12-15]. In these investigations, only optical metallography and electron microscopy of carbon replicas were used so that the precise nature of the precipitates developed could not be identified fully. More recently, these techniques have been supplemented by transmission electron microscopy in order that the ferrite morphologies and the carbide precipitate dispersions in this type of steel could be comprehensively examined after isothermal transformation at a number of subcritical temperatures in the ferrite and bainite ranges [16] and after the normalizing and tempering procedures typically employed with components for electricity generating plant [11].

The complex structures generally found in the normalised and tempered condition can be described most easily in terms of their development on cooling from the austenitizing temperature ($1238 \pm 15K$) to form the predominantly ferritic microstructures. It has been established that alloy carbides can precipitate concurrently with the direct transformation of austenite to equiaxed ferrite [16]. The ferrite phase develops from the original austenite boundaries. Periodic precipitation of carbide at the $\gamma - \alpha$ interface then leads to stepwise movement of the phase front and the generation of a banded dispersion of carbide particles in the ferrite regions. The appearance of this interphase vanadium carbide is shown in figure (3a). The carbon content of $\frac{1}{2}Cr\frac{1}{2}Mo\frac{1}{4}V$ steels is approximately twice that needed to precipitate all the vanadium as the carbide. Consequently, as precipitation of interphase vanadium carbide (VC_I) occurs, the carbon content of the remaining austenite increases. During isothermal transformation of these steels this effect eventually slows down the rate of transformation and the spacing of the bands of VC_I increases because the band spacing is related to the rate of movement of the $\gamma - \alpha$ interface [16]. A similar phenomenon has been found for the normalized and tempered condition [11] since the regions of VC_I gradually merged into zones with a more "random" vanadium carbide (VC_R) precipitate dispersion (Fig. 3b). This random VC_R precipitation usually appears as a transition zone between the VC_I (Fig.3a) and a more complex structure composed of vanadium carbide particles associated with dislocations (VC_D) in the ferrite regions obtained on transformation of the remaining austenite as cooling proceeded (Fig. 3c).

Whilst $\frac{1}{2}Cr\frac{1}{2}Mo\frac{1}{4}V$ ferritic steel in the normalised and tempered condition usually shows clear areas of VC_I, VC_R, and VC_D, the proportion of each constituent structure can vary from about 80% VC_I to almost 80% VC_D in any individual cast.

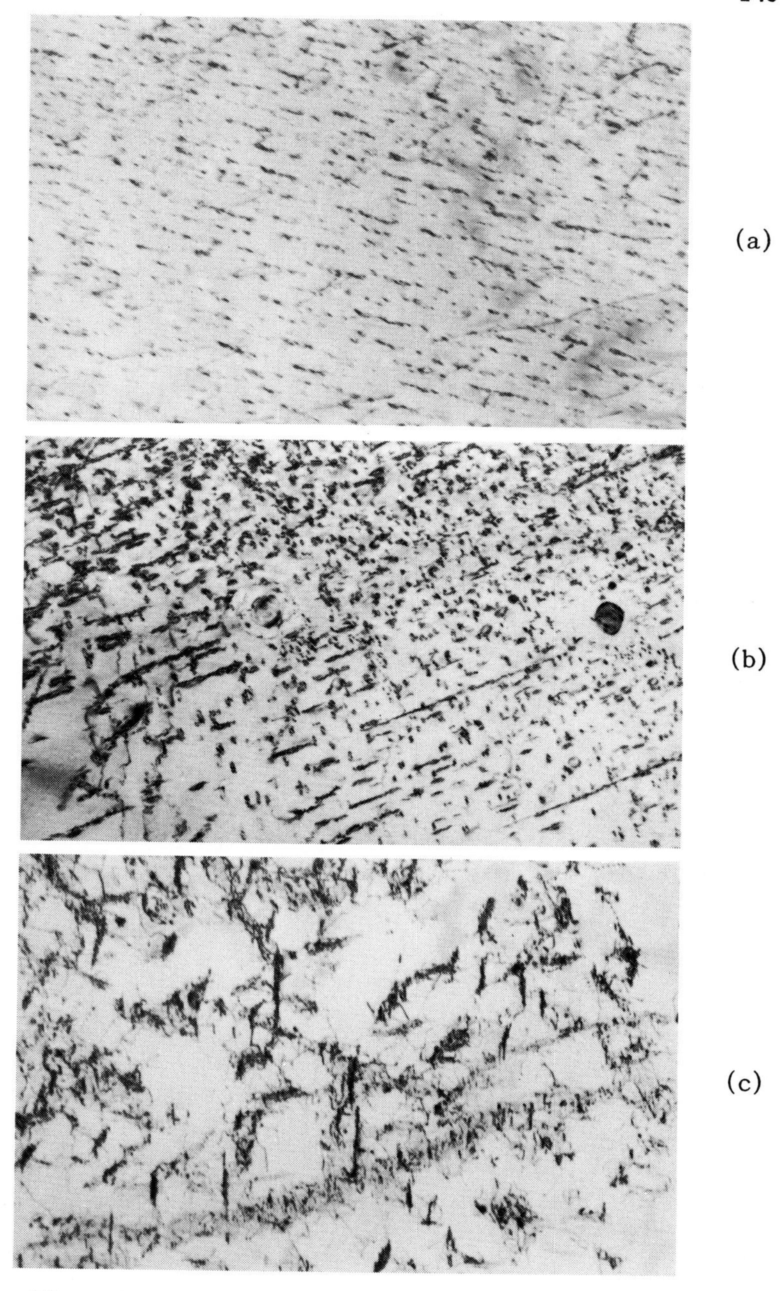

Figure 3. Microstructures present in $\frac{1}{2}$Cr$\frac{1}{2}$Mo$\frac{1}{4}$V ferritic steel in the normalised and tempered condition showing (a) interphase vanadium carbide, VC_I (x40,000), (b) random vanadium carbide VC_R (x40,000) and (c) vanadium carbide associated with dislocations, VC_D (x20,000). [11]

The proportion of each of these forms of precipitate observed in different batches of material appeared to depend primarily on minor variations in cooling rate during normalization [11]. This variation in microstructure could then be a major factor in accounting for the batch-to-batch scatter in the uniaxial creep and stress rupture data reported for this type of steel.

2.2 Microstructural changes during creep of $\frac{1}{2}Cr\frac{1}{2}Mo\frac{1}{4}V$ ferritic steel

In order to determine the extent to which the particle dispersions produced by normalizing and tempering are stable at elevated temperatures, a study has been made [11] of the microstructures of samples taken from components which had operated at nominally between 30 and 50 MNm^{-2} at 838K for periods of between 20,000 and 100,000 hours and from tensile creep specimens tested for similar periods under conditions simulating those encountered in service.

The size, spacing and type of carbides observed in the normalized and tempered conditions (Fig. 3) were found to have changed considerably (Fig. 4). The carbides had coarsened gradually in a manner so as to produce a more uniform structure. In particular, the interphase carbide regions which were easily discernible in the original materials became progressively more difficult to identify. The distinctly different carbide dispersions present initially in different regions of the normalized and tempered materials became increasingly less apparent with increasing service times. In addition to the general increase in size of the vanadium carbide particles, H-type precipitates became increasingly evident (Fig. 4b). These H-type carbides are composed of a central particle rich in vanadium with Mo_2C wings, suggesting that growth of Mo_2C occurs on pre-existing vanadium carbide particles. At temperatures below $\sim$875K, molybdenum diffuses to and dissolves in the vanadium carbide and, as the molybdenum content of the particle increases toward saturation, nucleation of molybdenum carbide (Mo_2C) becomes possible at the periphery of the vanadium carbide plate.

As changes in the size, spacing and form of the carbide particles occurred, the density of dislocations evidence within the grains increased very gradually. The individual dislocations observed within the grains were almost invariably found to be held up at carbides suggesting a gradual loss of creep resistance as particle coarsening takes place. Moreover, growth of grain boundary carbides leads to the progressive formation of particle-free zones adjacent to the boundaries (Fig. 4a). The high dislocation density generally observed within these zones demonstrates that creep of $\frac{1}{2}Cr\frac{1}{2}Mo\frac{1}{4}V$ steels during service at low stress levels occurs in

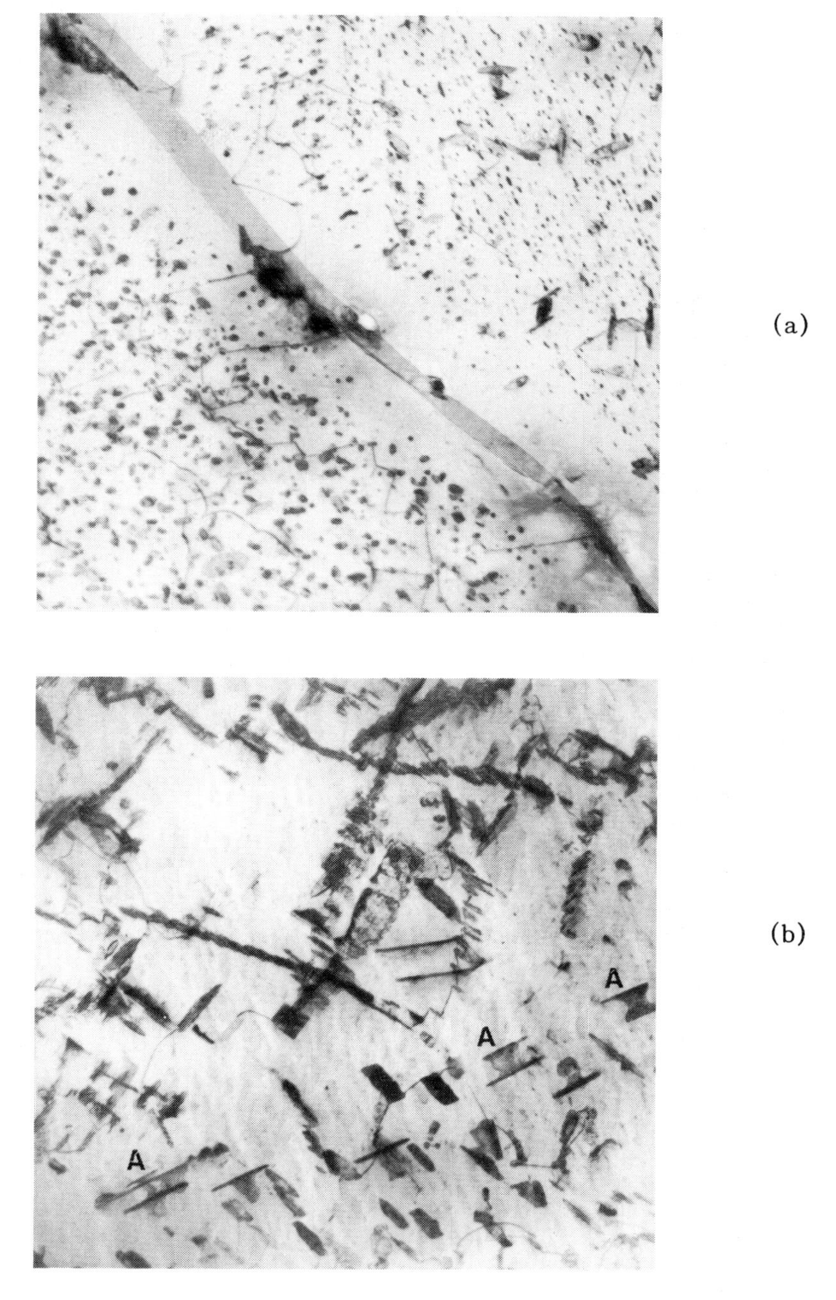

Figure 4. Microstructures developed in $\frac{1}{2}Cr\frac{1}{2}Mo\frac{1}{4}V$ ferritic steel during exposure at 838K in (a) a uniaxial creep test at 49 MNm^{-2} after 35,000 hours(x27,000) and (b) in a main steam pipe after 64,000 hours service at 40 MNm^{-2} (x27,000). [11]

a highly inhomogeneous manner (Fig. 5a). This inhomogeneity of deformation leads to the development of poorly formed subgrain boundaries in limited regions within the ferrite grains (Fig. 5b).

Few cavities were discernible by metallographic examination until the late stages of the creep life in tests carried out at low stress levels [11]. However, the transmission electron microscopy did reveal several examples of the isolation of small cavities within grains due to the occurrence of migration in the precipitate-free zones (Fig. 5c). Cavities so isolated will not continue to grow during subsequent creep. It is therefore suggested that, under conditions comparable with those encountered in service, intergranular cavities and cracks develop rapidly only late in the creep life when changes in particle dispersion have resulted in a significant loss of creep strength.

2.3 Tertiary creep of $\frac{1}{2}Cr\frac{1}{2}Mo\frac{1}{4}V$ ferritic steel

Microstructural examination of service components and samples subjected to long term uniaxial creep exposures therefore has established the carbide dispersion changes gradually with time under stress at elevated temperatures whereas grain boundary cavities and cracks develop in significant numbers only late in the creep life. Since the tertiary stages dominate the creep curves recorded in tests of long duration, the acceleration in creep rate at least during most of the tertiary curve must then reflect only the progressive loss of creep strength associated with microstructural instability.

In order to determine whether this conclusion is applicable under relatively short term test conditions, a metallographic and fractographic study has been carried out for samples tested at stress levels causing fracture at 838K in times up to a maximum of several thousand hours [10]. Over the stress range from 306 to 125 MNm^{-2}, fracture occurred in an intergranular manner. The development of the intergranular cracks was then studied by microstructural examination of testpieces which had been cooled under load after varying fractions of the creep life in a series of tests carried out at 306 MNm^{-2} and 208 MNm^{-2}. At these stresses, creep lives of 2 and 461 hours respectively were recorded. The testpieces were subsequently broken at several positions along the gauge length by impact at low temperatures. Scanning electron microscopy and optical metallography revealed that intergranular voids were absent during most of the creep life except very near to the specimen surface [10]. Whilst these voids could appear as general cavitation with samples having a small diameter, their development just at the specimen surface was clearly evident with the 8mm diameter testpieces used in the present study.

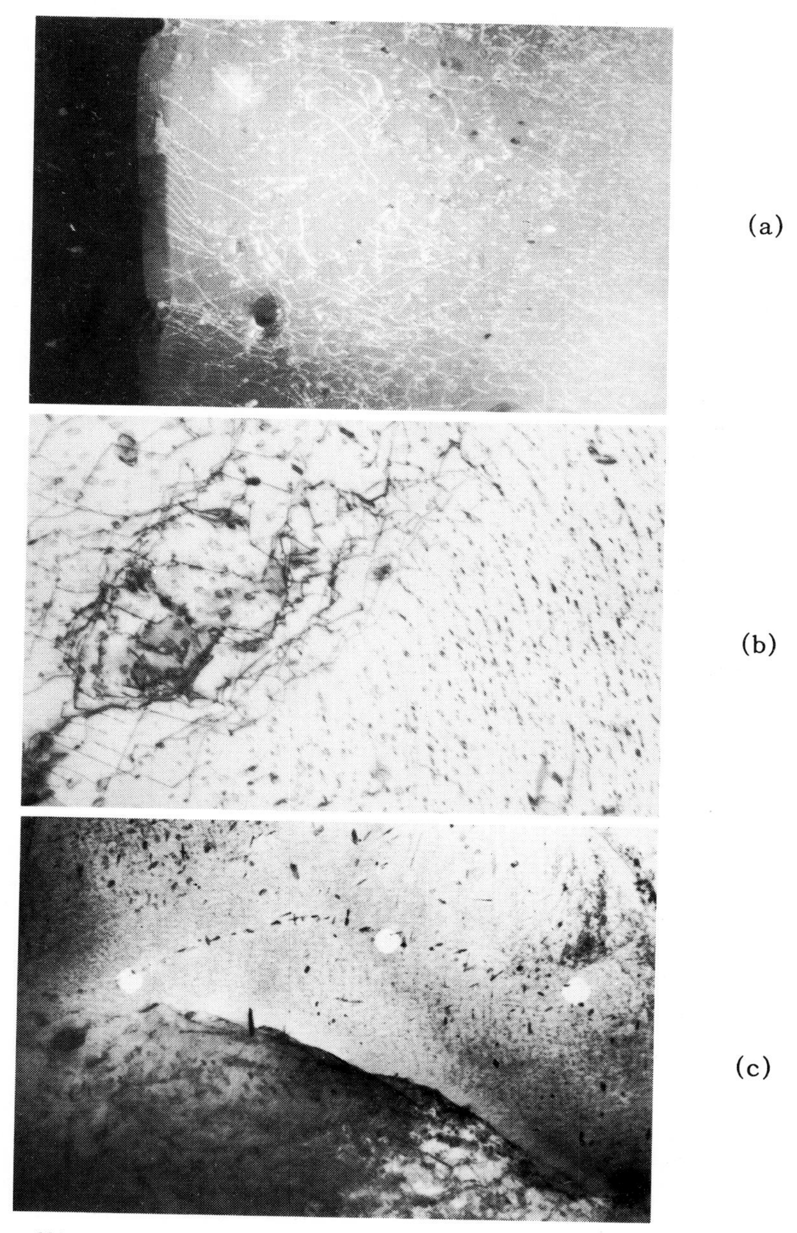

Figure 5. Microstructures developed in $\frac{1}{2}$Cr$\frac{1}{2}$Mo$\frac{1}{4}$V ferritic steel (a) after 50,000 hours service at $\sim$40 MNm^{-2} and 838K in a re-heat header (x30,000, dark field), (b) after 35,000 hours in a uniaxial creep test at 49 MNm^{-2} and 838K (x27,000) and (c) after 60,000 hours at 40 MNm^{-2} and 838K in a main steam pipe (x7,000). [11]

Metallographic examination of longitudinal sections of samples which had failed during creep established that the incidence of internal cavities and cracks decreased with distance from the fracture surface such that no internal cavities and cracks were detectable outside the region where necking was apparent. This conclusion was confirmed by fractographic studies in which the two portions of the failed testpieces were broken away from the necked region by impact at low temperatures. Except for the very limited surface voiding which was discernible, grain boundary cavity and crack development again appeared to be associated with neck formation which occurred only late in the creep life [10].

These observations indicate that, in short duration tests, grain boundary cavities and cracks are associated predominantly with neck formation. Consideration of the stages at which intergranular damage was discernible in relation to the form of the creep curve at each stress then suggests that the processes responsible for the commencement of tertiary creep differ over this high stress range. With tests lasting up to about 100 hours at 838K, the effects of microstructural instability can be neglected and since intergranular cracks developed primarily in the necked region formed late in the creep life, the onset of tertiary creep coincides with the onset of necking (Curve a in figure (2)). In tests lasting more than $\sim$100 hours, as the level of the applied stress is reduced the tertiary stage begins at a progressively earlier fraction of the creep life and certainly well before necking occurs (Curves b and c in figure (2)). Under these conditions, tertiary creep starts as a result of microstructural instability. The tertiary creep behaviour observed at 838K in tests at stresses below $\sim$250 MNm^{-2} therefore apperas to be representative of that encountered both in long duration tests and in service in electricity generating plant.

3. CREEP CURVE FITTING PROCEDURES FOR DATA EXTRAPOLATION

Many different equations have been used to describe the shapes of the creep curves recorded at elevated temperatures. In particular, several studies have established that the changes in creep strain, ε, with time for a wide range of materials can be accurately represented by an equation of the general form [17]

$$\varepsilon = \varepsilon_c - \varepsilon_o = \Phi_1 (1 - e^{-\Phi_2 t}) + \Phi_3 t + \Phi_4 e^{\Phi_5 (t - t_t)} \quad \ldots\ldots\ldots(5)$$

In this case, the creep strain, ε, is defined as the total strain less the initial extension on loading, ε_o, at the commencement of the creep test. The parameters Φ_1 and Φ_2 then describe the primary creep behaviour, Φ_3 , is the secondary creep rate and Φ_4 and Φ_5 quantify the tertiary creep curve. Describing the creep curve by means of equation (5) has then allowed the rate of change of creep rate with time throughout the primary [18,19] and tertiary stages [17,20] to be envisaged in terms of generalised mechanisms of creep and fracture.

In the present analysis, curve fitting procedures are developed which show that the behaviour of $\frac{1}{2}Cr\frac{1}{2}Mo\frac{1}{4}V$ ferritic steel in tests lasting less than one year is well represented by a modified form of equation (5). More importantly, the properties of the parameters defined in the modified equation are such as to allow the present short term data to be extrapolated to the time scales relevant to service in electricity generating plant. The predictions made not only account for the known trends in material behaviour as the creep lives increase but also appear to be fully compatible with the limited amount of long term data available to us on $\frac{1}{2}Cr\frac{1}{2}Mo\frac{1}{4}V$ steel. However, in order to allow the usefulness of the new procedures to be assessed in absolute terms, full predictions are given for all of the relevant creep and fracture properties for creep lives of up to 30 years and more. In this way, the validity of the present predictions are open to scrutiny by laboratories accumulating long term creep or even stress-rupture data for $\frac{1}{2}Cr\frac{1}{2}Mo\frac{1}{4}V$ ferritic steel.

3.1 Representation of creep curves

The form of equation (5) implies that the creep rate observed during the primary stage decreases gradually towards a definite 'steady-state' value, Φ_3, and then, after a period t_t, tertiary creep begins. It is debatable, however, whether a true 'steady-state' condition is ever achieved. For example, superpurity aluminium does not cavitate and, under test conditions such that the material is microstructurally stable, very large creep strains are necessary to cause discernible necking. Yet, in constant-stress tests at temperatures where neither recrystallization nor grain growth occur, superpurity aluminium exhibits a continuously decreasing creep rate even with creep strains in excess of 0.30. An alternative approach is therefore to consider normal creep curves merely as a consequence of two competing events, i.e. a primary creep process which is considered to decay throughout the entire creep life and a tertiary creep process which accelerates from the commencement of the test. The secondary or 'steady-state' rate is then the minimum rate which can appear to be approximately constant over the period of inflection.

With materials such as the present $\frac{1}{2}Cr\frac{1}{2}Mo\frac{1}{4}V$ ferritic steel, this approach has the distinct advantage that the overaging processes responsible for the acceleration in creep rate during the tertiary stage in tests exceeding $\sim$100 hours duration must occur throughout the creep life. Even with materials and test conditions such that the tertiary stage is attributable to grain boundary cavity and crack development, density measurements have shown that cavities are present very early in the creep life and will therefore exert a progressively greater effect as the number and size of cavities increase with increasing creep strain [20]. Only when tertiary creep occurs by necking could it be considered possible to define a time, t_t, when tertiary creep begins. Yet, with necking, the associated acceleration in apparent creep rate at the end of the creep life is rapid so that the values of the parameters defining the sudden tertiary stage would ensure that their contribution to the overall creep rate would not significantly affect the early stages of the creep curve. On this basis, attempts to define a 'time to the onset of tertiary creep', t_t, simply identify the stage at which the acceleration due to the tertiary component is sufficient to cause a discernible deviation away from the apparently constant or minimum rate during the period of inflection.

It is therefore suggested that normal creep curves should be described using an equation of the form

$$\varepsilon = f\{\boldsymbol{\theta},t\} = \theta_1(1 - e^{-\theta_2 t}) + \theta_3(e^{\theta_4 t} - 1) \qquad \text{.......(6)}$$

where $\boldsymbol{\theta} = [\theta_1 , \theta_2 , \theta_3 , \theta_4]^T$

This equation sees the creep curve as an addition of decaying and accelerating exponential functions of time in which $\boldsymbol{\theta}$ plays the role of governing creep curve shape. This can be most readily seen by differentiating equation (6) to give

$$\frac{d\varepsilon}{dt} = \theta_1\theta_2\, e^{-\theta_2 t} + \theta_3\theta_4\, e^{\theta_4 t} \qquad \text{........(7)}$$

and

$$\frac{d^2\varepsilon}{dt^2} = -\,\theta_1\theta_2^{\,2}\, e^{-\theta_2 t} + \theta_3\theta_4^{\,2}\, e^{\theta_4 t} \qquad \text{........(8)}$$

θ_1 and θ_3 act as scaling parameters which control the extent with respect to strain of the primary and tertiary stages of creep. θ_2 and θ_4 have an important effect on curve shape and equation (8) shows that they sensitively affect the curvature of the primary and tertiary stages. Increasing θ_2 and θ_4 rapidly increases the deceleration of creep rate in primary and acceleration in tertiary. The effect is illustrated in figure (6) where θ_2 and θ_4 values increase from curve 1 to curve 5 for constant creep life and rupture strain. It is interesting to see that this variation qualitatively mirrors the changes in creep curve shape described in Section 2.

Equation (8) indicates that the function $f\{\boldsymbol{\theta},t\}$ has a minimum gradient for positive values of θ_i and that this minimum creep rate occurs at time t_m given by

$$t_m = \frac{1}{\theta_2 + \theta_4} \ln \frac{\theta_1 \theta_2{}^2}{\theta_3 \theta_4{}^2} \qquad \text{........(9)}$$

Thus if $\boldsymbol{\theta}$ is known at any set of testing conditions it is possible to calculate t_m and hence through equation (7) the minimum creep rate, $\dot{\varepsilon}_m$. This gives a useful check on the accuracy of creep curve shape description since the calculated value can be directly compared to those observed experimentally. Further, if suitable descriptions of the variation of $\boldsymbol{\theta}$ with stress and temperature can be found then the application of equations (6) to (9) will provide a useful means of extrapolating creep data to provide low stress minimum creep rates.

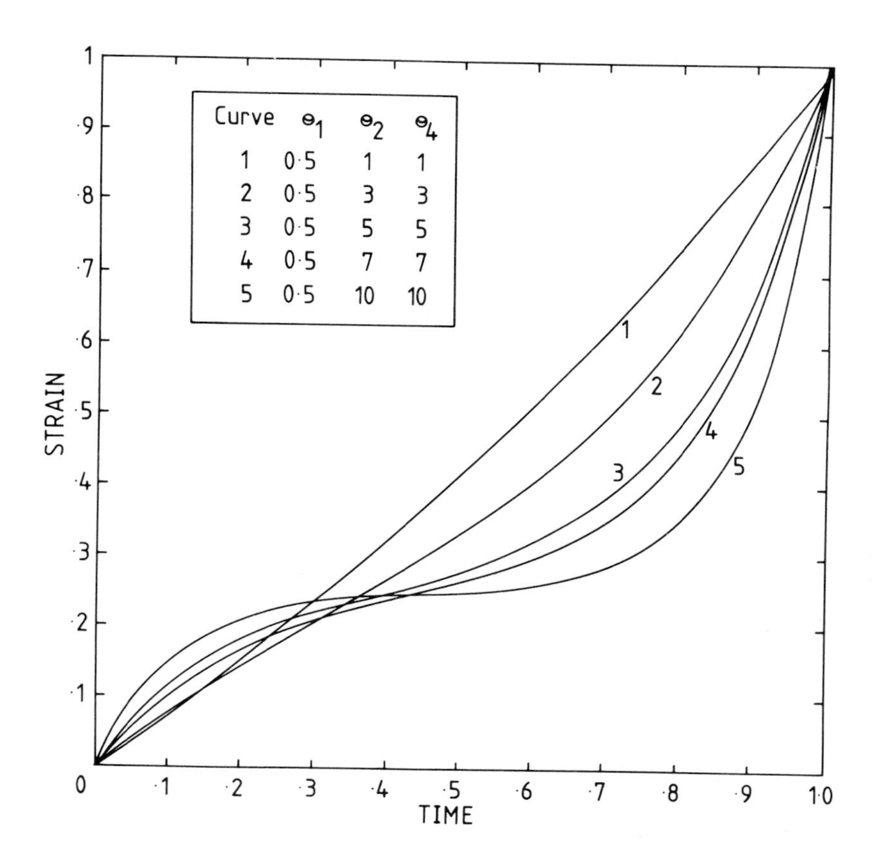

Figure 6. Variation of curve shape with various $\boldsymbol{\theta}$ values.

3.2 Analysis of creep curves

Between one and two hundred strain/time readings were available for each of the constant stress creep curves determined over the stress range 306 to 125 MNm^{-2} at 838K. Each curve has been analysed using equation (6) as an object function. Rough estimates of $\boldsymbol{\theta}$ can be obtained graphically but good estimates require the use of a suitable numerical parameter estimation procedure. In the present case, a least-squares method has been used and the Appendix contains a detailed description. The method uses a guarded Newton-Gauss procedure for minimising the least- squares function and it has been found to converge reasonably well in the present circumstances. The least-squares method has the advantage that confidence intervals for $\boldsymbol{\theta}$ can be obtained in a straight-forward way during the analysis. In all cases analysed in the present work, the final standard error on the strains was always less than 0.4×10^{-3}.

Figure (7) shows the variation of $\boldsymbol{\theta}$ with stress at 838K over the stress range 306 to 125 MNm^{-2}. In all cases, these variations could always be well represented by a linear change of the logarithm of θ_i with stress and the least-squares

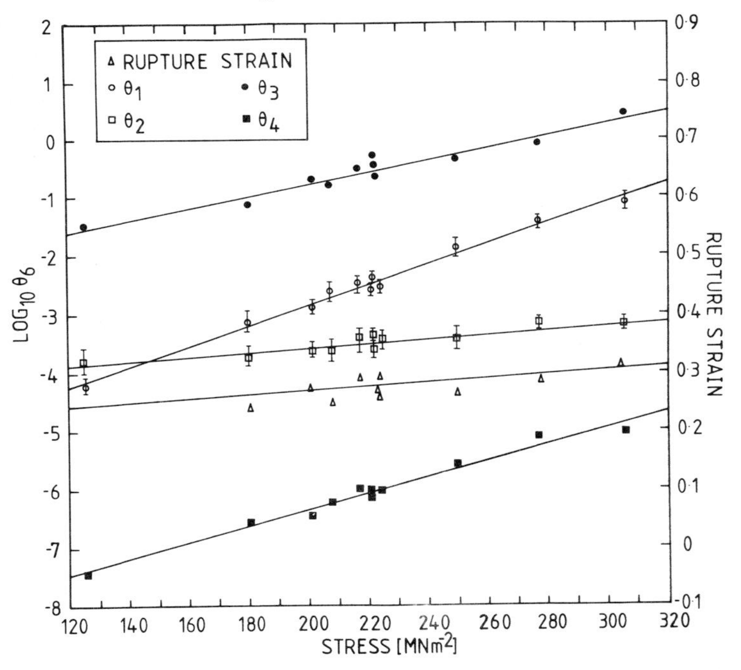

Figure 7. Variation of $\boldsymbol{\theta}$ and rupture strains with stress at 838K. Error bars for θ_i are 95% confidence intervals. The confidence intervals for θ_3 and θ_4 are too small to be represented on the present scale.

estimates of the parameters were

$$g_1(\sigma) = \quad = \exp(-14.5 + 0.040\sigma)$$
$$g_2(\sigma) = \quad = \exp(-9.85 + 0.0081\sigma)$$
$$g_3(\sigma) = \quad = \exp(-6.51 + 0.024\sigma)$$
$$g_4(\sigma) = \quad = \exp(-21.03 + 0.032\sigma) \qquad \text{........(10)}$$

The above description of creep curves is a good representation of deformation behaviour but there is no reference to rupture processes. In order to complete the description, it is necessary to know how the creep curves terminate. This may be done either by a knowledge of rupture times or of rupture strains but both parameters need not be specified since they are related through equations (6) and (10). The quantity which shows the smallest variation with stress over the stress range used in the present work was rupture strain and these strains are shown in figure (7). It is clear that the variation is small but that rupture strain decreases slightly with decreasing stress.

3.3 The $\boldsymbol{\theta}$ projection concept

The well-behaved linear variation of ℓn ($\boldsymbol{\theta}$) with stress suggests that functions (6) and (10) may be a useful means of extrapolating creep behaviour to low stresses. If this is to be done, it is important to establish whether the creep properties predicted agree reasonably well with the low stress-long time tests available. One useful quantity for such a comparison is the minimum creep rate. The minimum creep rate can be readily measured at low stresses since the tertiary accelerating portion of the creep curve commences at a small fraction of the total creep life. The extrapolated minimum creep rates have been calculated from equations (7), (9) and (10) and are shown in figure (8) together with known experimental points from the literature [1,21]. There is good agreement between extrapolated and measured values.

The approach which has become almost universally adopted both for presentation and analysis of creep data obtained at constant temperature involves consideration of the stress dependence of this minimum or 'secondary' rate, $\dot{\varepsilon}_m$ as

$$\dot{\varepsilon}_m = f(\sigma) \qquad \text{........(11)}$$

Of the various functions employed, Norton's law i.e. the power function

$$\dot{\varepsilon}_m \quad \alpha \quad \sigma^n \qquad \text{........(12)}$$

has been most widely adopted. In common with virtually all reported work involving long term test data, the present graph of log ($\dot{\varepsilon}_m$) against log (σ) shows that the gradient, n, varies depending on the stress range. It has then been usual to represent experimental plots by means of straight line segments which, for the $\frac{1}{2}Cr\frac{1}{2}Mo\frac{1}{4}V$ steel, n would be about 8 to 9 at high stresses decreasing towards values of unity at low stresses. However, the present analysis suggests that such representations are merely convenient approximations to the creep rate function of figure (8). Thus, extrapolation through variations in with stress is more satisfactory than trying to extrapolate minimum creep rate/stress plots using 'n' values whose range of applicability can only be guessed at from the few low stress tests usually available.

Extrapolations carried out by the above procedures allows the estimation of much more than minimum creep rates. The complete constant stress curves can be predicted for any stress level and, in particular, can be predicted for low stress levels where detailed experimental creep curves cannot be obtained. Figure (9) shows some predicted low stress creep curves calculated with equations (6) and (10). It is noticeable that they are virtually entirely 'tertiary curves' and that the steepness of the curves at near fracture is such that the rupture lives are only slightly dependent on the final rupture strains.

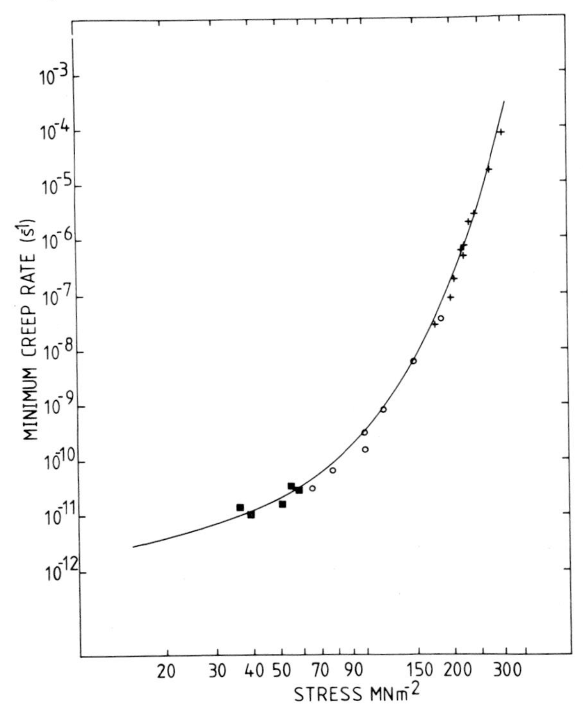

Figure 8. Variation of the calculated minimum creep rate with stress for $\frac{1}{2}Cr\frac{1}{2}Mo\frac{1}{4}V$ steel at 838K. Experimental points are as follows:- + ref. 10, o ref. 1, ■ ref. 21.

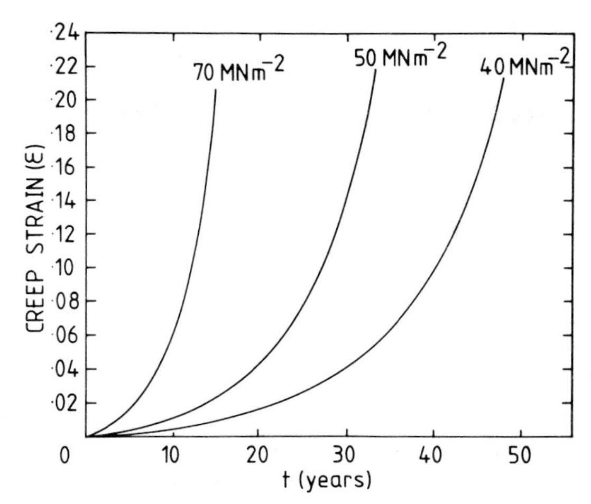

Figure 9. Calculated constant stress creep curves for $\frac{1}{2}Cr\frac{1}{2}Mo\frac{1}{4}V$ ferritic steel at 838K at stress levels approaching those encountered in service in electricity generating plant.

4. IMPLICATIONS OF THE θ PROJECTION CONCEPT FOR DATA EXTRAPOLATION

When the minimum creep rate is considered to vary as a power function of the applied stress (eqn. (12)), it has become widely accepted that different values of the stress exponent, n, are evidence of a change in creep mechanism [5]. With most pure metals, single phase alloys and commercial creep-resistant materials, it is acknowledged that creep at high stresses (when n is greater than about 4) occurs by diffusion-controlled generation and movement of dislocations. In contrast, n values close to unity are almost always discussed in terms of mechanisms involving stress-directed vacancy flow without dislocation movement. However, with $\frac{1}{2}Cr\frac{1}{2}Mo\frac{1}{4}V$ ferritic steel, dislocations were clearly evident even in samples taken from service after up to 100,000 hour exposure at the very low stress levels encountered in electricity generating plant (Fig. 5). More importantly, the dislocation configurations observed in the service material deformed in the $n \simeq 1$ stress regime were not dissimilar from those present after creep under high stresses suggesting that creep of $\frac{1}{2}Cr\frac{1}{2}Mo\frac{1}{4}V$ steel occurs by diffusion-controlled generation and movement of dislocations over the entire stress range down to the levels encountered in service, irrespective of the changes in the apparent 'n' value 'characterizing' the creep behaviour.

The microstructural and fractographic studies (Section 2) demonstrate that not only the deformation mechanism but also the processes governing tertiary creep and fracture do not change below $\sim$250 MNm^{-2}. At high stresses, the tertiary stage becomes apparent late in the creep life and appears to coincide with the onset of necking, with cavities and cracks developing in the necked region. Although neck formation and the associated development of voids was always detectable late in the creep life even in the tests carried out at relatively low stress levels, a clear acceleration in creep rate was evident well before the onset of necking as a result of the progressive loss of creep strength accompanying particle coarsening.

At stresses sufficiently high that dislocations can bow easily around the carbide particles present, pronounced primary stages are recorded and, since the effects of microstructural instability can be ignored under these conditions, the tertiary stage coincides with the onset of discernible necking. As the stress level decreases, the extent of the primary stage (defined by θ_1 and θ_2) decreases to become almost trivial at low stresses. Conversely, the loss of creep strength as overaging proceeds (represented by θ_3 and θ_4) begins to dominate the overall deformation behaviour as the test duration increases, this loss of strength resulting in an acceleration in creep rate which is apparent well before strains sufficient to cause necking are attained. The changes in creep curve shape with decreasing stress (Fig. 2) then simply reflect changes in the dominance of these effects which can be quantified directly in terms of the stress dependences of $\boldsymbol{\theta}$ (eqn. 10). In line with the microstructural evidence, no change in creep mechanism need therefore be invoked to explain the decrease in apparent n value (Fig. 8). Instead, the varying stress dependence of the minimum creep rate is then an inevitable consequence of the changing form of the creep curves which can be expressed precisely in terms of the $\boldsymbol{\theta}$ functions.

The predictions obtained using the $\boldsymbol{\theta}$ projection concept (Fig. 8) therefore appear to be free from problems of mechanism change over the range of extrapolation required. It is also a source of confidence that projections based on high stress data where 'n' is constant have forecast the exact curvature change towards n values of about unity for long-term experimental studies. However the facilities used to generate the present creep curves are of a level of sophistication well in advance of the standard test equipment employed in conventional experimental determinations of long term properties. For example, the creep curves used for extrapolation were obtained under 'constant stress' conditions since the equipment was designed to compensate exactly for the decrease in cross sectional area and the consequent increase in stress occurring as the specimen extends whereas virtually all long term data on commercial alloys is for 'constant load' conditions in which the stress increases progressively throughout the test. In order that the current predictions can be assessed with less uncertainty in relation to actual results from other laboratories, the extent to which conventional constant load behaviour would be unduly pessimistic has therefore been derived from the present constant-stress properties.

Although high precision facilities would undoubtedly be required in order to derive the full benefits which would accrue if the θ projection concept is validated, the increased equipment and operating costs would be trivial in relation to the elimination of the inherent objections to other projection methods or the very considerable expense and problems of long-term testing. In addition, the θ projection concept is ideally suited to the derivation of the exact form of data required to take advantage of modern design methods.

4.1 Consequences of using conventional constant load procedures

The θ prediction method relies on the analysis of creep data obtained from high accuracy constant stress creep curves. In most cases, information is available only in the form of constant load stress-rupture data. Even when full creep tests are conducted, the curves are usually obtained at constant load. In the case of uniaxial tensile deformation, constant load conditions imply a continuous increase in stress with strain so that the shape of the creep curves can be severely distorted as well as the rupture lives shortened. It is important to realise that the use of such data for prediction purposes may lead to considerable error.

The magnitude of these effects can be estimated by calculating the constant load creep curves from the constant stress data summarised in equations (6) and (10). As the specimen strain changes, the increase in stress can be readily calculated and hence changes in θ (which govern the instantaneous creep rate) can be calculated from equation (10). Suppose that a uniaxial tensile creep specimen is loaded at an initial stress σ_o in a constant load test. Then, provided that extension is uniform, at some time t the true strain will be ε and the corresponding stress will be σ' where

$$\sigma' = \sigma_o e^{\varepsilon} \qquad \text{........(13)}$$

The shape of the constant load creep curve can be constructed by integrating equation (7) with appropriate initial conditions provided that variations in θ as a result of stress increases during straining can be adequately represented. Two extreme schemes are possible. In the first scheme, the material is considered to be 'time hardening'. This implies that for any change in stress, the creep rate can be calculated from changes in θ with time remaining constant in equation (7). Alternatively, the material may be assumed to be 'strain hardening' so that changes in creep rate accompanying stress increases are calculated on the basis of strain remaining constant.

For a time hardening rule, the differential equation governing the constant load creep curve at the point (t,ε) is

$$\frac{d\varepsilon}{dt} = g_1(\sigma')g_2(\sigma')\exp\{-g_2(\sigma')t\} + g_3(\sigma')g_4(\sigma')\exp\{g_4(\sigma')t\} \quad (14)$$

for $\varepsilon = 0$ at $t = 0$, ' given by equation (13) and g_i by equation (10).

For a strain hardening rule the corresponding differential equation at (t,ε) is

$$\frac{d\varepsilon}{dt} = g_1(\sigma')g_2(\sigma')\exp\{-g_2(\sigma')t'\} + g_3(\sigma')g_4(\sigma')\{\exp g_4(\sigma')t'\} (15)$$

for $\varepsilon = 0$, $t = 0$. In this case t' is the time given by the solution of the equation

$$g_1(\sigma')\{1 - \exp(-g_2(\sigma')t)\} + g_2(\sigma')\{\exp(g_4(\sigma')t) - 1\} - \varepsilon = 0 \quad (16)$$

For the two hardening rules, equations (14) and (15) have been integrated by a fourth order Runge-Kutta process with numerical values of g_i given by equation (10) and equation (16) being solved by Newton's method. Figure (10) shows the resulting constant load creep curves for high and low

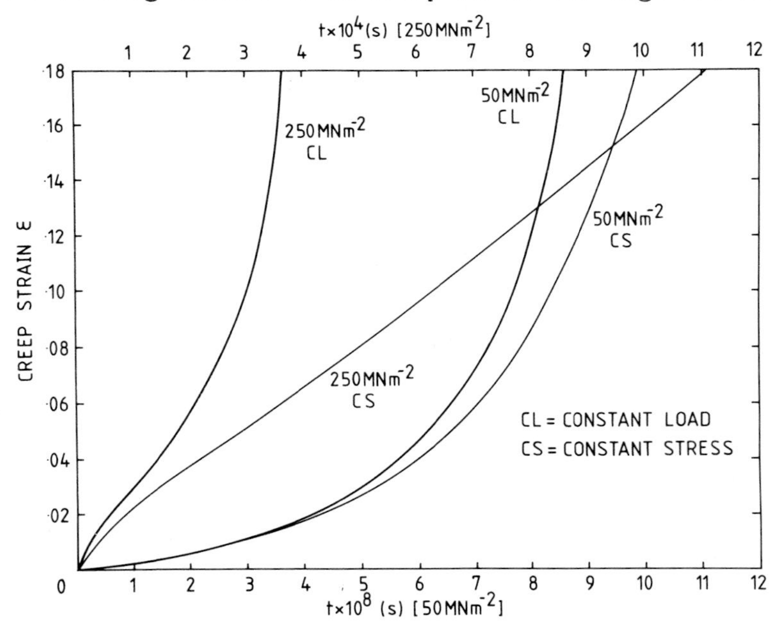

Figure 10. Calculated constant load creep curves in comparison with the corresponding constant stress curves at 250 and 50 MNm^{-2} and 838K for $\frac{1}{2}Cr\frac{1}{2}Mo\frac{1}{4}V$ steel.

stresses as well as the basic constant stress creep curves at the same initial stresses. The constant load curves are the mean of those produced by the two hardening rules. It is clear that the effects of constant load testing are particularly severe at high stresses where not only is the rupture time curtailed but the entire shape of the creep curve is altered. Table 2 gives a more detailed summary of the constant stress/constant load comparisons. At 50 MNm^{-2} the 'time to a given strain' ratios for constant load and constant stress are not very different from 1 even at a strain of 0.2. On the other hand at 300 MNm^{-2} the ratios are small, the constant load life at a strain of 0.2 being less than 1/5 the life at constant stress. The difference in the constant load/constant stress behaviour at high and low stress levels is clearly a result of the change of creep curve shape with stress. At high stresses much of the strain occurs in primary creep so that stress intensification (and hence creep rate acceleration) occurs early in the test leading to much reduced life. On the other hand, at low stresses the creep curve shape is such that the major proportion of strain takes place late on in the tertiary stage so that stress intensification also occurs late and has little effect on creep life.

Table 2

Ratio of the time to a given strain (ε) at constant load to that at constant stress for various initial stresses (σ_0). The upper figures represent time hardening and the lower represent strain hardening rules.

σ_0 \ ε	0.01	0.02	0.05	0.10	0.20
50 MNm^{-2}	0.986 0.964	0.976 0.959	0.945 0.953	0.897 0.937	0.819 0.908
100 MNm^{-2}	0.973 0.963	0.949 0.947	0.886 0.912	0.793 0.866	0.671 0.797
150 MNm^{-2}	0.957 0.955	0.921 0.923	0.824 0.848	0.698 0.760	0.533 0.641
200 MNm^{-2}	0.936 0.959	0.886 0.900	0.761 0.778	0.618 0.654	0.414 0.471
250 MNm^{-2}	0.932 0.923	0.818 0.779	0.638 0.606	0.483 0.476	0.299 0.304
300 MNm^{-2}	0.926 0.925	0.807 0.770	0.620 0.600	0.432 0.407	0.195 0.178

It is clear that constant load creep data must be treated with caution if it is to be used for low stress prediction since the distortions caused by this testing method vary with stress level. If extrapolations are to be made from high stress levels then quantities such as minimum creep rate and rupture times will be considerably in error since the high stress range is where distortions are most severe. It has been common to attempt some creep extrapolations on the basis of constant load creep-rupture data alone. In particular, Norton's n values have been estimated from the stress-time to rupture graphs. Table 2 indicates that any n value obtained in such a way must be treated with considerable suspicion.

4.2 Derivation of long-term iso-strain data

Modern plant design and assessment procedures require a greater knowledge of long term creep behaviour than simple creep-rupture data. Finite element procedures for calculating stresses and strains in components require to know the variation of creep strain rate at all creep strains as input parameters. Similarly any design code which uses as operational lifetime the time to a specified strain (say 1%) must have a precise knowledge of the shape of creep curves at low stresses. The present θ projection method allows the straightforward calculation of the contours in stress-time space for various total accumulated strains.

In order to calculate the time, t', required to attain a strain at an initial stress σ, it is only necessary to solve (numerically) the equation

$$g_1(\sigma)\ \{1 - \exp(-g_2(\sigma)t)\} + g_3(\sigma)\{\exp(g_4(\sigma)t) - 1\} - \varepsilon = 0 \quad (17)$$

with g_i being given by equation (10).Figure (11) shows the results of such calculations for constant stress tests in the range 10 to 320 MNm^{-2}. The technique allows the calculation of the strain/time curves for any stress in the range but, before rupture life can be estimated, it is necessary to have some knowledge of rupture strains. At high stresses, these strains are only slowly varying functions of stress (Fig. 7) but this may not be so at low stresses and this would appear to raise difficulties in the estimation of low stress rupture lives. However, the form of the curves shown in Figure (11) is such that the exact choice of rupture strain for life estimation becomes less critical as stress decreases. The curves reflect the fact that the curvature of the tertiary stage increases rapidly as stress decreases so that the exact rupture strain has only a small influence on rupture time. For instance, at 200 MNm^{-2}, the estimated rupture times for the extreme choice of rupture strains of 0.01 and 0.30 are in the ratio of 1 : 21 whereas for the same assumed rupture strains at 40 MNm^{-2} the rupture life ratio is only 1 : 3.

Thus choices of the assumed rupture strains become less and less important in estimating life as the stresses decrease toward those areas of particular practical importance.

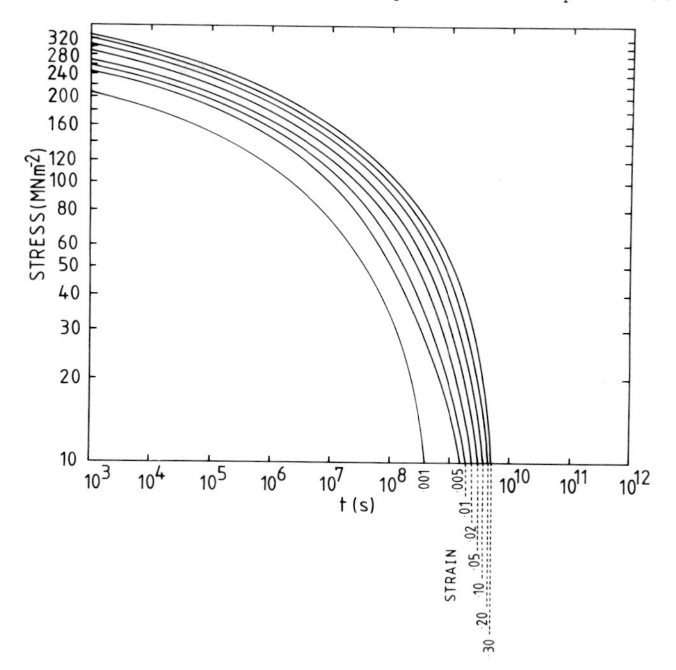

Figure 11. Calculated iso-strain data for $\frac{1}{2}$Cr$\frac{1}{2}$Mo$\frac{1}{4}$V ferritic steel at 838K. The curves are for constant true strains between 0.001 and 0.3 under constant stress conditions.

4.3 Implications of the θ projection concept

The θ projection concept predicts the exact curvature of the experimentally determined long term $\log\sigma/\log\dot{\varepsilon}_m$ relationships (Fig. 8) from the gradual changes in creep curve shape which are already apparent at stresses well above the levels at which the pronounced decrease in 'n' values becomes evident. In a similar manner, the projected form of the iso-strain data (Fig. 11) provides a direct interpretation for the curvature of the standard $\log\sigma/\log t_f$ plots (Fig. 1). The serious overestimation of long term performance caused by conventional methods of extrapolating $\log\sigma/\log\dot{\varepsilon}_m$ and $\log\sigma/\log t_f$ data from high to low stress ranges is then attributable entirely to a failure to recognise that the minimum creep rate, the rupture life and many other standard quantities are determined only by the form of the creep curves. When the θ parameters are defined in a manner which allows the changing form of the creep curve to be quantified with respect to the applied stress, a consistent pattern of behaviour emerges.

It must be emphasised that all of the present predictions, which extend even beyond the low stress ranges appropriate to electricity generating plant are made on the basis of analysis of creep curves with a maximum rupture life of about 3 months. The range of projection is therefore well over an order of magnitude greater than the factor of three in creep life considered reliable with ISO procedures [7,9]. With any forecasting procedure, the accuracy of the predictions must inevitably improve as the extent of the extrapolation decreases. Since forecasts are generally made on the basis of continuity of trends, the shorter the extent of extrapolation, the lower is the risk of unforeseen events influencing the predictions. Yet, the present analysis (Section 3) and the microstructural studies quoted in support (Section 2) suggest that no change in either creep or fracture mechanism will cause modification of the current projections even though the marked decrease in n value occurring over the range of extrapolation (Fig. 8) is usually taken as indicative of mechanism changes. Furthermore, the confidence intervals established for the θ functions (eqn.(10)) are sufficiently small to suggest that only trivial improvements in accuracy of the predictions would result from extending the actual test base beyond the present limited range. Despite the fact that the predicted properties are for creep lives up to two orders of magnitude greater than the maximum test data available to us, the predictions given are presented for assessment without reservation.

If the θ projection concept can be validated for materials of the $\frac{1}{2}Cr\frac{1}{2}Mo\frac{1}{4}V$ type, the approach developed would appear to offer numerous advantages over other methods of creep and fracture extrapolation. Indeed, in order to provide the fullest possible basis for assessment of the current procedures and their predictive capabilities, the implications of the θ projection analysis for other extrapolation methods are now discussed.

Consideration of the gradually changing forms of the creep curves suggests that attempts to base long term predictions on extrapolation of the quantitites normally derived from high stress data can introduce problems. For example, the curvature of the $\log\sigma/\log\dot{\varepsilon}_m$ and $\log\sigma/\log t_f$ plots are sufficient to limit their use to very short extrapolation ranges. However, in the high stress range, the product of the minimum creep rate and the time to rupture has been found to be approximately constant for a range of pure metals, single phase alloys and commercial materials [22]. If the $\dot{\varepsilon}_m.t_f$ product is essentially stress-independent, a method of long term life estimation from short duration test programmes would appear possible since the minimum creep rate is attained in relatively short times even at low stresses (Fig. 2). However, this simplistic approach would seem to result in overestimation of long term performance almost as serious as that associated with direct extrapolation of short term stress rupture data. The present analysis shows that the changing form of the creep curves affects differentially the stress dependences of the minimum creep rate and the rupture life (or time to a specified strain) causing the product $\dot{\varepsilon}_m . t_f$ to decrease rapidly as the stress decreases towards service levels (Table 3). Although the product becomes less sensitive to variations in rupture strain as the test duration increases, the product is not fully stress independent even at high stresses and is certainly affected by variations in rupture ductility in these short duration tests.

An alternative extrapolation procedure entails carrying out tests at the service stress which are accelerated by choosing significantly higher temperatures. The iso-stress data may then be extrapolated to the appropriate service temperature. Unless these tests are performed under vacuum or in inert atmospheres, problems can be encountered due to progressive loss of cross-section of the small specimens by oxidation over what are still relatively long times. Moreover, with steels of the $\frac{1}{2}Cr\frac{1}{2}Mo\frac{1}{4}V$ type, the form and dispersion of the carbides present can be critically dependent on the test temperature so that time dependent changes in carbide distribution occurring in accelerated tests at high temperatures may not be representative of plant conditions.

Table 3
Calculated values of the product $\dot{\varepsilon}_m t_f$ at various stresses (σ) for different assumed rupture strains (ε)

σ \ ε	0.30	0.25	0.20	0.15	0.10
250 MNm^{-2}	0.2452	0.2092	0.1709	0.1302	0.0865
200 MNm^{-2}	0.1779	0.1579	0.1354	0.1097	0.0786
150 MNm^{-2}	0.1019	0.0936	0.0838	0.0720	0.00567
100 MNm^{-2}	0.0484	0.0456	0.0422	0.0379	0.0321
50 MNm^{-2}	0.0203	0.0195	0.0184	0.0170	0.0151

This type of problem has resulted in the adoption by ISO of long-term multi laboratory test programmes which aim to establish internationally agreed property levels for design applications. Unfortunately, the multi laboratory data is associated with wide scatter bands [7,9]. Since the lower limits of the scatter bands are used for design, the long term properties selected may be unnecessarily conservative.

With the narrowness of the composition range for any element in the material specification, variations within the permitted limits seems unlikely to account for the rupture lives varying by more than an order of magnitude at any applied stress and temperature but, even for the same cast of material, some scatter will be introduced by variations in microstructure caused primarily by differences in cooling rate from the austenitizing temperature. Yet, examination of the microstructures after long term exposure show that the marked differences initially obvious in the carbide dispersion (Fig. 3) gradually become less pronounced as the test duration increases (Figs. 4 and 5). This suggests that scatter due to heat treatment variations should decrease as the applied stress decreases towards service levels. For these reasons, it would appear that much of the scatter recorded is attributable to laboratory to laboratory variations.

The θ projection concept, using high-precision constant-stress equipment in conjunction with curve analysis procedures, should largely eliminate test scatter. The gradual approach towards microstructural stability with time should result in property extrapolations, based on high stress data with some scatter due to compositional and heat-treatment variations, converging to narrower limits as the predictions approach service stresses, The test facilities are more expensive than standard creep and certainly stress-rupture machines and the operational care necessary is considerably increased. Yet, if the present forecasting procedure can be validated, these cost increases are offset completely by drastic reductions in the test times necessary and in the extent of test duplication required. By obviating the need for very long test times or enhanced temperatures for test acceleration, θ projections based only on relatively short duration creep curves obtained for the actual service temperature ranges overcome problems associated with specimen oxidation without the necessity for expensive vacuum or inert atmosphere protection. The most spectacular feature of the present approach is, however, the provision of a range of uniaxial properties rather than just rupture lives or other limited quantities. Moreover, these properties are derived for 'constant stress' rather than for less informative 'constant load' test conditions. The θ projection concept therefore offers full characterization of the uniaxial creep and fracture behaviour of materials for exploitation of modern design methods.

5. CONCLUSIONS

(i) Tensile creep tests have been carried out over a range of stresses from 306 to 125 MNm^{-2} at 838K for $\frac{1}{2}Cr\frac{1}{2}Mo\frac{1}{4}V$ ferritic steel using high-precision constant-stress machines. The creep curves obtained are shown to be described accurately using the equation

$$\varepsilon = \theta_1(1 - e^{-\theta_2 t}) + \theta_3(e^{\theta_4 t} - 1)$$

The numerical procedures, full computer programme and operating instructions needed to derive the θ parameters are presented in the Appendix.

(ii) The well behaved variations of the θ functions with respect to stress allow not only a quantitative description of the gradually changing form of the creep curves as the stress decreases but also prediction of the full creep curves for the stress levels encountered in service in electricity generating plant.

(iii) Derivations of the minimum creep rate as a function of stress predict that the stress exponent, n, (i.e. the slope of the log σ/log $\dot{\varepsilon}_m$ relationship) should decrease from $\sim$8 to 9 over the stress range studied towards values approaching unity at low stresses. The predicted minimum creep rates coincide exactly with the published values determined experimentally at low stresses.

(iv) The analysis demonstrates that the decrease in n value, which is normally interpreted in terms of a distinct change in creep mechanism, results simply from the change in form of the creep curve as the stress decreases. The present analysis also accounts for the curvature of log σ/log t_f plots which causes estimates obtained by direct extrapolation of short term rupture data to seriously overestimate actual performance at low stresses.

(v) Microstructural studies undertaken for short duration test samples and for material withdrawn from service up to 100,000 hour exposure in electricity plant support the view that no signficant change in the mechanism of either creep or fracture occurs over the stress range involved in extrapolation.

(vi) Although the maximum creep life of the constant stress tests used for the curve analysis was only about three months, entire creep curves having lives up to 30 years and more have been derived. The small confidence intervals on the θ parameters and the conclusion that no mechanism change occurs down to the stress levels relevant to service at 838K suggest that no unforeseen events should affect these long term creep forecasts. The detailed creep and fracture properties predicted using the **θ** projection concept are therefore presented for scrutiny without reservation.

(vii) If the **θ** projection concept is validated as long term data becomes available for $\frac{1}{2}Cr\frac{1}{2}Mo\frac{1}{4}V$ ferritic steel, numerous advantages will follow. These include dramatic reductions in the current scatter associated with multi laboratory test data, elimination of concern over the possible effects of specimen oxidation during long term test programmes at service temperatures or during accelerated testing using higher temperatures and the provision of 'constant stress' rather than less informative 'constant load' data. More importantly, the **θ** projection concept offers full characterization of the long term uniaxial creep and fracture behaviour needed in order to exploit modern design methods.

REFERENCES

1. BROWNE, R.J., CANE, B.J., PARKER, J.D. and WALTERS, D.J., - 'Creep and Fracture of Engineering Materials and Structures', (Eds. B. Wilshire and D.R.J. Owen), Pineridge Press, Swansea, 1981, p. 645.

2. JOHNSON, R.F., and GLEN, J., - Proc. Conf. on the Creep Strength of Steels and High Temperature Alloys, Metals Society, London, 1973, p. 37.

3. LARSON, F.R., and MILLER, J., - Trans. ASME, 1952, 74, 765.

4. MANSON, S.S., and HAFERD, A.V., - 1953, NACA - TN 2890

5. ASHBY, M.F., - Acta Metall., 1972, 20, 887.

6. ASHBY, M.F., GANDHI, C. and TAPLIN, D.M.R., Acta Metall., 1979, 27, 699.

7. TAYLOR, R.R., and JOHNSON, R.F., - JISI, 1971, 209,714.

8. MENDELSON, A., ROBERTS, E. and MANSON, S.S. - 1965, NASA - TN - D - 2975.

9. GOODMAN, A.M., ORR, J. and DRAPER, J.H.M., Proc. Inst. Conf. on Engineering Aspects of Creep, Inst. Mech. Eng., London, 1980, p. 159.

10. PARKER, J.D. and WILSHIRE, B. - Mech. Materials, 1982. To be published.

11. WILLIAMS, K.R. and WILSHIRE, B., - Mater. Sci. Eng., 1981, 47, 151.

12. WERNER, F.W., EICHELBURGER, T.W. and MANN, E.K. - Trans. ASM, 1960, 52, 376.

13. MURPHY, M.C. and BRANCH, G.D., - JISI, 1969, 207, 1347.

14. NORTON, J.F. and STRONG, A. - JISI, 1969, 207, 193.

15. TIPLER, H.R., TAYLOR, L.H., THOMAS, G.B. WILLIAMSON, J., BRANCH, G.D. and HOPKINS, B.E. - Met. Techol. 1975, 2, 206.

16. DUNLOP, G.L.and HONEYCOMBE, R.W.K., - Metal Sci., 1976, 10, 124.

17. DAVIES, P.W. WILLIAMS, K.R., EVANS, W.J., and WILSHIRE,B., - Scripta Metall., 1969, 3, 671.

18. GAROFALO, F., RICHMOND, C., DOMIS, W.F. and von GEMMINGEN, F. Joint Int. conf. on Creep. Inst. Mech. Eng., London, 1963, pp 1.

19. WEBSTER, G.A., COX. A.P.D. and DORN, J.E., - Metal Sci. J. 1969, 3, 221.

20. EVANS, R.W. and WILSHIRE, B., - Creep and Fracture of Engineering Materials and Structures, (Eds. B. Wilshire and D.R.J. Owen), Pineridge Press, Swansea, 1981, p303.

21. PLASTOW, B. and DAVISON, J., - 1982. To be published.

22. MONKMAN, F.C., and GRANT, N.J., - Proc. ASTM, 1956, 56, 593.

APPENDIX - PARAMETER ESTIMATION

A.1. Variable List

$\mathbf{Q}$	Linear approximation to Hessian of Φ
$\mathbf{D}$	Diag $(Q_{11}, Q_{22}, \ldots, Q_{nn})$
$\mathbf{B}$	$= \mathbf{Q} + \lambda\mathbf{D}$
$\boldsymbol{\theta}$	Parameter Vector
$\mathbf{x}$	Parameter estimator vector
$\mathbf{g}$	Gradient vector of Φ
$\mathbf{s}$	Experiential creep strain vector
$\mathbf{t}$	Time vector
$\mathbf{r}$	Residuals vector $r_j = f(\mathbf{x},t_j) - s_j$
$\mathbf{p}$	Vector of iteration step direction $= -\mathbf{B}^{-1}\mathbf{g}$
$\mathbf{x^*}$	$\mathbf{x}$ which minimises Φ
χ	Theoretical creep strain
$f(\boldsymbol{\theta},t)$	Theoretical creep strain function $= \chi$
ε	Estimated creep strain
$f(\mathbf{x},t)$	Estimated creep curve function $= \varepsilon$
σ_j	Standard deviation of errors in creep strain s_j
Φ	Least square function $\mathbf{r}^T\mathbf{r}$
α	Step length
n	Number of components in $\boldsymbol{\theta}$ and $\mathbf{x}$
m	Number of components in $\mathbf{s}$ and $\mathbf{t}$
i	Counter for $\mathbf{x},\boldsymbol{\theta}$ components $(i = 1,n)$
j	Counter for $\mathbf{s}$, $\mathbf{t}$ components $(j = 1,m)$
λ	Marquardt damping factor

A.2. Introduction

This appendix describes a computational technique for the estimation of the parameters which govern creep curve shape. The programme presented deals with creep strain (χ), time (t) curves given by the equation

$$\chi = \theta_1(1 - e^{-\theta_2 t}) + \theta_3(e^{\theta_4 t} - 1) \qquad \text{............(A.1)}$$

where the various θ are the parameters which are to be estimated. However, the programme is readily adapted to other object functions and, in order to facilitate this, the analysis is presented in terms of a general creep curve function.

A.3. General Statement of the Problem

Suppose the theoretical creep curve is described by the general relationship

$$\chi = f(\boldsymbol{\theta},t) \qquad \text{............(A.2)}$$

where $\boldsymbol{\theta} = [\theta_1, \theta_2, \ldots, \theta_i, \ldots, \theta_n]^T$ is a vector of the n unknown parameters which describe the curve shape. In order to make estimates of $\boldsymbol{\theta}$ there will be available m data points consisting of the experimental variation of the creep strain with time. It is assumed that time can be measured with a very high precision but that the creep strains measured will not co-incide exactly with the theoretical strains, χ, due to the presence of measurement errors arising from a wide variety of sources. It will hardly ever be possible to know $\boldsymbol{\theta}$ exactly but it will be hoped to determine a vector $\mathbf{x} = [x_1, x_2, \ldots, x_i, \ldots, x_n]^T$ which is a sufficiently good estimator of $\boldsymbol{\theta}$ to allow the function

$$\varepsilon = f(\mathbf{x}, t) \qquad \text{.........(A.3)}$$

to adequately represent the creep strain. The measured strains available for analysis will be represented by the vector $\mathbf{s} = [s_1, s_2, \ldots, s_j, \ldots s_m]^T$ and the corresponding times by the vector $\mathbf{t} = [t_1, t_2, \ldots, t_j, \ldots, t_m]^T$. It should be noted that $\mathbf{x}$ is essentially a random variable since for a different set of measurements taken under identical conditions a different $\mathbf{x}$ may be derived by whatever scheme is developed for estimation.

A.4 Estimation

The question immediately arises as to how to ensure that the procedure adopted for choosing $\mathbf{x}$ makes it a good estimator of $\boldsymbol{\theta}$. The question can only be answered satisfactorily if there is information available on the nature of the errors which are present in the measured strains $\mathbf{s}$. Errors can arise from many sources. They may be inherently connected with the sensitivity of the transducer monitoring devices or the linearity of the transducers themselves. They may arise from inadequacies in the extensometer design and manufacture. They will contain a component due to the small changes which are inevitable in specimen and furnace temperature and to random fluctuations in room temperature and draughts. The important point to note is that these errors are unlikely to be systematic for the long times involved in creep testing. It is more probable that each source will contribute its own fraction, on an independent additive basis, to overall error in s_j such that its expected value is zero and its variance is independent of the value of s_j. If it is further assumed that the errors are normally distributed with constant variance then the choice of estimating procedure can be made on a meaningful statistical basis.

For any observed value of strain, s_j, the mean error is zero and its variance is σ_j so that its probability density function is

$$p(s_j) = \frac{1}{\sqrt{2}\pi\sigma j} \exp\left(- \tfrac{1}{2}\left(\frac{s_j - f_j}{\sigma j} \right)^2 \right) \qquad \text{.......(A.4)}$$

where f_j is the expected value of s_j. This will be $f(\boldsymbol{\theta}, t_j)$. For the set of observations $\mathbf{s}$ the conditional probability of obtaining this set for given $\boldsymbol{\theta}$ is

$$p(\mathbf{s}|\boldsymbol{\theta}) = \left(\frac{1}{\sqrt{2}\pi} \right)^m \prod_{j=1}^{m} \frac{1}{\sigma_j} \exp\left(-\tfrac{1}{2} \left| \frac{s_j - f_j}{\sigma_j} \right|^2 \right) \qquad (A.5)$$

Since all the errors are assumed to have the same variances, σ^2, this may be written as

$$\ln p\,(\mathbf{s}|\boldsymbol{\theta}) = -\frac{m}{2} \ln 2\pi - m \ln \sigma - \frac{1}{2\sigma^2} \sum_{j=1}^{m} (s_j - f_j)^2 \quad ..(A.6)$$

The conditional probability, $p\,(\mathbf{s}|\boldsymbol{\theta})$, is the probability of obtaining the observation set **s** for a given **θ**. However it may be regarded as a likelihood function if the **s** are considered fixed and **θ** is regarded as a particular value of the random variable **x**. It is necessary to maximize this probability of **x** being equal to **θ**. The log-likelihood function given in (A.6) contains two terms which are independent of **θ** so that the log-function can be maximized by minimizing the third term on the right hand side. Thus the estimator, **x**, for **θ** is best in a maximum likelihood sense if we minimize the scalar, Φ, given by

$$\Phi = \sum_{j=1}^{m} (s_j - f(\mathbf{x}, t_j))^2 \quad \text{...........}(A.7)$$

The estimation procedure thus requires us to find a value of **x**, equal to **x***, which makes the sum of the squares function, Φ, a minimum. If the residuals $\mathbf{r} = [r_1, r_2, \ldots, r_j, \ldots, r_m]^T$ are defined as the difference between calculated values of strain and observed values then

$$r_j = f(\mathbf{x}, t_j) - s_j \quad \text{...........}(A.8)$$

and $$\Phi = \mathbf{r}^T\mathbf{r} = \sum_{j=1}^{m} r_j^2 \quad \text{...........}(A.9)$$

A.5 Minimization of Φ

Some methods for minimizing Φ operate directly on equation (A.9). These direct search methods merely evaluate Φ for predetermined values of x_i and hence search for minimizers, x^*_i. Although these methods can be reasonably quick for $n \leq 2$, they quickly become very inefficient for large dimensioned **x** because of the enormous number of Φ evaluations required. It is more usual to operate through the equations

$$\frac{\partial \Phi}{\partial x_i} = 0, \quad i = 1, n \quad \text{.........}(A.10)$$

These equations may be written as

$$\sum_{j=1}^{m} (s_j - f_j) \frac{\partial f_j}{\partial x_i} = 0 \qquad i = 1,n \qquad \text{........(A.11)}$$

where $f_j = f(\mathbf{x},t_j)$. If the fundamental equation is linear in $\mathbf{x}$ then the partial derivatives $\frac{\partial f_j}{\partial x_i}$ do not contain x_i so that the equations are linear. In other cases, $\frac{\partial f_j}{\partial x_i}$ contain x_i so that the equations are non-linear and an iterative scheme is required. A method which has proved reasonably successful for creep curve equations is a guarded Gauss-Newton procedure. At the minimizing point, $\mathbf{x}^*$, equation A.11 becomes

$$\sum_{j=1}^{m} (s_j - f_j^*) \frac{\partial f_j^*}{\partial x_i^*} = 0 \qquad i = 1,n \qquad \text{....... (A.12)}$$

where $f_j^* = f(\mathbf{x}^*, t_j)$. We proceed by choosing a value of $\mathbf{x}$ which is not far from $\mathbf{x}^*$. As an approximation to equation A.12 we replace $\frac{\partial f_j^*}{\partial x_j^*}$ by $\frac{\partial f_j}{\partial x_i}$ and replace f_j^* by a Taylor series expansion about $\mathbf{x}$ ignoring items of order higher than one.

i.e. $$f(\mathbf{x}^*,t_j) = f(\mathbf{x},t_j) - \sum_{k=1}^{n} (x_k^* - x_k) \frac{\partial f_j}{\partial x_k} \qquad \text{...(A.13)}$$

so that the equation set A.12 becomes

$$\sum_{j=1}^{m} s_j - f(\mathbf{x},t_j) - \sum_{k=1}^{n} (x_k^* - x_k) \frac{\partial f_j}{\partial x_k} \frac{\partial f_j}{\partial x_i} = 0$$

$$i = 1,n \qquad \text{.......(A.14)}$$

.These n equations are linear in $\mathbf{x^*} - \mathbf{x}$ and can be represented by

$$\mathbf{Q(x^* - x) + g} = 0 \qquad \text{............(A.15)}$$

where $\mathbf{Q}$ is a linear approximation to the Hessian of Φ and $\mathbf{g}$ is its gradient vector.

Gathering terms in equations A.14 yields

$$\mathbf{Q} = [Q_{ik}] = [\sum_{j=1}^{m} \frac{\partial f_j}{\partial x_i} \cdot \frac{\partial f_j}{\partial x_k}]_{nxn} \quad ....(A.16)$$

and $$\mathbf{g} = [g_i] = [\sum_{j=1}^{m} - r_j \frac{\partial f_j}{\partial x_i}] \quad ....(A.17)$$

so that $\mathbf{x}^* = \mathbf{x} - \mathbf{Q}^{-1}\mathbf{g} = \mathbf{x} + \mathbf{p}$ ....(A.18)

where $\mathbf{p} = -\mathbf{Q}^{-1}\mathbf{g}$ is the iterative step direction. In view of the approximations involved in deriving A.14 it is not to be expected that the value of $\mathbf{x}^*$ will be exactly equal to the minimizer of Φ except in the case where $f(\mathbf{x},t)$ is linear in parameters. However, it may be used to continue the iterative process so that the equation can be written in the iterative form

$$\mathbf{x}^{(u+1)} = \mathbf{x}^{(u)} + \mathbf{p}^{(u)} \quad ....(A.19)$$

such that $\mathbf{x}^{(u)} \rightarrow \mathbf{x}^*$ as $u \rightarrow \infty$.

A.6 Implementation of the Method

In practice the implementation of equation A.19 can frequently give convergence failure and the method has to be safeguarded in several ways. These safeguards will be described but it must be emphasised that, as with all non-linear iterations, adequate convergence cannot always be assured in certain difficult numerical situations.

Some of the computation difficulties are associated with the approximations involved in equation A.14. $\mathbf{x}^{(o)}$ may be so far from $\mathbf{x}^*$ that the direction of the step in A.19 is not downhill in Φ at any point. This can be tested for by evaluating $\mathbf{g}^T\mathbf{p}$, for the scalar must be negative for downhill directions. If it is not then the step is unacceptable and either a new $\mathbf{x}^{(o)}$ must be chosen or an alternative temporary iterative scheme used.

Even if $\mathbf{g}^T\mathbf{p} < 0$, $\Phi^{(u+1)}$ may still be greater than $\Phi^{(u)}$. This will generally be because the linear approximation is not accurate enough over the whole step length of equation A.19. In this case, since Φ is known to decrease initially, equation A.19 is modified to

$$\mathbf{x}^{(u+1)} = \mathbf{x}^{(u)} + \alpha^{(u)}\mathbf{p}^{(u)} \quad ....(A.20)$$

where $\alpha^{(u)}$ (the step length) is a scalar between 0 and 1. A number of methods for choosing α are available.

The programmes described subsequently use the method of Meyer and Roth [1] and α is chosen so that $\alpha^{(u)} = 2^{-\nu}$ where ν is the smallest possible integer such that

$$\Phi^{(u+1)} \leq \Phi^{(u)} + 2.10^{-5}.\ \alpha^{(u)}\ \mathbf{g}^{(u)T}\ \mathbf{p}^{(u)} \qquad \text{....(A.21)}$$

During an iteration it frequently happens that **Q** becomes difficult to invert. There are various numerical difficulties which give rise to this condition and it is important to devise a scheme which will make **Q** better conditioned while still maintaining its effectiveness in the iterative scheme. The scheme proposed by Marquardt [2] makes **Q** sufficiently well conditioned by transforming to the matrix **B** by

$$\mathbf{B} = \mathbf{Q} + \lambda\mathbf{D} \qquad \text{.........(A.22)}$$

where **D** is a positive definite matrix and λ a scalar of sufficient size. A good choice for **D** is

$$\mathbf{D} = \text{Diag}\ (Q_{11},\ Q_{22},\ \dots\ Q_{nn}) \qquad \text{.........(A.23)}$$

For large λ, the second term in A.22 dominates **B** so that the iterative step $(x_i^{(u+1)} - x_i^{(u)})$ becomes $-\lambda^{-1} D_{ii}^{-1} g_i$ which is a very small step in the downward direction. If λ is small the method reduces to Gauss-Newton iteration. The value of λ is thus adjusted depending on the degree of ill-conditioning of **Q**.

A further danger is that A.20 may not converge to a local minimum but to a saddle point in Φ. At such a point equations A.11 are satisfied so that no further progress can be made by iteration. In these circumstances it is wise to perform a small grid search around $\mathbf{x}^{(u+1)}$ to ensure that there is not a lower Φ in the immediate neighbourhood. If such a point is found, iteration can recommence at this point.

A.7 Programming

A flow chart for the computation procedures is shown in Fig.A1. The programme reads the m data points of time and strain and calculates the strain rates at each point by calculating the gradient of the least squares line through the point and the two adjacent points. The data is scaled so that strain and time have the range 0 → 1. Initial estimates of the parameters which are reasonably close to the final values are required and these are estimated by a subroutine working on the strain rates previously calculated.

The programme allows up to 100 iterations for convergence and if no satisfactory solution is found in this time, the most recent values of **x** are reported. For the first iteration,

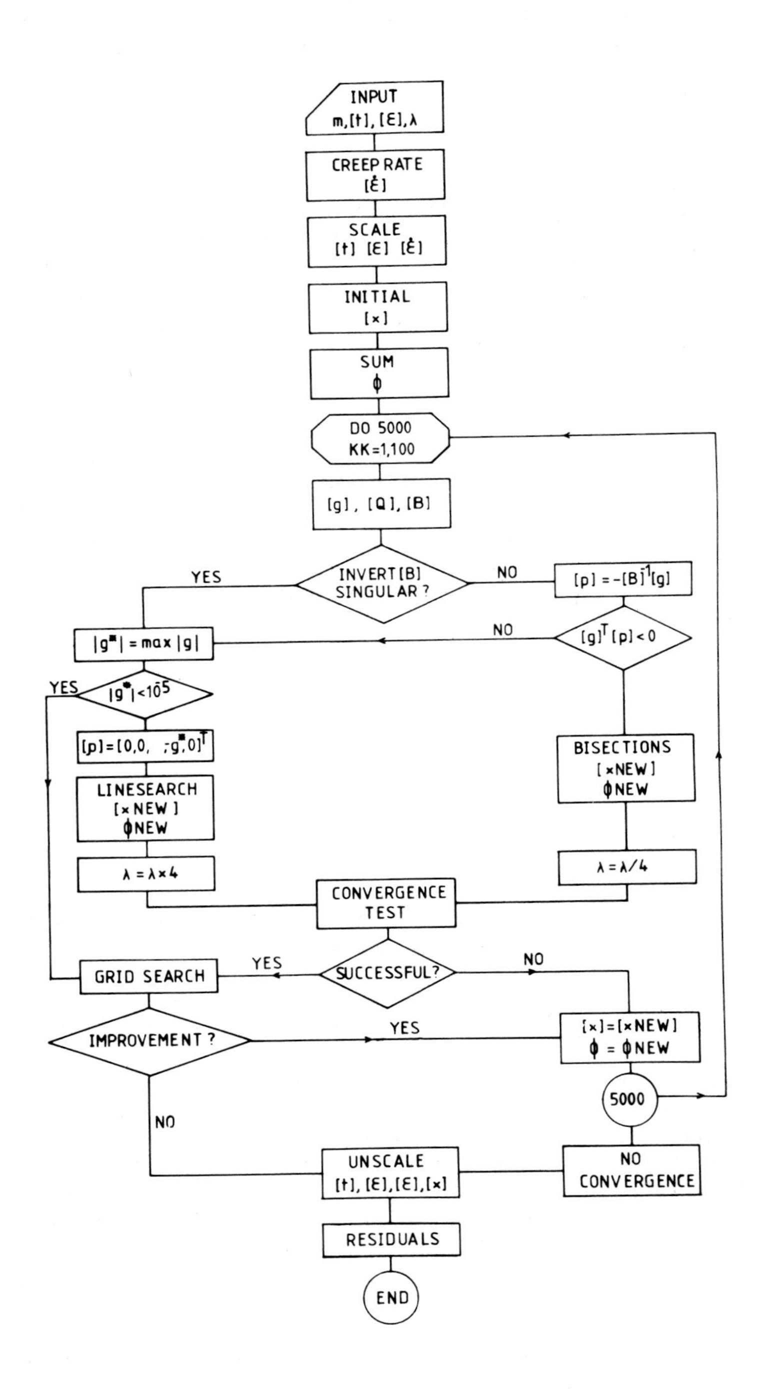

Figure A.1. Flow Chart

λ is set to 0.01 and **Φ**, **g**, **Q**, **D** and **B** are calculated followed by an attempt to invert **B**. If this fails the matrix is assumed singular and the programme proceeds down the left hand side. A new estimate of **x** is obtained by proceeding along a direction $[0,0, \ldots, g_p, \ldots, 0]^T$ (where g_p is the maximum absolute value of **g**) by a distance α chosen by line search routine. λ is increased by 4 times to avoid future difficulties with inversion. The new values of **x** and **Φ** are checked for convergence. If **B** has been inverted successfully, a check is made on the acceptability of the step by evaluation of $\mathbf{g}^T\mathbf{p}$. If it is not acceptable, progress is made by reverting to the technique described above. Otherwise a suitable α is chosen by means of equation A.21 and new estimates of **x** and **Φ** are sent to the convergence check routine.

Convergence is checked by determining whether **x** and **Φ** have ceased to change significantly between iterations. If they have not, iteration is continued. If they have, a final grid search is made and only if this is not successful in finding lower **Φ** is total convergence assumed. Otherwise iteration is re-commenced. Finally, s_j, t_j, and x_i are unscaled to their initial values and the residuals r_j are calculated.

A.8 Use

Data must be inputed in the following FORMAT

Card 1 : M (I3) No. of data points

Card 2 →M + 1 : TIME, STRAIN (2F 10.5)

The programme has been found to perform well for creep curves but some care must be taken to avoid using unsuitable functions. For instance, if primary strains are very small in equation A.1, $\theta_1 \simeq 0$, $\frac{\partial f_j}{\partial x_2} = 0$ and the rows and columns in **Q** containing the term will all be zero. Thus **Q** will be singular and the Marquardt technique will not help much. Convergence via successive line searches will be very slow. It would be better to accept the absence of primary and use a different equation recognizing this fact. The programme has been written to run in FORTRAN IV on an ICL 1904S machine with a 48 bit word length and may require slight modification for other machines.

A.9 References

1. MEYER, R.R. & ROTH, P.M.: J. of Math. and Its applications, 1972, 9, 218.
2. MARQUARDT, D.W., J. of Soc. of Industrial and Applied Math. 1963, 11, 431.

PROGRAMME LISTING

```
      PROGRAM (RELS)
      INPUT 5 = CR0
      OUTPUT 6= LP0
      COMPACT DATA
      END
C
C
C
C
      MASTER REDATA
      DIMENSION EXTIM(250),STRAN(250),RES(250),CRATE(250)
      DIMENSION RESN(250),SCAL(250)
      DIMENSION G(5),P(5),X(5),XN(5),B(5,5),BI(5,5)
      AMDA=0.01
      READ(5,1000)M
 1000 FORMAT(I3)
      DO 1010 J=1,M
 1010 READ(5,1011)EXTIM(J),STRAN(J)
 1011 FORMAT(2E15.5)
      CALL CRRATE(EXTIM,STRAN,CRATE,M,NMIN)
      FACTOR1= STRAN(M)
      FACTOR2= EXTIM(M)
      DO 1030 J=1,M
      EXTIM(J)=EXTIM(J)/FACTOR2
      STRAN(J)=STRAN(J)/FACTOR1
 1030 CRATE(J)=CRATE(J)/FACTOR1*FACTOR2
      CALL INITIAL(STRAN,EXTIM,CRATE,X,M,N,NMIN,IAB)
      IF(IAB.EQ.1)GO TO 7000
      CALL SUM(EXTIM,STRAN,RES,X,M,PHI)
      DO 5000 KK=1,100
      WRITE(6,1300) KK
 1300 FORMAT(/1X,18HITERATION NUMBER  ,I3)
      WRITE(6,1301)
 1301 FORMAT(1X,27HINITIAL SCALED THETA VECTOR)
      WRITE(6,1302)(X(I),I=1,N)
 1302 FORMAT(1X,5E13.5)
      WRITE(6,1303)PHI
 1303 FORMAT(1X,19HINITIAL PHI VALUE  ,E15.6)
      CALL GB(EXTIM,STRAN,RES,X,M,N,AMDA,B,G)
      CALL INVERT(B,BI,N,G,P,DET,IPIVOT)
      IF(IPIVOT.NE.1)GO TO 2000
 1210 GL=0.
      L=0
```

```
      DO 1201 I=1,N
      IF(ABS(G(I)).LE.GL)GO TO 1201
      L=I
      GL=ABS(G(I))
 1201 CONTINUE
      IF(GL.LT..1E-4)GO TO 3500
      DO 1202 I=1,N
 1202 P(I)=0.
      P(L)=-G(L)
      CALL LINSEARCH(PHI,P,N,STRAN,EXTIM,M,XN,X,PHIN,RESN,AMDA)
      GO TO 3000
 2000 GTP=0.
      DO 2100 I=1,N
 2100 GTP= GTP+ G(I)*P(I)
      IF(GTP.GT.0.)GO TO 1201
      CALL BISECT(X,XN,P,STRAN,EXTIM,M,N,G,PHI,PHIN,KK,AMDA,RESN)
 3000 CALL CONVERGE(X,XN,PHI,PHIN,ICON,N)
      IF(ICON.NE.1)GO TO 4000
 3499 GO TO 3501
 3500 DO 3502 I=1,N
 3502 XN(I)=X(I)
      PHIN=PHI
 3501 CALL GRID(XN,PHIN,STRAN,EXTIM,M,N,IGRID,RESN)
      IF(IGRID.NE.1) GO TO 5001
 4000 PHI=PHIN
      DO 4001 I=1,N
 4001 X(I)=XN(I)
      DO 4002 J=1,M
 4002 RES(J)=RESN(J)
 5000 CONTINUE
      WRITE(6,5005)
 5005 FORMAT(/1X,35HNO CONVERGENCE AFTER 100 ITERATIONS)
 5001 WRITE(6,5006)
 5006 FORMAT(1X,25HFINAL SCALED THETA VECTOR)
      WRITE(6,1302)(XN(I),I=1,N)
      CALL DESCALE(EXTIM,STRAN,RES,XN,M,N,FACTOR1,FACTOR2,CRATE)
      CALL SUM(EXTIM,STRAN,RES,XN,M,PHIN)
      WRITE(6,5010)
 5010 FORMAT(/1X,27HFINAL UNSCALED THETA VECTOR)
      WRITE(6,1302)(XN(I),I=1,N)
      WRITE(6,6303)PHIN
 6303 FORMAT(1X,20HFINAL PHI VALUE IS  ,E15.6)
      S=1./(M-1)*SQRT(PHIN)
      WRITE(6,5100)S
 5100 FORMAT(1X,19HSTANDARD ERROR IS  ,E15.6)
      WRITE(6,6000)
 6000 FORMAT(/1X,60H         TIME(S)     EXP. STRAIN    CALC. STRAIN
     1RESIDUAL)
      DO 6001 J=1,M
      SCAL(J)=STRAN(J)-RES(J)
 6001 WRITE(6,6002)EXTIM(J),STRAN(J),SCAL(J),RES(J)
```

```
 6002 FORMAT(1X,4E15.5)
 7000 STOP
      END
C
C
C
C
      SUBROUTINE INITIAL(STRAN,EXTIM,CRATE,X,M,N,NMIN,IAB)
      DIMENSION STRAN(250),EXTIM(250),CRATE(250),X(5)
      N=4
      IAB=0
      WRITE(6,1)
    1 FORMAT(/////1X,22HTHE OBJECT FUNCTION IS)
      WRITE(6,3)
    3 FORMAT(1X,44HE=X(1)*(1-EXP(-X(2)*T))+X(3)*(EXP(X(4)*T)-1))
      WRITE(6,2)M
    2 FORMAT(1X,11HTHERE ARE  ,I3,17H DATA POINTS WITH)
      WRITE(6,4)NMIN
    4 FORMAT(1X,29HMINIMUM CREEP RATE AT POINT  ,I3)
      IF(NMIN.LT.6)GO TO 500
      NM=M-NMIN
      IF(NM.LT.6)GO TO 500
      NM=M-NMIN
      IF(NM.LT.6) GO TO  600
      N1=NMIN/5+1
      N2=NMIN/2+1
      X(2)=(ALOG(CRATE(N2))-ALOG(CRATE(N1)))/(EXTIM(N1)-
     1EXTIM(N2))
      X(1)=CRATE(N2)/(X(2)*EXP(-X(2)*EXTIM(N2)))
      N1= NMIN+(M-NMIN)/2
      N2= NMIN+(M-NMIN)*3/4
      X(4)=(ALOG(CRATE(N2))-ALOG(CRATE(N1)))/(EXTIM(N2)-
     1EXTIM(N1))
      X(3)=CRATE(N2)/(X(4)*EXP(X(4)*EXTIM(N2)))
      GO TO 700
  500 IAB=1
      WRITE(6,501)
  501 FORMAT(1X,29HISUFFICIENT POINTS IN PRIMARY)
      GO TO 700
  600 IAB=1
      WRITE(6,601)
  601 FORMAT(1X,30HISUFFICIENT POINTS IN TERTIARY)
  700 RETURN
      END
C
C
C
C
      SUBROUTINE  CRRATE(EXTIM,STRAN,CRATE,M,NMIN)
      DIMENSION EXTIM(250),STRAN(250),CRATE(250)
      N=M-1
      CRATE(1)=(STRAN(2)-STRAN(1))/(EXTIM(2)-EXTIM(1))
```

```
      CRATE(M)=(STRAN(M)-STRAN(N))/(EXTIM(M)-EXTIM(N))
      DO 100 J=2,N
      N1=J-1
      N3=J+1
      XY=0.
      XX=0.
      X=0.
      Y=0.
      DO 200 I=N1,N3
      XY=XY+STRAN(I)*EXTIM(I)
      XX=XX+EXTIM(I)**2
      X=X+EXTIM(I)
  200 Y=Y+STRAN(I)
  100 CRATE(J)=(3.*XY-X*Y)/(3.*XX-X*X)
      NMIN=1
      TEMP=1000.
      DO 300 J= 1,M
      IF(CRATE(J).GT.TEMP) GO TO 300
      TEMP=CRATE(J)
      NMIN=J
  300 CONTINUE
      RETURN
      END
C
C
C
C
      SUBROUTINE GB(EXTIM,STRAN,RES,X,M,N,AMDA,B,G)
      DIMENSION STRAN(250),EXTIM(250),RES(250),X(5),C(5),B(5,5)
      DIMENSION G(5)
      DO 10 I=1,N
      G(I)=0.
      DO 10 K=1,N
   10 B(I,K)=0.
      DO 20 J=1,M
      C(1)=1.-EXP(-X(2)*EXTIM(J))
      C(2)= X(1)*EXTIM(J)*EXP(-X(2)*EXTIM(J))
      C(3)= EXP(X(4)*EXTIM(J))-1.
      C(4)= X(3)*EXTIM(J)*EXP(X(4)*EXTIM(J))
      DO 20 I=1,N
      G(I)= G(I)-RES(J)*C(I)
      DO 20 K=1,N
   20 B(I,K)=B(I,K)+C(I)*C(K)
      DO 30 I=1,N
      DO 30 K=1,N
   30 IF(I.EQ.K)B(I,K)=B(I,K)+ AMDA*B(I,K)
      RETURN
      END
C
C
C
C
```

```
      SUBROUTINE SUM(EXTIM,STRAN,RES,X,M,PHI)
      DIMENSION EXTIM(250),STRAN(250),RES(250),X(5)
      PHI=0.
      DO 10 J=1,M
      A1=-X(2)*EXTIM(J)
      IF(A1.GT.5.)GO TO 9
      A1=X(1)*(1.-EXP(A1))
      A2=X(4)*EXTIM(J)
      IF(A2.GT.5.)GO TO 9
      A2=X(3)*(EXP(A2)-1.)
      RES(J)=STRAN(J)-A1-A2
      GO TO 10
    9 RES(J)=200
   10 PHI=PHI+ RES(J)*RES(J)
      RETURN
      END
C
C
C
C
      SUBROUTINE DESCALE(EXTIM,STRAN,RES,XN,M,N,FACTOR1,FACTOR2
     1,CRATE)
      DIMENSION EXTIM(250),STRAN(250),RES(250),XN(5)
      DIMENSION CRATE(250)
      DO 10 J=1,M
      STRAN(J)=STRAN(J)*FACTOR1
      EXTIM(J)=EXTIM(J)*FACTOR2
      CRATE(J)=CRATE(J)*FACTOR1/FACTOR2
   10 RES(J)= RES(J)*FACTOR1
      XN(1)= XN(1)*FACTOR1
      XN(2)= XN(2)/FACTOR2
      XN(3)=XN(3)*FACTOR1
      XN(4)= XN(4)/FACTOR2
      RETURN
      END
C
C
C
C
      SUBROUTINE CONVERGE(X,XN,PHI,PHIN,ICON,N)
      DIMENSION X(5),XN(5),D(5)
      WRITE(6,9000)
 9000 FORMAT(1X,14HENTER CONVERGE)
      ICON=0
      DO 10  I=1,N
   10 D(I)=(XN(I)-X(I))/X(I)
      DTD=0.
      DO 20 I=1,N
   20 DTD= DTD + D(I)*D(I)
      IF(SQRT(DTD).GT.1.0E-2)GO TO 31
      IF(ABS((PHIN-PHI)/PHI).GT.5.E-3)GO TO 31
```

```
   30 ICON=1
   31 RETURN
      END
C
C
C
C
      SUBROUTINE GRID(XN,PHIN,STRAN,EXTIM,M,N,IGRID,RESN)
      DIMENSION EXTIM(250),STRAN(250),XN(5),RESN(250),GAM(4)
      DIMENSION XG(5)
      DIMENSION RT(250)
      WRITE(6,9000)
 9000 FORMAT(1X,10HENTER GRID)
      IGRID=0
      GAM(1)=.90
      GAM(2)=.99
      GAM(3)=1.01
      GAM(4)=1.10
      DO 10 I=1,N
      DO 20 IG=1,N
   20 XG(IG)=XN(IG)
      DO 30 L=1,4
      XG(I) = XG(I)*GAM(L)
      CALL SUM(EXTIM,STRAN,RT,XG,M,S)
      IF((PHIN-S)/PHIN.LT.1.0E-3)GO TO 30
      IGRID=1
      PHIN=S
      DO 50 IJ=1,N
   50 XN(IJ)=XG(IJ)
      DO 60 JJ=1,M
   60 RESN(JJ)=RT(JJ)
   30 CONTINUE
   10 CONTINUE
      RETURN
      END
C
C
C
C
      SUBROUTINE INVERT(C,B,N,G,P,DEL,IPIVOT)
      DIMENSION A(5,5),B(5,5),GG(5),P(5),G(5),C(5,5)
      WRITE(6,9000)
 9000 FORMAT(1X,12HENTER INVERT)
      DO 3 I=1,N
    3 GG(I)=-G(I)
      DO 4 I=1,N
      DO 4 J=1,N
    4 A(I,J)=C(I,J)
      IPIVOT =0
      DO 6 I=1,N
      DO 6 J=1,N
      B(I,J)=0.
```

```
    6 IF(I.EQ.J)B(I,J)=1.0
      DEL=1.
      DO 45 K=1,N
      IF(K-N.GE.0) GO TO 30
      IMAX=K
      AMAX= ABS(A(K,K))
      KP1= K+1
      DO 20 I=KP1,N
      IF(AMAX-ABS(A(I,K)).GE.0.)GO TO 20
      IMAX=I
      AMAX=ABS(A(I,K))
   20 CONTINUE
      IF(IMAX.EQ.K)GO TO 30
      DO 29 J=1,N
      ATMP= A(IMAX,J)
      A(IMAX,J)= A(K,J)
      A(K,J)=ATMP
      BTMP= B(IMAX,J)
      B(IMAX,J)= B(K,J)
   29 B(K,J)=BTMP
      DEL=-DEL
   30 CONTINUE
      IF(ABS(A(K,K)).LT.1.E-20)GO TO 90
      DEL= A(K,K)*DEL
      DIV= A(K,K)
      DO 38 J=1,N
      A(K,J)=A(K,J)/DIV
   38 B(K,J)=B(K,J)/DIV
      DO 43 I=1,N
      AM=A(I,K)
      IF(I.EQ.K)GO TO 43
      DO 42 J=1,N
      A(I,J)=A(I,J)-AM*A(K,J)
   42 B(I,J)=B(I,J)-AM*B(K,J)
   43 CONTINUE
   45 CONTINUE
      IF(ABS(DEL).LT..00001)GO TO 90
      DO 47 I=1,N
      P(I)=0.
      DO 47 J=1,N
   47 P(I)=P(I)+B(I,J)*GG(J)
      GO TO 91
   90 IPIVOT =1
   91 RETURN
      END
C
C
C
C
      SUBROUTINE BISECT(X,XS,P,STRAN,EXTIM,M,N,G,S,SS,KK,AMDA
     1,RESN)
```

```
      DIMENSION X(5),XS(5),STRAN(250),EXTIM(250),RESN(250),P(5)
      DIMENSION G(5)
      WRITE(6,9000)
 9000 FORMAT(1X,12HENTER BISECT)
      K=-1
      ALPHA=ALPHA*32.
      IF(ALPHA.GT.1.)ALPHA=1.
      IF(KK.EQ.1)ALPHA=1.
    1 K=K+1
      DO 10 I=1,N
   10 XS(I)=X(I)+ALPHA*P(I)
      CALL SUM(EXTIM,STRAN,RESN,XS,M,SS)
      GTP=0.
      DO 20 I=1,N
   20 GTP= GTP+ G(I)*P(I)
      A=S+ 2.*1.0E-5*ALPHA*GTP
      IF(SS.LE.A) GO TO 1000
      ALPHA=ALPHA/2.
      GO TO 1
 1000 AMDA=AMDA/4.
      RETURN
      END
C
C
C
C
      SUBROUTINE LINSEARCH(S,P,N,STRAN,EXTIM,M,XS,X,SS,RESN,AMDA)
      DIMENSION P(5),STRAN(250),EXTIM(250),XS(5),X(5),RESN(250)
      WRITE(6,9000)
 9000 FORMAT(1X,15HENTER LINSEARCH)
      SE=1.E-3
      SB=S
      DO 10 I=1,N
   10 PTP= PTP+ P(I)*P(I)
      A= (S-SE)/PTP
      B= 1./SQRT(PTP)
      IF(A.LT.B)H=A
      IF(B.LE.A)H=B
    1 CONTINUE
      DO 200 I=1,N
  200 XS(I)=X(I)+ P(I)*H
      CALL SUM(EXTIM,STRAN,RESN,XS,M,SS)
      IF(SS.LT.SB) GO TO 11
      GO TO 2
   11 SB=SS
      H= 2.*H
      GO TO 1
    2 CONTINUE
    3 ALPHA=H*H*PTP/(SS-S+2.*H*PTP)
      IF(H/4.LE.ALPHA.AND.ALPHA.LE.3.*H/4.)GO TO 22
      IF(ALPHA.LT.3.*H/4.)A=ALPHA
```

```
      IF(3.*H/4..LE.ALPHA)A=3.*H/4.
      IF(H/4..GT.A)H=H/4.
      IF(A.GE.H/4.)H=A
      DO  23 I=1,N
   23 XS(I)=X(I)+H*P(I)
      CALL SUM(EXTIM,STRAN,RESN,XS,M,SS)
      IF(SS.LT.S)GO TO 1000
      GO TO 3
   22 DO 25 I=1,N
   25 XS(I)=X(I)+ ALPHA*P(I)
      CALL SUM(EXTIM,STRAN,RESN,XS,M,SS)
      IF(SS.LT.S)GO TO 1000
      H=ALPHA
      GO TO 3
 1000 AMDA=AMDA*4.
      RETURN
      END
      FINISH
****
```

Chapter 4

THE BREE DIAGRAM - ORIGINS AND LITERATURE: SOME RECENT ADVANCES CONCERNING EXPERIMENTAL VERIFICATION AND STRAIN-HARDENING MATERIALS

H. W. Ng and D. N. Moreton.

Department of Mechanical Engineering, University of Liverpool,

Liverpool L69 3BX, England.

SUMMARY

The logic of Bree [1] which lead to the formulation of the Bree diagram is presented. A survey of the literature following from this work is given. Three diagrams, similar to that of Bree have been produced for alternative loading conditions. A method of experimentation which models the Bree loading is described and results from these experiments are presented for both elastic/perfectly-plastic and strain-hardening materials. Finally, a computer based study undertaken to predict the ratchetting behaviour of strain-hardening materials is presented.

1. INTRODUCTION

In recent years much effort has been directed towards an understanding of the mechanism of cyclic inelastic deformation of structures. Whilst it has been realised that purely elastic design of structures is both wasteful of material and often impractical, the presence of cyclic inelastic strains may lead to the failure of structures by excessive distortion or low cycle fatigue.

The power generating industry has provided much of the recent motivation for achieving both efficient design and freedom from cyclic inelastic deformation (ratchetting) and alternating plasticity. It is in power generating plant, both conventional and nuclear, that situations of high cyclic thermal gradients and mechanical loads often occur together. As will be demonstrated later, it is this type of loading which can give rise to severe accumulations of ratchet strain, although either "loading" acting alone may not be damaging. Present design codes (see e.g. [2]) require that a design at a specified operating condition shall be free from ratchetting, or if ratchetting does occur, that the strain accumulation shall be less than a specified limit during the expected operating lifetime.

Two broad areas of research concerning ratchetting can be identified. Firstly, that area dealing with the behaviour of structures under one or more cyclic mechanical load (e.g. pressure vessels subjected to cyclic pressure variations) and secondly, structures subjected to one mechanical load together with a cyclic thermal loading. It is the latter case which this paper reviews although the former is of considerable importance. References [3] - [10] are included to provide a starting point for the reader interested in pursuing this class of loading.

Structures which are subjected to continuous mechanical loading together with cyclic thermal loading have received considerable attention in recent years. One particularly severe case of this type of loading occurs in the liquid metal fast breeder reactor (LMFBR). Here, the fuel cans are subjected to what may be regarded as a continuous internal pressure together with an alternating thermal gradient through the wall of the can. This class of loading was studied by Bree [1] and through a particularly skillful analysis the strain behaviour of the can was identified for varying combinations of mechanical and thermal load. This analysis will be discussed at length later together with a survey of contributions to the literature which followed from the work of Bree.

As had been the case for much of the earlier work on shakedown (e.g. [3], [4], [7], [8] and [9]),Bree idealised the fuel can material as elastic/perfectly-plastic. This assumption permitted a closed form solution to be obtained but was an unrealistic approximation of the type of materials proposed for LMFBR's. Such materials exhibit strain-hardening characteristics and thus the model of Bree will give an unrealistic assessment of the ratchet strains, although will give nearly correct bounds to the shakedown regime. This was demonstrated for kinematic hardening by Mulcahy [11], [12]. Consequently, the need for a solution of the Bree problem for realistic material behaviour exists and the reported work which has attempted to satisfy this need is surveyed later. In addition the authors have experimentally obtained strain behaviour data for the Bree problem for both elastic/perfectly-plastic and strain-hardening materials. The technique used and results obtained are reported in the final section of this paper. A numerical solution of the problem using a hardening model which is in essence kinematic is included.

2. THE BREE DIAGRAM

For a thin walled cylinder subjected to a constant internal pressure and cyclic through wall thermal gradient Bree obtained a closed form solution for the strain behaviour for any combination of pressure and thermal gradient. This solution was presented as a plot of σ_t/σ_y against σ_p/σ_y, where σ_t is the maximum fictitious elastic thermal stress and σ_p the pressure stress. This plot depicted zones of elastic, shakedown, alternating plasticity and ratchetting strain behaviour and has become known as the Bree diagram.

For an elastic/perfectly-plastic material the basis of the derivation of the diagram is now given.

At the core of a fast-nuclear-reactor the fuel is contained within thin-walled cylinders - the fuel cans. During the life of the reactor the fuel is burnt up and produces both a change in volume of the fuel and some gaseous products. The effect of this is to subject the fuel cans to an internal pressure, which may be considered constant throughout the life of the can. In addition, at start-up and shut-down of the reactor the fuel cans experience a high thermal gradient across the can wall. The effect of the pressure is to produce a constant bi-axial elastic stress whilst the temperature gradient causes a temporary thermal stress sufficient to cause plastic deformation. Since start-up and shut-down operations are likely to be cyclic, then the thermal stress will be cyclical in nature.

2.1 Elastic Analysis

(a) Pressure Stresses. From equilibrium for a thin wall cylinder subjected to an internal pressure we have in polar form

$$\int_{-t/2}^{+t/2} \sigma_\theta dx = 2 \int_{-t/2}^{+t/2} \sigma_z dx = \frac{pd}{2t} \qquad (1)$$

Where p is the pressure, σ_θ the hoop stress, σ_z the axial stress, t the thickness and x the radial co-ordinate measure + VE outwards from the mid-wall of the cylinder.

Assuming that σ_θ and σ_z have mean values σ_p and $\sigma_p/2$ respectively we have

$$\sigma_p = \frac{pd}{2t} \qquad (2)$$

(b) Thermal Stresses. The temperature distribution measured relative to the can mid-wall is given by

$$T = - \Delta T x/t \qquad (3)$$

If an element of the can is considered as a flat plate which is prevented from bending in both the axial and hoop directions then the temperature gradient gives rise to thermal stresses (see e.g. [13])

$$\sigma_t = E\alpha\Delta T/2(1 - \nu) \tag{4}$$

in both the axial and hoop directions. This stress is minimum ($= -\sigma_t$) at the inside surface and maximum ($= +\sigma_t$) at the outside surface.

Thus the total elastic stresses are:

$$\begin{aligned} \sigma_\theta &= \sigma_p + \frac{2x\sigma_t}{t} \\ \sigma_z &= \frac{\sigma_p}{2} + \frac{2x\sigma_t}{t} \end{aligned} \tag{5}$$

It should be noted that these are elastic stresses and may thus be fictitious since we have not yet introduced a yield criterion.

2.2 The Uniaxial Model

An elastic/plastic analysis of the stresses given by (5) can only be achieved by a numerical method. Bree, therefore, attempted to simplify the analysis by considering only the stresses in the hoop direction (which are the dominant stresses).

By assuming that

$$\Delta T' = \Delta T/(1 - \nu) \tag{6}$$

where $\Delta T'$ denotes the temperature difference for the uniaxial model, the hoop stress could be written as

$$\sigma = \sigma_p + \frac{2x\sigma_t}{t} \tag{7}$$

Consequently the model contains all the physical features of the biaxial stress system but yields simpler analytical solutions. Thus, Bree argued, a better qualitative understanding of the problem could be gained than from a detailed numerical solution.

2.3 Statement of the Simplified Problem

We now have to consider the deformation of an element of the can which is subjected to a steady hoop stress, σ_p, together with an alternating thermal stress due to the alternating temperature distribution

$$T = -\frac{\Delta T'x}{t} \rightarrow \text{constant} \rightarrow -\frac{\Delta T'x}{t} \rightarrow \cdots\cdots \quad (8)$$

the element being restrained from bending due to the temperature gradient.

If the total strain in the hoop direction is denoted by ε_T then the condition the bending is prevented can be written as

$$\varepsilon_T = \text{constant} \quad (9)$$

and from equilibrium we have

$$\int_{-t/2}^{+t/2} \sigma dx = \sigma_p t \quad (10)$$

The yield condition may be written

$$\begin{aligned} |\sigma| &= \sigma_y \ \text{(in plastic regions)} \\ |\sigma| &< \sigma_y \ \text{(in elastic regions)} \end{aligned} \quad (11)$$

where σ_y is the yield stress, and finally the uniaxial stress/strain relationship may be written as

$$\varepsilon_T = \frac{\sigma}{E} + \varepsilon_p + \alpha T \quad (12)$$

where ε_p is the plastic strain, and αT is the thermal strain.

<u>1st Half Cycle</u>: During this period the can is subjected to the pressure loading together with the temperature gradient. On substituting for αT in (12) the stress strain relationship becomes

$$\varepsilon_T = \frac{\sigma}{E} + \frac{2x\sigma_t}{Et} + \varepsilon_p \quad (13)$$

<u>2nd Half Cycle</u>: During this period the can is subjected to the pressure loading only hence (12) becomes

$$\varepsilon_T = \frac{\sigma}{E} + \varepsilon_p \quad (14)$$

We have therefore to solve (13) and (14) subject to the restraints of (9), (10) and (11).

It can be noted that for both the 1st and 2nd half cycles, ε_T must be constant through the thickness of the can wall. Hence from (13), i.e. for the first half of the cycle $(\frac{\sigma}{E} + \varepsilon_p)$ must be linear in x with slope $\frac{2\sigma_t}{Et}$

$$\text{i.e.} \quad \left(\frac{\sigma}{E} + \varepsilon_p\right) = \frac{-2\sigma_t}{Et} x + \varepsilon_T \qquad (15)$$

Thus in regions of the can where ε_p is constant $\frac{\sigma}{E}$ has slope $\frac{2\sigma_t}{Et}$.

For the second half of each cycle, equation (14) applies and since ε_T is constant $\frac{\sigma}{E} + \varepsilon_p$ is uniform across the can wall. Thus $\frac{\sigma}{E}$ and ε_p have equal and opposite slopes at each point in the can.

2.4 The Condition of Ratchetting

Consider firstly Fig. 1(a). This shows the stress distribution through the wall of the cylinder due to the pressure and the thermal stress. The pressure stress alone

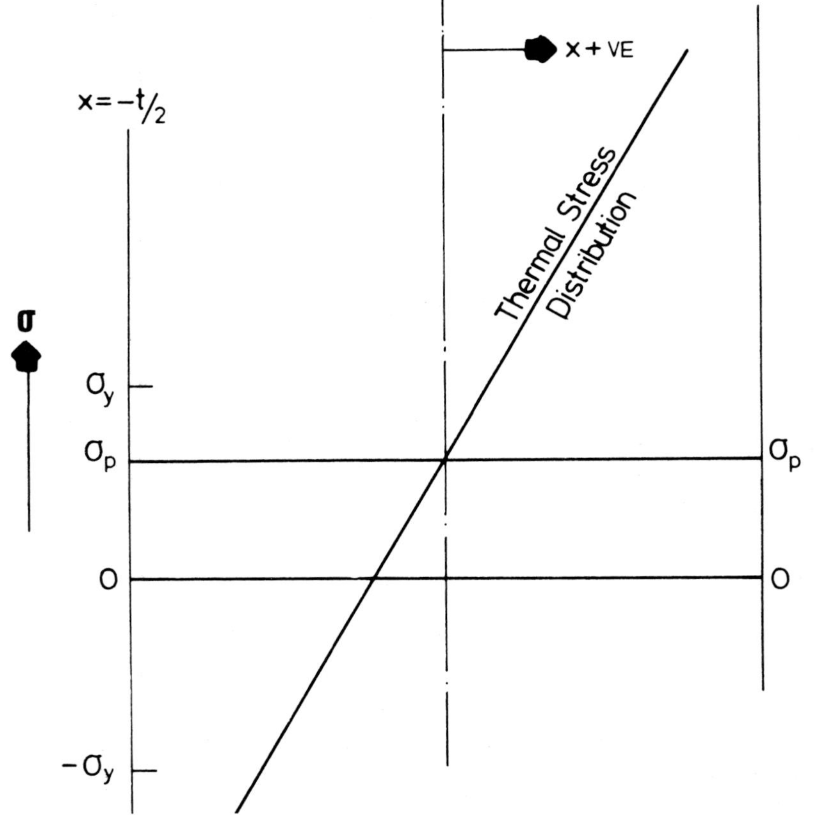

FIG.1(a) Summation of Stresses during the First Half Cycle but Violating the Conditions of Equilibrium & Yield.

will always be less than the yield stress and to this is added the thermal stress. However, this summation then violates both of equations (10) and (11). In order to satisfy (10) and (11) the stress distribution must take the form of Fig. 1(b) and the value of x = a may be found from the equilibrium condition - i.e. equation (10), and results in

$$x = a = - \frac{1}{2} \cdot \frac{t}{E} [1 - 2 \sqrt{(\sigma_y - \sigma_p)/\sigma_t}] \tag{16}$$

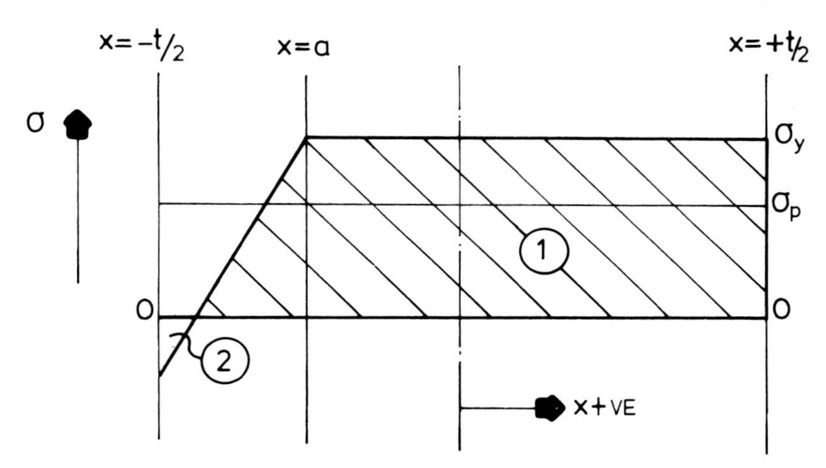

FIG.1(b) Stress Distribution during 1st Half Cycle

Referring now to Fig. 2, A, B and C show the contributions to the total strain D for the first half cycle. We know that the total strain must be uniform through the can wall and the summation of $(\sigma/E) + \varepsilon_p + T$ shows this to be so. Thus the left hand column of Fig. 2 shows the contributions to the total strain for the first half cycle - i.e. pressure load + thermal gradient.

The centre column of Fig. 2 shows the strain changes required to reach the second half cycle, i.e. pressure only. The strains at this stage are shown in the right hand column of Fig. 2.

Working across the top row of Fig. 2, E shows the fictitious elastic stresses to be added to A to remove the thermal stress. This results in H (dotted) which again violates (10) and (11). In order to satisfy (10) and (11) the full line H is required. This incurs no additional plastic strain in the zone x = b to x = +t/2, but in the zone

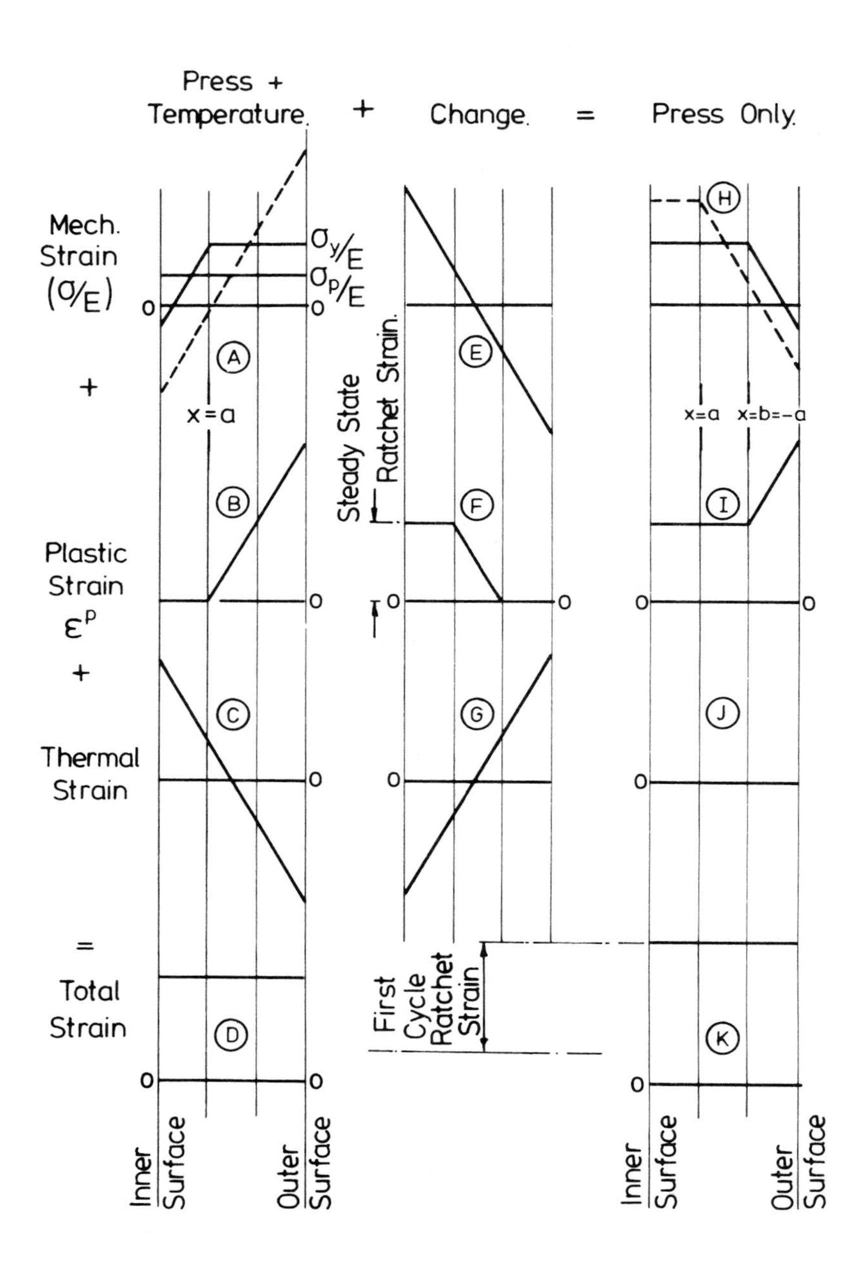

FIG.2 Ratchetting Regime R1

x = -t/2 to x = b the additional plastic strain of F is incurred. The total plastic strain is then I which is B + F. The thermal strains C and G add to give zero at J. For the second half cycle the total strain is now the sum H + I + J giving K which is uniform as required. The ratchet strain accumulated over this period is shown between D and K.

After the first cycle the strain accumulated in each cycle will be that shown in F. This may be written as

$$\delta = 2.a.\frac{2\sigma_t}{t}$$

and using (16) this becomes

$$\delta = \frac{2\sigma_t}{E}\left[1 - 2\sqrt{(\sigma_y - \sigma_p)/\sigma_t}\right] \qquad (17)$$

Bree introduced the quantity σ_e being

$$\sigma_e = \sigma_p + \frac{1}{4}\sigma_t - \sigma_y$$

So that (17) may be written

$$\frac{E\delta}{2\sigma_t} = 1 - \left(1 - \frac{4\sigma_e}{\sigma_t}\right)^{\frac{1}{2}} \qquad (18)$$

and if $\sigma_e << \frac{1}{4}\sigma_t$ then (18) reduces to

$$\delta = \frac{4\sigma_e}{E} \qquad (19)$$

Fig. 3 shows a plot of ratchet strain against σ_e/σ_t as given by equations (18) and (19).

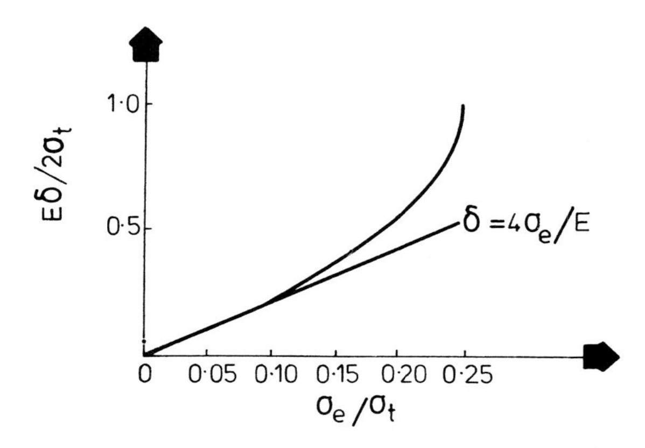

FIG. 3 Ratchet Strain per Cycle for the Stress Regime R_1 of Fig. 5.

For differing combinations of σ_p and σ_t the behaviour of the can material will be different. We have seen here the conditions which lead to ratchetting but other stress conditions may lead to shakedown, alternating plasticity or purely elastic performance.

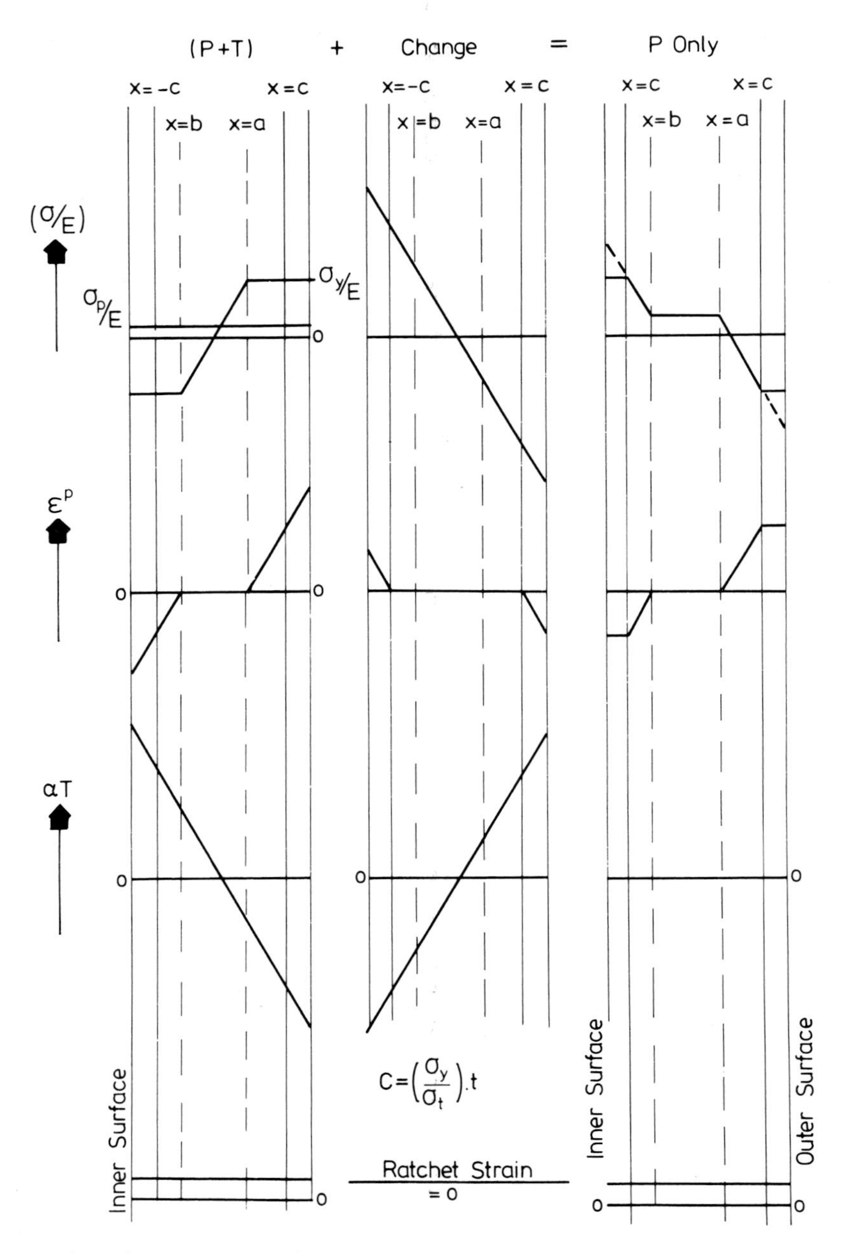

FIG. 4 Alternating Plasticity Regime P.

For example Fig. 4 shows a combination of σ_p and σ_t in the same manner as Fig. 2. The reader may work through this to establish that although yielding occurs repeatedly on both surfaces of the can no strain is accumulated. This condition is termed alternating plasticity.

Thus we have identified two distinct types of strain behaviour occurring for different combinations of σ_p and σ_t. Clearly if neither σ_p nor σ_t violate the yield criterion then a condition of purely elastic behaviour will exist. A fourth type of strain behaviour occurs when after the first half cycle, during which the yield criterion is violated, subsequent deformation is entirely elastic. This condition is termed shakedown.

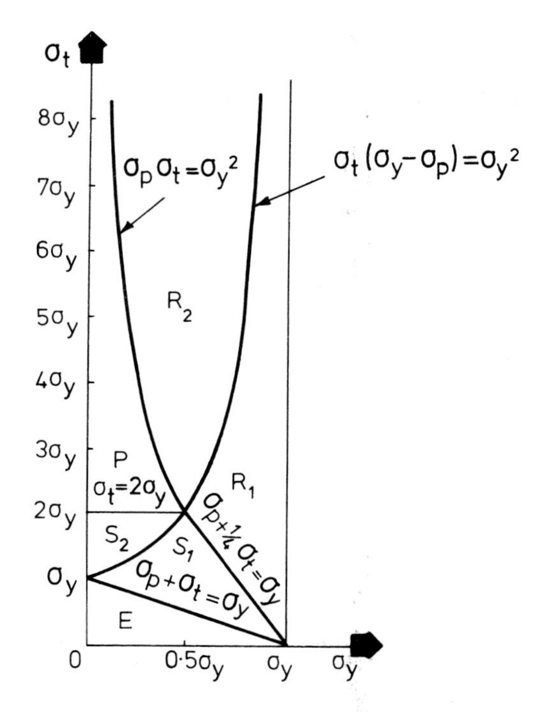

STRESS REGIME	CAN BEHAVIOUR
R_1 and R_2	RATCHETTING
S_1 and S_2	SHAKEDOWN AFTER FIRST HALF CYCLE
P	PLASTIC CYCLING
E	ELASTIC

FIG.5 Stress Regimes for Non-Work-Hardening Materials.

Bree was able to identify the boundaries separating these types of strain behaviour. These are shown in Fig. 5 which is the Bree diagram.

In using the Bree diagram it should be noted that σ_t is a fictitious elastic stress, which is therefore calculated assuming the material to have an infinite yield stress. On dividing this stress by the actual yield stress of the material the location on the σ_t/σ_y axis is found. Since the σ_p/σ_y axis is bounded at unity (the collapse condition) the concept of fictitious elastic stresses is not important. However, for strain-hardening materials the σ_p/σ_y axis can extend beyond unity in which case σ_p must be calculated assuming purely elastic behaviour of the material.

3. A SURVEY OF CONTRIBUTIONS TO THE LITERATURE FOLLOWING THE WORK OF BREE[1]

The phenomenon of ratchetting (incremental growth) in structures subjected to cyclic loading has received much theoretical and experimental attention over the past decade or more. A survey of the literature is now undertaken to trace the development of the subject to the present investigation undertaken by the authors.

The ratchetting phenomenon had been observed and reported by many workers often as the results of investigations into pressure vessel behaviour. Weil and Rapasky [14] while investigating pressure vessels subjected to repeated thermal stresses observed progressive distortion. Coffin [15] in his work on fatigue under repeated straining of uniaxial specimens also observed incremental growth under combined steady and cyclic loads. Moreton and Moffat [6] among others reported persistent cyclic strain increments in stainless steel pressure vessel components when the repeated pressure load exceeded the shakedown pressure. The work by Parkes [16] which involved the study on the problem of incremental collapse of an aircraft wing with aerodynamic skin heating superimposed on the normal wing loading was perhaps the first to seriously study the subject.

The early literature in this field is mainly confined to studies in cyclic loading of structures made of elastic/perfectly-plastic materials with time independent properties. Studies on such structures have the advantage of simplicity when time dependence is avoided. Therefore the states from one cycle to the next can be assumed to be quasi static, thus load controlled deformation or thermal strains assume stepped values. The occurrence of ratchetting effects due to the interaction of elastic and plastic strains have been noted and simple analytical examples have been given by Prager [17], Gill [18], Ruiz [19] and Hill [20]. More examples of greater complexity and practical value are works carried out by

Miller [21], Edmunds and Beer [22], Bree [1] and Burgreen [23]. The approach used by these authors were basically the same and each have obtained suitable solutions to simplified versions of their own specific problems. The approach used was later modified and improved by others to include the effects due to time dependence such as creep, fatigue and rapid transient type loading.

From this point, the discussion will center on the model by which Miller [21] analysed the reactor fuel can and the further detailed analyses conducted by others. Miller's model [21] is one of the elements of the tube or fuel can of radius R, thickness d and subjected to pressure. The problem was reduced to one in which an element of tube was subjected to an axial stress $\sigma_z = pR/2d$ and a circumferential stress $\sigma_\theta = pR/d$ with a linear or parabolic temperature distribution through the thickness. This idealisation gives rise to a uniaxial beam model which is subjected to a primary stress of pR/d and a secondary stress due to the temperature difference T. From this model, Miller was able to obtain solutions for the loadings which give rise to shakedown or ratchetting. These solutions now form the basis of the ASME III [24] restrictions on primary stress in the presence of thermal loadings of this type.

Burgreen [23] firstly studied the problem by considering the behaviour of a two-bar structure under thermal loading. The structure consisted of two bars of equal lengths rigidly attached together at the ends and subjected to a constant axial load and variable temperature at one of the bars. Analytical formulae were obtained to calculate ratchetting strains with the relative area of the two bars being variables. The analysis was later extended in a subsequent publication [25] in which a multi-element assembly was proposed to study the same problem. Such an assembly, containing a large number of elements, is anologous to the two bar model. However, each bar experiences a slightly different temperature from its nearest counterpart. By having a large number of elements, a homogenous structural state was approached and by subjecting the structure to a temperature variation across its width, a solution similar to Miller's was obtained.

At this point, it is of interest to deviate briefly to other structures used in the study of ratchetting. Two or three bar structures are structural models which are also currently being used for the experimental and theoretical study of ratchetting under constant primary and cyclic thermal loadings. Conceptually these structures simulate the condition in a real structure where the main part operates at a constant stress whilst, smaller parts of it operate at higher stress levels which are caused by cyclic thermal loads or stress concentrations. Loading conditions are varied by means of the imposed constant axial load and the cyclic temperature variations of the bars. The length and cross-sectional areas of the bars are varied to vary the relative interacting volumes of the structure. The two or three bar structures are in fact similar in all respects if one were to imagine that one of the bars in a two bar structure is split into two halves lengthwise and replaced on both sides of the remaining bar to form a three bar structure. The three bar structures have the experimental advantage that no lateral restraints are required to keep the bars straight during the temperature changes on either the centre or the two side bars. This is a necessary requirement for the full imposition of the thermal stresses caused by the variable temperature loading. These structural models also exhibit the same Bree-like responses under the imposed loadings. Calculations using various geometrical parameters, loading conditions and material assumptions have been performed by Parkes [26] and Megahed [27] and in the experimental area Ainsworth [28], Megahed et al. [29] and Uga [30] have undertaken interesting studies. As the two, or three bar structures form a separate research area which is outside the scope of this article, the references above will not be examined.

Bree [1] extended Miller's analysis by using the uniaxial beam model and extended the $\frac{\sigma_t}{\sigma_y}$ vs. $\frac{\sigma_p}{\sigma_y}$ plot to include the by now familiar categories of structural response which are referred to as R1, R2, P, S1, S2 and E regimes of behaviour for an elastic-plastic material with time independent material properties.

It is later shown that a small incursion into the R region can cause large deformation by ratchetting after a few cycles. Incursion into the P region indicates alternating plasticity which may ultimately lead to failure by low cycle fatigue. By using the relationship derived by Coffin [31], the number of cycles to failure can be predicted. Bree also calculated the effect on the diagram if there is a complete relaxation of residual stresses by creep during the hot part of the cycle. Residual stresses are set up by plastic deformation and are beneficial in that the elastic range of stresses are increased and consequently a greater opportunity for shakedown exists. As a consequence of creep, residual strains are converted to creep strains and without the beneficial residual stress distribution, ratchetting will occur in regions where shakedown would have otherwise taken place. When relaxation is complete, the S1-R1 and P-R2 boundaries fall to the elastic boundary. Next, Bree assessed the effect when the yield stress is temperature dependent by assuming a higher yield stress σ'_y for the low temperature part of the cycle and a uniform lower value σ_y for the linear temperature distribution. It was shown that the effect is not significant when compared with a Bree diagram calculated using a yield stress equal to $\frac{\sigma'_y + \sigma_y}{2}$. When work hardening is considered, it is found that the ratchetting criterion is not greatly affected. However, work-hardening causes the increment of strain per cycle to decrease asymptotically to zero after an infinite number of cycles. Work-hardening is defined in this context as the increase in the stress in the plastic range with increasing plastic strain.

On the basis of the assumption of complete relaxation of residual stresses due to creep, the ratchet strain per cycle is increased greatly compared to the case without creep. This assumption was seen to be unduly pessimistic and was later reconsidered. Bree in the second publication [32] extended his work to the case where creep causes only partial relaxation of internal stresses and evaluated the ratchet strains. Creep was considered to occur only in the high temperature excursion and in accordance with a secondary creep law of the form $\dot{\eta} = A\sigma^n$ where $\dot{\eta}$ is the creep rate, σ the applied stress and A and n are constants for the material. An additional time parameter was introduced to account for the on power (high temperature) duration of the cycle during which creep occurs. The solution was based on the construction of a steady state stress cycle for the uniaxial model which enabled the ultimate behaviour of the can to be determined without tracing the detailed history of its deformation. This procedure is based on the theorem by Frederick and Armstrong [33] which states that two bodies differing only in their initial patterns of internal stress will develop identical stress patterns in regions of creep and plasticity if they are subjected to the same variations of temperature and load. The total strain increment is considered to be a sum of two components, δ_1 and δ_2, δ_1 being the plastic ratchet strain and δ_2 the increase in creep strain due to the steady stress state. As the on power duration increases, the incremental strains tends to an upper limit corresponding to complete relaxation of internal stress. When the on power duration is small, creep is insignificant and on the incremental growth criterion becomes that of the Bree Diagram without creep. It was noted that the creep law described was of a simple form which took no account of the effect of temperature. Because of the complexity of the results, it was impossible to present them in the simple form of a Bree diagram.

Mulcahy [12] extended the analysis of Bree by assuming for the material properties an idealised Bauschinger effect. The material model was different from the work-hardening model described by Bree [1] which was of linear isotropic hardening while Mulcahy's model was one of linear kinematic hardening. The result was presented as a Bree diagram with the contours of upper limit accumulated ratchet strains. The hardening coefficient was chosen at the value K = 4.62 which corresponds to a hardening coefficient fitted to Type 316 stainless steel in the annealed state at room temperature. It was seen that the contour of zero accumulated strain approximated very closely to the perfect-plasticity line. The rate of strain accumulation per cycle was found to be a decreasing function of the cycle numbers. For a given strain-hardening coefficient, it was found that the maximum rate of strain accumulation occurred for the largest value of the temperature and is relatively insensitive to the pressure load. It was expected that 90% of the strain accumulation for an infinite number of cycles would occur within the first 25 cycles. The material model used by Mulcahy did not include creep effects because of the complexity of the analysis.

The material assumption of non-linear kinematic hardening have been investigated by Moreton and Ng [34]. The uniaxial stress-strain curve was modelled by an Osgood-Ramberg [35] power law, using the uniaxial stress-strain data of a Type 316 stainless steel tested at room temperature. A computer program was written to calculate the accumulated strain, cycle by cycle. The results obtained showed the cycle numbers necessary to reach assymptotic strain accumulation is dependent on the thermal and pressure stress. These range from 5 to 20 cycles for a thermal stress range from $\frac{\sigma_t}{\sigma_y} = 3.0$ to $\frac{\sigma_t}{\sigma_y} = 4.5$.

The results of Mulcahy [12] using the linear kinematic hardening material shows that greater than 27 cycles are required. The smaller number of cycles required by a non-linear kinematic model is due to the conservatism of the linear model.

At present, no work has been published which investigates the effect of cyclic hardening or softening on the Bree beam model. There is however a broad knowledge base of uniaxial cyclic stress/strain experiments and theoretical models. The authors have attempted to simulate the cyclic hardening effect by using a cyclic hardening model proposed by Jhansale [36] and preliminary results indicate that cyclic hardening has little effect.

At this juncture, it is useful to review the work of O'Donnell and Porowski [37] since it uses a simple approximate method to estimate creep and plastic strains. The generic name for these methods is simplified inelastic analysis methods. These methods are very useful in initial design studies since they provide an approximate evaluation of structural behaviour without expensive detailed computer inelastic analysis. A review by Leckie [38] introduces and discusses some of these methods used today. O'Donnell and Porowski's approach in bounding creep strain offer an alternative to Bree's partial relaxation method [32]. They also considered the case of the thin tube as in the Bree analysis and used the same assumptions employed by Bree including the stress distribution he devised when the cylinder reaches a steady cyclic state. The method applies if the stresses are within the S1, S2 or P regions of the elastic-plastic Bree diagram and uses the maximum stress that occurs at any point in the shell wall which remain below the yield strength to bound the creep strains. The maximum stress value, called the creep stress σ_c was used to obtain the maximum creep strain that could be accumulated during the expected life time of the component. The maximum creep strain or upper bound creep strain is obtained from creep properties of the material or from isochronous stress strain curves. The contours of stresses, denoted by σ_c/σ_y to be used in obtaining upper bound strains where plotted on the Bree diagram. It should be emphasised that the bounds of O'Donnell and Porowski, bound only the uniform creep strain which accumulates during the elevated temperature part of the cycle. In order to evaluate the total inelastic strain (creep plus plastic) the plastic ratchet strain from Bree's analysis is added to the upper bound creep strain obtained by O'Donnell and Porowski. This additive operation may be questionable due to the possible micro-structural interaction between plasticity and creep. However, due to the lack of information regarding this complex situation the assumption was not regarded as unreasonable. The procedures have been tested and verified against the results of a series of detailed inelastic analysis [39] and is now being incorporated into the ASME Code Case 1592 [40], now the N-47-12 [41].

The simplified methods for bounding creep strain was extended by O'Donnell et al. [42] to include biaxial stresses in shells. The method considers the simultaneous asymmetric loading on a shell of revolution where an overall expansion and thermal shock result in any combination of biaxial thermal stress components which are superimposed on the biaxial membrane stresses. A biaxial model was developed around a biaxially stressed element using the assumption of elastic/perfect-plasticity and the Tresca yield condition. The two orthogonal faces of the element were subjected to membrane stresses of the same or opposite signs and thermal stresses of positive or negative gradients through the thickness. Three different load situations were investigated. The analysis which is omitted here involved finding the elastic core stresses σ_{c1}, and σ_{c2} in the two principle directions at the shell midwall. From these two stresses, the effective stress $\bar{\sigma}$ was determined. The effective stress is related to the effective core strain for creep by the isochronous curves of the ASME Code Case N-47 within the specified component life-time. The separation of the effective core creep strain into its components in the principal directions was achieved by means of the Prandtl-Reuss flow relations. As a comparison, the biaxial and the uniaxial model were used to bound creep strains in the case of the pressurised cylinder. Assuming a constant effective core stress, the maximum of the strain components calculated was 15% less than that given by the bounds of the uniaxial model. For the case of a pressurised sphere the value was 50%. These values indicated that the uniaxial model is conservative for pressurised spheres and cylinders and that the bounds can be improved by a biaxial model. For its simplicity, the uniaxial model can be advantageously used in shells where the biaxial membrane stresses are of the same signs and the biaxial thermal stresses on one face are both higher than on the other.

One of the major problems encountered in inelastic analysis is that of thermal transients in Fast Breeder Reactor Cores. These transients are caused by the rapid changes in temperature within the structures during shutdown or startup either in emergencies or in normal operation. The review up to this point has been concerned with steady cycling where the steady state temperature distribution is assumed to be allowed to develop. The effects of rapid thermal transients give rise to non-linear temperature distribution leading to more complex behaviour of the Bree beam model. Goodman [43] analysed the effects of these transient loadings on the Bree problem and used the same assumptions that Bree adopted, i.e. elastic-plastic material and temperature independent yield stress without creep. He considered the effect of two forms of non-linear temperature distribution on the Bree diagram and therefore illustrate the types of problems to which the Bree diagram cannot be applied directly. Two cases are considered, the first considers the effect of a thermal downshock (drop in temperature) on only one surface and the second case considers the effect on both surfaces simultaneously. To quantify the properties of a transient, two non-dimensional parameters were introduced, the Biot number B and the Fourier number F. The Biot number B is defined by $B = \frac{hd}{k}$ where h = heat transfer coefficient of fluid/surface interface, d = thickness of plate and k = thermal conductivity. This quantity defines the relative ability of the interface and the plate thickness to transmit the same amount of heat across a unit area. The Fourier number is defined by:

$$F = {}^{\tau}/\bar{\tau} \quad \text{where} \quad \bar{\tau} = {}^{\rho c d^2}/k$$

ρ = density and c = thermal capacitance. $\bar{\tau}$ is the characteristic time and it measures the rate of decay in temperature in a transient. F is therefore a measure of the relative value of τ (the ramp time) and $\bar{\tau}$ (the characteristic decay time). The ramp time is the duration taken by the drop or rise in temperature from the normal to the minimum or maximum temperature in a transient. From these two dimensionless groups, a large B means a high heat transfer rate to the surface and vice versa and a large F means a very slow change in temperature and vice versa. The results calculated numerically by a computer program showed that for case 1 where the downshock is on a single side, there is a considerable reduction in allowable thermal loading for small primary loading but the effect is less significant for case 2. For the largest values of F (nearly quasi-static) and B, the ratchet strain per cycle coincides with the quasi-static case and is hence in accord with Bree's quasi-static model. Another important result comes from the study of plate thickness in the case of the double sided downshock. For a prescribed value of the primary and secondary loading, the ratchetting strain per cycle decrease with the plate thickness, therefore there exists a critical plate thickness at which ratchetting will not occur. It has to be noted that although these results show that transients do have a severe effect upon ratchetting, the assumptions have however ignored the important effects of creep and the variation in yield stress due to temperature.

A theoretical analysis given by Phillips [44] extended the work of Goodman by considering the accumulation of creep and fatigue damage on the Bree plate. The calculations examine the dependence of creep and fatigue damage on such factors as the applied loads, the degree of hardening of the material, the length of the creep dwell periods and the severity of the transient thermal loadings as measured by the Fourier and Biot numbers. The procedure for calculating the stress and stress distribution as well as the ratchet strain is similar to that of Goodman [27]. The assumptions used for the material properties were that of kinematic hardening, and included creep. The creep law used represented the creep behaviour by

$$\dot{\varepsilon}_c = A_1 \left(\frac{\sigma}{\sigma_y}\right)^n e^{-Q/RT}$$ where A_1, σ_y, n, Q, R are constant.

The creep rupture lifetime t_r is represented by the equation

$$\left(\frac{|\sigma|}{\sigma_y}\right)^m t_r = A_2\, e^{Q/RT}$$ where A_2, m are constants.

The values of the constants are chosen to represent as well as possible the behaviour of Type 316 stainless steel at 600°C. The creep damage is accumulated according to the rule of Robinson [45] given by

$$D_c(t_2) - D_c(t_1) = \int_{t_1}^{t_2} \frac{dt}{t_r(\sigma(t),T(t))}$$

where $\sigma(t)$ and $T(t)$ are the stress and temperature histories of the material under consideration. The fatigue damage is accumulated according to the rule of Miner [46] given by

$$D_f(N_2) - D_f(N_1) = \sum_{r=N_1}^{r=N_2} \left[\frac{\Delta\varepsilon_p . r}{A_3} \right]^{1/\beta}$$

where N_2 and N_1 are the cycle numbers, $\Delta\varepsilon_p$ is the plastic strain range and A_3 and β are material constants taken from low strain-rate fatigue data for Type 316 stainless steel at 600°C. The results of this analysis show that for an elastic-plastic material, creep and fatigue damage accumulates rapidly under large compressive stresses at the hot surface which leads to local failure of the surface fibres of the plate. For a hardening material, the strain and creep damage accumulation becomes very rapidly deaccelerated.

Creep damage accumulation was shown to increase with increase in dwell time. The effect was most significant on the hot surface and at large dwell times. However a larger dwell time meant a lesser degree of fatigue damage accumulation in the same period since the cycle per unit time is proportionally decreased. The realistic values of F and B in problems of this type are in the range of 0.01 to 1.0 for the former and large (>>1) for the latter. When $B >> 1$, creep damage accumulation was found to be affected less compared with fatigue damage accumulation under variation of F. Fatigue damage was therefore significantly related to large B values. The overall deformation of the beam was increased by introducing rapid transient loadings as opposed to quasi-static thermal loadings and this occurred more significantly near the ratchetting boundary on the Bree diagram. The assumption of kinematic hardening lead to reduced accumulation in both deformation and damage when compared to an elastic/perfectly-plastic material. Again, it is necessary to note that simple models have been used in this study to obtain the above results.

As for experimental studies of the Bree beam problem, the first tests were performed by Goodall and Cook [47] on En58J stainless steel (18 Cr:8 Ni:Mo) beam specimens in the annealed condition at 600°C. The experimental procedure used a mechanical loading technique consisting of two hydraulic rams which were independently controlled to exert axial and bending load on a beam type specimen. In these tests the cycle time was 4 cycles/min. which was assumed to be short enough for creep to be neglected. Three tests were conducted under varying axial loads each with a fixed bending load. The ratchet strains were plotted against cycle number for both theory and experiment. The theoretical ratchet strains were calculated on the elastic/perfectly-plastic model of Bree, the yield stress being selected from an elastic-plastic fit to the stress strain curve of the test material. It was seen that the results accorded very well with the theory for the initial cycles. The accumulated strain thereafter decreased as is expected of a strain-hardening material.

About 10 cycles of loading were applied to each specimen.

The second set of tests were carried out by Anderson [48] using an experimental set-up which was discussed in a previous publication [49]. The loading method used to simulate the thermal loading was again by bending a specimen to a fixed curvature. In the test procedure, the specimens were subjected to constant axial stress and to represent the thermal strains were bent alternatively around two mandrels of opposite curvature at a temperature of 649°C. It is noted that the tests simulate a particular type of loading where the temperature gradient alternates in the sign of its slope every other cycle. Such a case had not been considered in Bree's analysis where the gradient is only of one sign. The tests which were conducted on Type 304 stainless steel, were 20 in number and the number of cycles applied varied from 10 to 16. A number of testing points were selected on the Bree diagram. The experimental results again showed that the first few cycles of accumulated strain correlated very well with the prediction of Bree. The tests were all completed within 24 hours but because the temperature was so high, the magnitude of the creep strain contribution to the total accumulated strain was uncertain.

A series of tests were carried out by authors [34] using an experimental rig which is described elsewhere in this article. The material used was mild steel and the tests conducted at room temperature. Mild steel was selected for its near elastic/perfectly-plastic characteristic. This material was thus ideally suited for verifying the Bree diagram. The experimental results showed that like the results of Goodall and Cook [39] and Anderson [41], the point of onset of ratchetting, is in good agreement with Bree, the rate of strain accumulation, however, was not in agreement with Bree.

The last and most relevant series of tests have been conducted by Yamamoto, Kano and Yoshitoshi [50]. Tests were conducted on hot rolled Type 304 stainless steel pipes in a solution treated condition by passing through it alternatively hot and cold sodium. This alternate flow of sodium subjected the inside of the pipe to an alternate increase and decrease of temperature varying from 550 to 250°C in one of the tests. The pipes were each subjected to increments of axial load and a total of 5000 thermal cycles. 140 thermal cycles were applied at each increment of axial load. The temperature difference (between hot and cold sodium) was maintained constant all the time. The cycle time was about 10 mins. A ratchetting boundary on the Bree diagram was plotted based on the experimental results. By calculating backwards, this boundary was found to coincide with theoretical prediction if the yield stress was 20% higher than the 0.2% proof stress. This test was carried out without interruption except for the increase in the axial load after every 140 cycles of thermal cycling. It was found that at the initial 5 increments of load (low values of axial load), no strain was accumulated indicating shakedown or alternating plasticity behaviour. At the 6th increment, the strain was seen to accumulate over the first few cycles and gradually drop to a rate which appeared to be constant. The comparison between the predicted ratchet strain increments per cycle using Bree's equation for non-workhardening and the experimental results shows that Bree may be over conservative.

4. ALTERNATIVE DIAGRAMS

The Bree diagram has come into popular use as a design tool. However it has, in the past, been used incorrectly for applications in which the loading (thermal and mechanical) was not as prescribed by Bree.

As an example, consider a heat exchanger tube in which is flowing a hot liquid at high pressure. At start-up conditions the pressure and thermal gradient are both zero and rise to the operating conditions. At shut-down both pressure and thermal gradient return to zero. Thus the thermal and mechanical loads are in phase. In such a situation the Bree diagram gives an incorrect solution for the various types of strain behaviour except for the elastic zone.

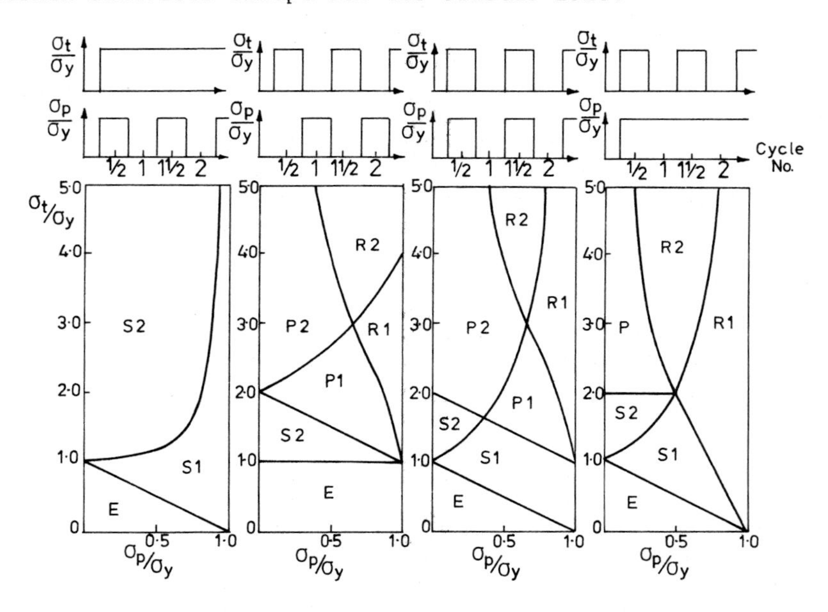

FIG.6 Alternative Bree Diagrams

The authors have constructed alternative diagrams for the following three loading conditions,

(a) Continuous thermal gradient with alternating pressure,
(b) Alternating thermal gradient and alternating pressure with these loadings in phase and
(c) Out of phase.

These are shown together with the original Bree diagram in Fig. 6. Thus for the example of the heat exchanger tube given above the "in phase" diagram is applicable. It is of interest to note that the Bree diagram is the most "conservative" of the four diagrams shown in Fig. 6.

5. EXPERIMENTAL VERIFICATION OF THE BREE DIAGRAM

In order to verify the Bree diagram experimentally for the case of a cylinder under steady internal pressure together with a cyclic through wall thermal gradient two severe problems of experimentation need to be overcome. Firstly, that of strain measurement at very high temperatures and secondly, the problem of supplying and removing sufficient heat to create the necessary thermal gradients.

The authors devoted some effort to overcoming the first of these problems but could find no method of strain measurement of sufficient accuracy for shakedown tests at temperatures above about 200°C. Similarly the second problem seems to be insoluble with only normal laboratory equipment available. It was thus concluded that experiments must be conducted using an "artificial" method of producing the stress system in a test specimen. Bree [1] gives a comparison between the state of stress in the cylinder and that in a beam element which is subjected to a steady axial load together with a cyclic deformation to a prescribed radius of curvature. Referring to Fig. 7 the analogy between the stress states of such a beam element and the Bree cylinder is easily understood. The axial stress in the beam is analogous with the pressure stress of the cylinder whilst the bending stress is analogous to the thermal stress of the cylinder. It is seen from Fig. 7 that no thermal strain exists in this case so that the total strain varies through the thickness. However the steady state ratchet strain is found to be the same as that of Fig. 3.

By this argument experiments have been conducted using long thin strips of steel as "test specimens" and cyclically deforming these to a prescribed radius of curvature whilst applying a steady axial load. The authors have developed a wheel rig, shown schematically in Fig. 8 for performing such experiments. In this rig the strip was formed over the wheel and the ends of the strip were attached to chains. These chains were passed over pulleys and carried deadweights at their ends. The wheel was drived by an electric motor (with thyristor control) through a rack and pinion and crank mechanism. The wheel was thus rotated clockwise through

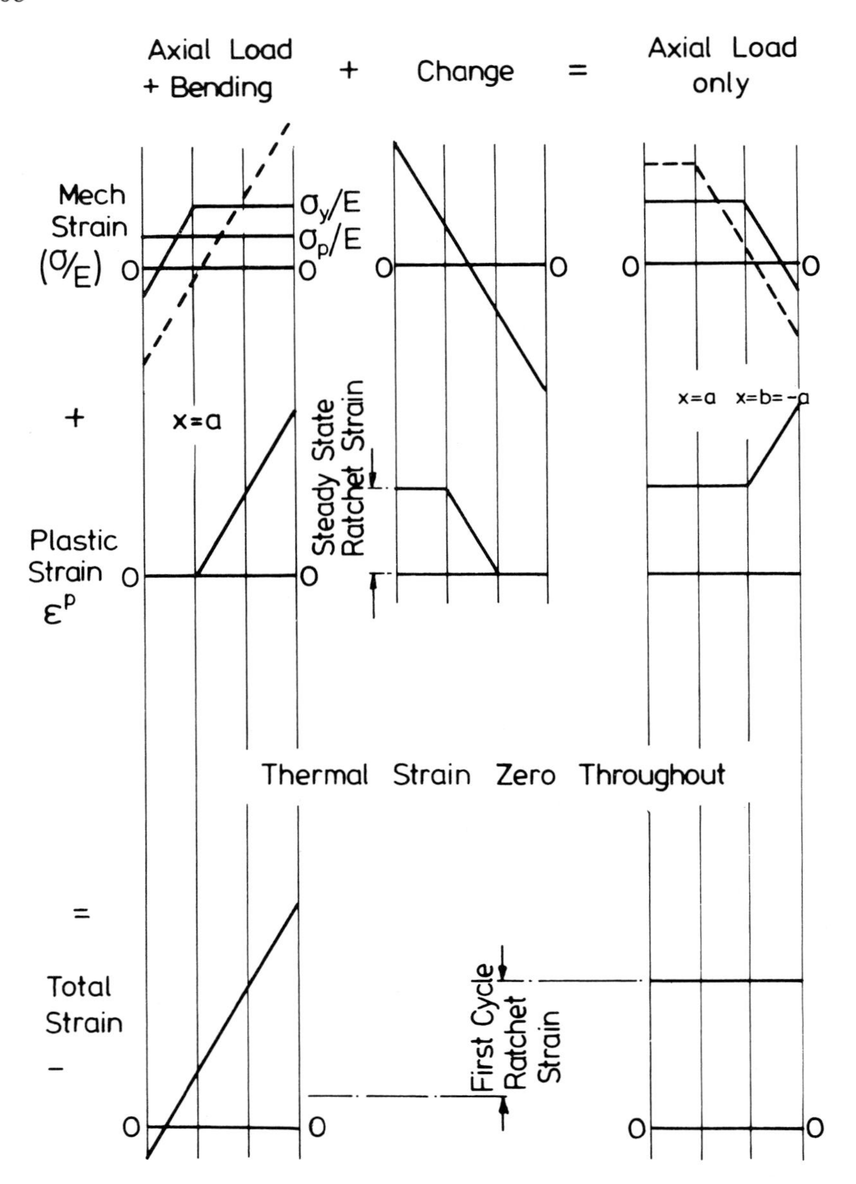

FIG. 7 Ratchetting with Dead-Weight and Bending Stresses

about 108° before the motion was reversed by the crank. By this means, the strip was brought into contact with the wheel (taking its curvature) and returned to its initial position. Thus a test section of the strip (about 250 mm) was cyclically deformed between the radius of curvature of the wheel and zero curvature whilst carrying a constant axial stress due to the dead-weight loads. The rate of cycling was initially fixed at 42s/cycle though subsequently the effect of cycle rate has been investigated.

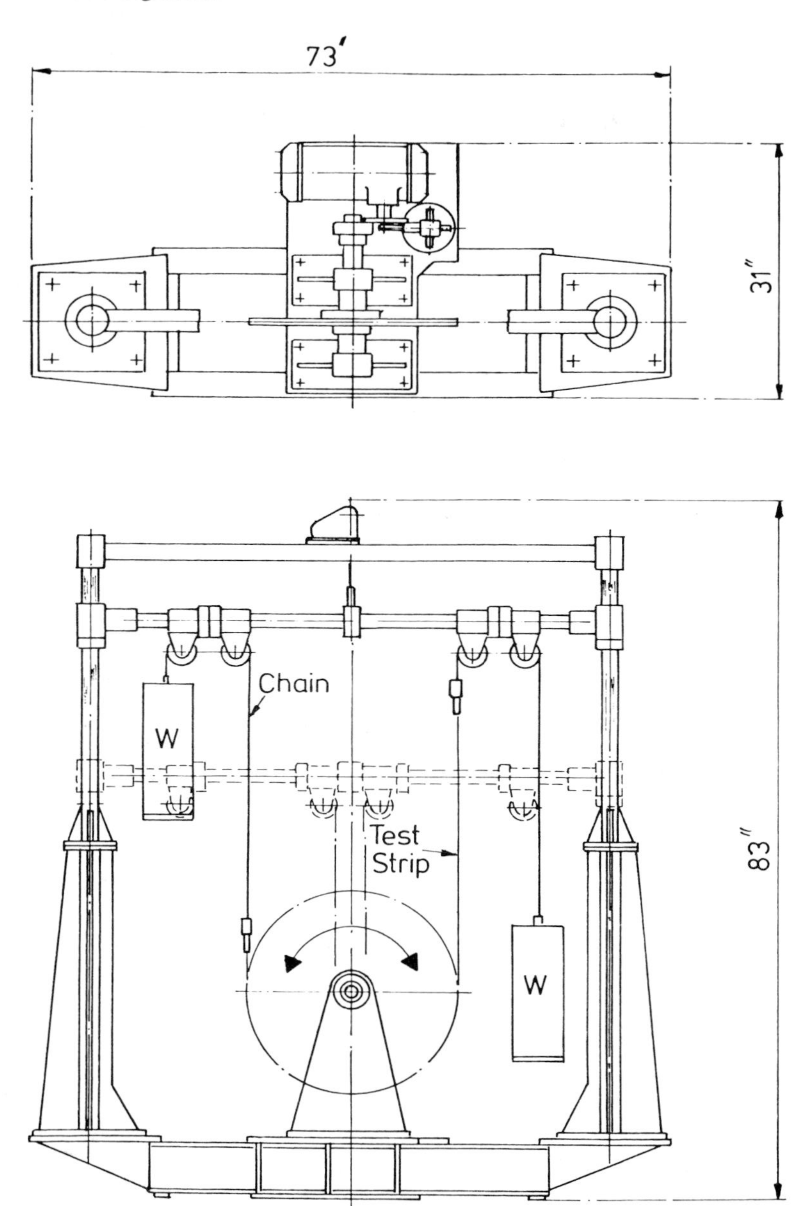

FIG.8 The Experimental "Wheel Rig"

Any chosen location on the Bree diagram could be investigated by choosing the wheel size and end loads to give the desired σ_t/σ_y and σ_p/σ_y ratios. Fig. 9 shows the locations chosen when investigating the strain behaviour of an elastic/perfectly-plastic strip. The deformation of the strip was measured by strain gauges on the strip and by strain gauge displacement transducers at the ends of the strip. The latter were found to give more repeatable results than the former and thus all results presented here were obtained by this means. The transducers had a resolution equivalent to an axial strain of 1 μ on the strip.

σ_t/σ_y	σ_p/σ_y						
5·050	·071	·142	·213	·283	·354	·425	–
3·710	·213	·283	·354	·425	·496	·567	–
3·328	·213	·283	·354	·425	·496	·567	·638
2·996	·283	·354	·425	·496	·567	·638	–
1·943	·425	·496	·567	·638	·708	·779	–

FIG. 9 Experimental Locations on the Bree Diagram.

Fig. 10 shows typical results from the ratchetting regime for a wheel size giving a σ_t/σ_y ratio of 3.710. Accumulated strains for various dead-weights (i.e. σ_p/σ_y ratios) are shown. Fig. 11 gives the results of all such tests conducted on the mild steel. These have been obtained using five different wheel sizes and various dead-weights. The accumulated strain per cycle is plotted against σ_p/σ_y ratios for each wheel size, i.e. for specific σ_t/σ_y ratios. Shown on Fig. 11 for comparison is the predicted strain accumulation according to Bree's model. It can be seen from Fig. 11 that whilst the Bree prediction for the onset of ratchetting is in good agreement with the experiment, the rates of strain accumulation do not agree well. It should be noted that these experimental results were obtained with a finite cycle rate and therefore include the effects of cold creep; Bree made no provision for this phenomena in his analysis.

Fig. 12 shows the experimentally obtained boundary of the ratchet regime plotted on the Bree diagram. Lines of constant cyclic strain accumulation are also shown. From this illustration it may be concluded that the Bree diagram is a most useful tool with which to locate the stress ratios which give rise to ratchetting. However, it should be used with some caution when assessing the rates of ratchetting.

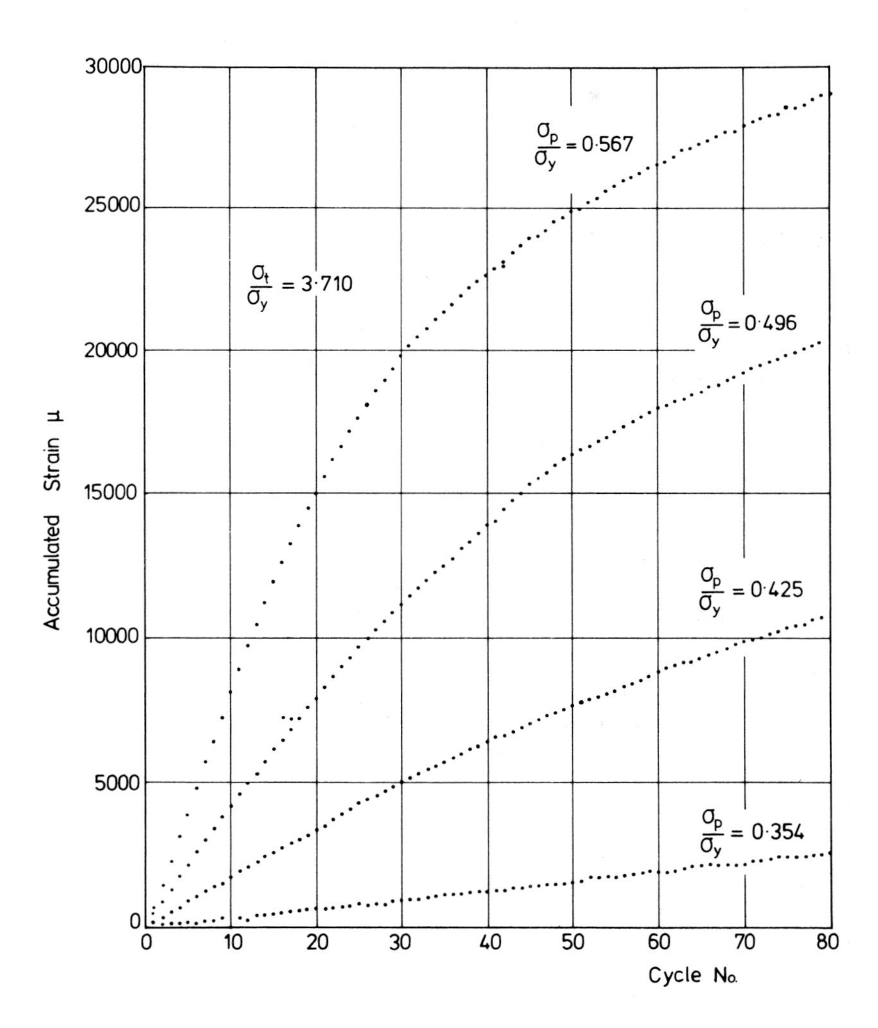

FIG. 10 Experimental Strain Accumulation with Cycling for Elastic/"Perfectly"-Plastic Mild Steel.

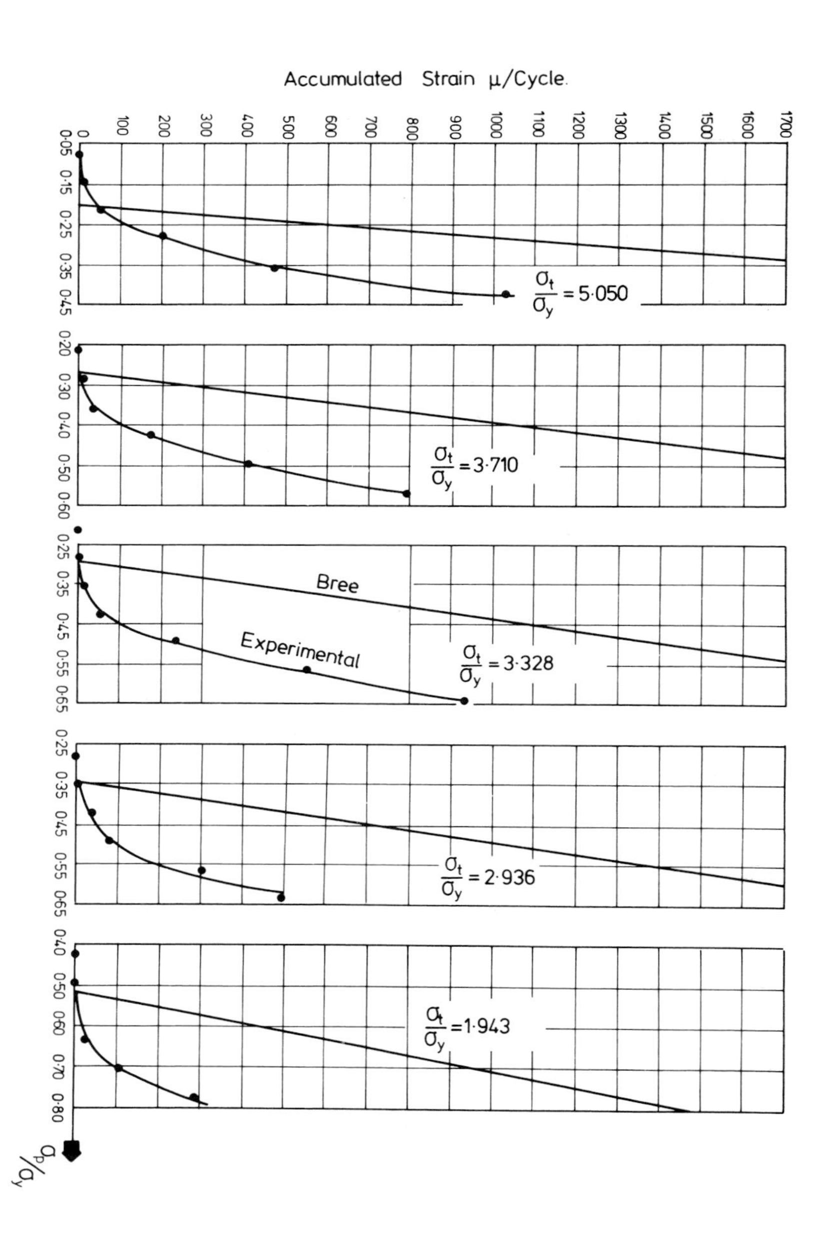

FIG. 11 Experimental Ratchet Strain/Cycle Compared with Bree [1]

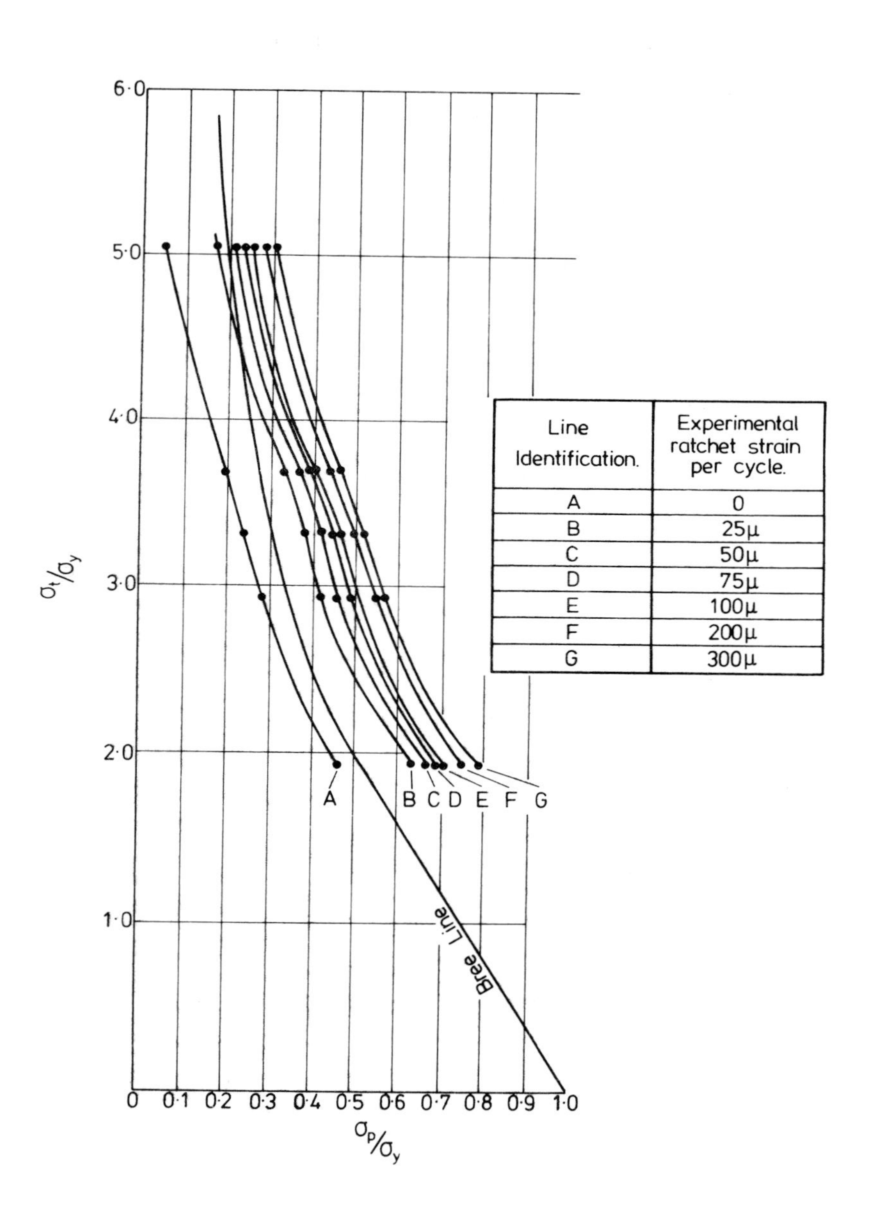

Line Identification.	Experimental ratchet strain per cycle.
A	0
B	25μ
C	50μ
D	75μ
E	100μ
F	200μ
G	300μ

FIG. 12 Lines of Constant Ratchet Strain/Cycle Compared with Bree [1]

6. RATCHETTING TESTS ON S/S 316

Ratchetting tests were conducted on stainless steel type 316 strip in the same manner as that used to verify the Bree diagram using mild steel strips. The S/S 316 (purchased as rolled strip) was subjected to a solution heat treatment at 1050℃.

The strip material was subjected to a "slow" tensile test which showed a yield stress (departure from linearity) at 75.4 MN/m². The stress/strain characteristic for the material is shown in Fig. 13. This illustration has been produced by fitting an Osgood/Ramberg [35] curve to the experimental data (range shown) and by using the Osgood/Ramberg descriptors the experimental curve has been extended to 35% strain. Materials properties to this order of magnitude of strain were necessary for the numerical analysis which follows. However, the experimental data was limited by the capability of the strain gauges, and thus extension of the curve was necessary.

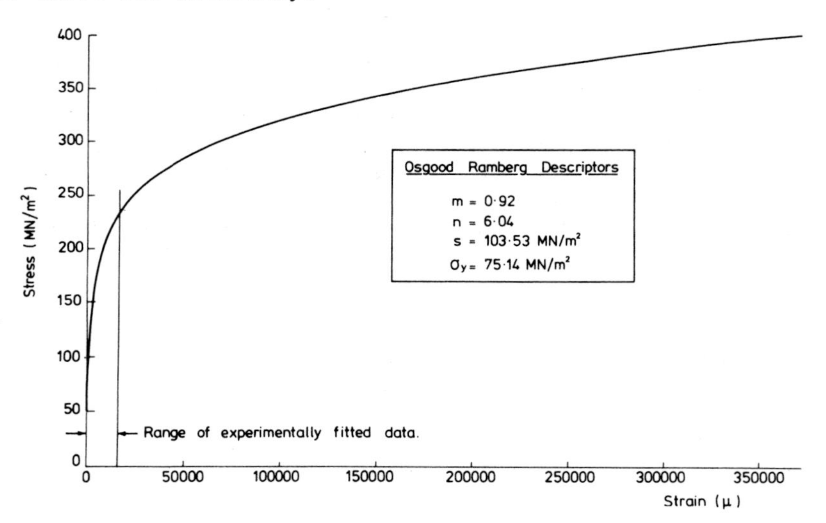

FIG.13 Uniaxial Stress Strain Curve for Stainless Steel Type 316

The authors were particularly interested in studying the ratchet behaviour of this steel over a large number of cycles. The wheel rig proved ideal for this purpose and several tests were conducted up to 3000 cycles. However, it was found that the rate of accumulation of strain change very little after about 1000 cycles and so all subsequent tests were conducted over this number of cycles. The rate of cycling was fixed at 41s/complete cycle although the effect of varying this rate is reported below.

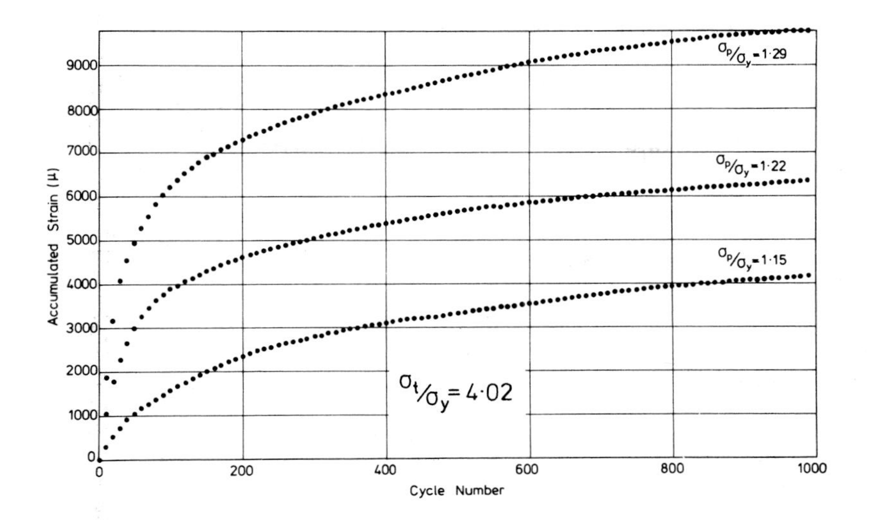

FIG. 14 Accumulated Strain over 1000 Cycles S/S 316

Fig. 14 shows some typical results up to 1000 cycles for $\sigma_t/\sigma_y = 4.02$ with three diferent σ_p/σ_y values. This illustration suggests that the ratchet strains will eventually reduce to zero. However, by plotting the incremental strain (at any cycle) against the natural log of the cycle number - Fig. 15 - it is seen that this decay will be very slow. (Fig. 15 shows a further set of data obtained at $\sigma_t/\sigma_y = 8.0$.)

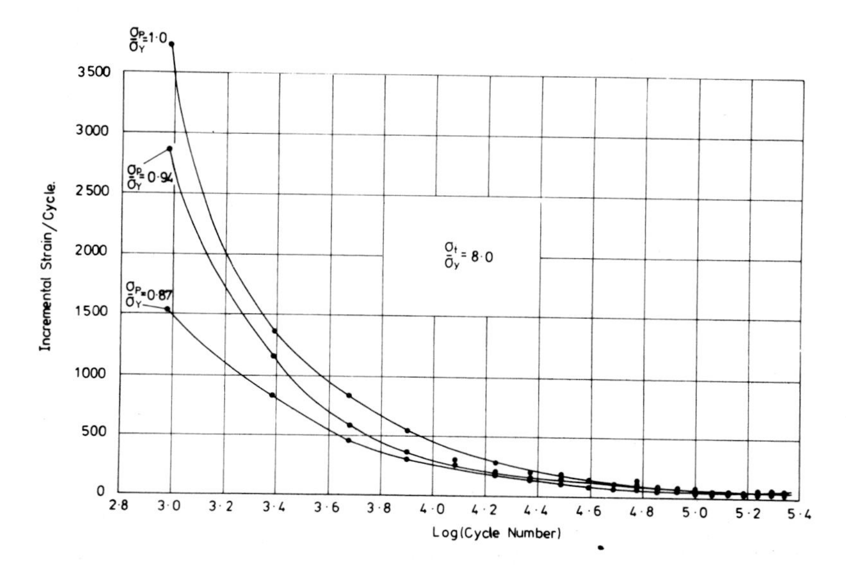

FIG. 15 Incremental Strain against Natural Log of Cycle Number

This aspect is relatively unimportant since the major contribution to the total accumulation of strain is seen to occur in the first say 500 cycles. However, at other σ_p/σ_y and σ_t/σ_y combinations it may take a greater or less number of cycles to accumulate this. The Bree diagram for a strain-hardening material is not bounded at $\sigma_p/\sigma_y = 1$ as is the diagram for an elastic/perfectly-plastic material provided that σ_y is defined as some stress below the tensile stress. The choice of σ_y for this purpose is quite arbitrary and is at present the centre of debate amongst workers in this field. It is natural to want to choose σ_y as some specific proof stress although no justification (except convenience) exists for doing so. Fig. 16 shows the results of wheel tests on S/S 316 in the form of a Bree diagram. For tests conducted at various σ_t/σ_y and σ_p/σ_y ratios the boundary of the ratchetting regime has been defined as those locations where shakedown[I] is/is not achieved within 100 cycles. Shown on Fig. 16 are the locations where "shakedown" was achieved within 100 cycles and those where it was not achieved even after 1000 cycles. Two sets of axes are drawn, firstly that using σ_y as the departure from linearity and secondly that defining σ_y as

$$\sigma_y = 1.35.P_{0.2}$$ (Where $P_{0.2}$ signifies the 0.2 proof stress.)

which is the ASME [2] recommendation. Also shown on Fig. 16 is a Bree ratchetting boundary assuming an elastic/perfectly-plastic idealisation based upon 200 MN/m^2. This figure is chosen by the equal area criterion using the experimental stress/strain curve up to 20,000 μ.

All of the above results were obtained at a cycle rate of 41s/complete cycle. The effect of varying this rate was investigated. The results shown in Fig. 17 indicate that the primary effect is to change the rate of accumulation of strain in the first 200 cycles. After this, the rates of accumulation of strain are reasonably consistent for 8, 13, 22 and 41 s/cycle. It is uncertain at this stage if the boundary of ratchetting is affected by cycle rate.

7. COMPUTER STUDIES ON THE BREE MODEL

Present design codes (e.g. [24]) permit ratchetting provided that the total accumulation of strain, within the life of the structure, is less than a specified limit. It is therefore important to be able to calculate, with

I. This method of experimentation identifies only two regimes, i.e. ratchetting and non-ratchetting. Within non-ratchetting is included shakedown, alternating plasticity and elastic.

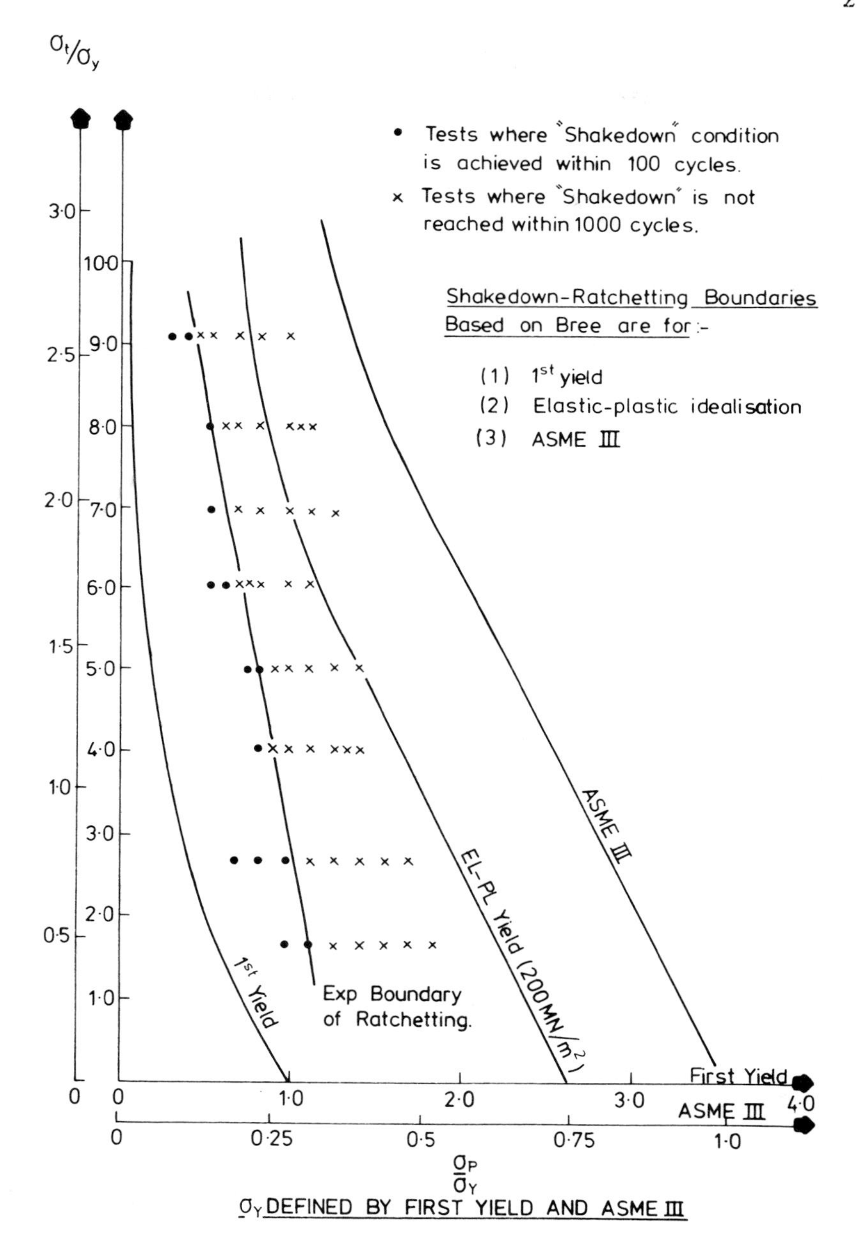

FIG. 16 Ratchetting Tests on Bree Beam Element for 316 S/S at Room Temperature

confidence, the degree of strain accumulation. With this in mind the previously discussed experiments have been conducted to give data with which calculation techniques can be compared.

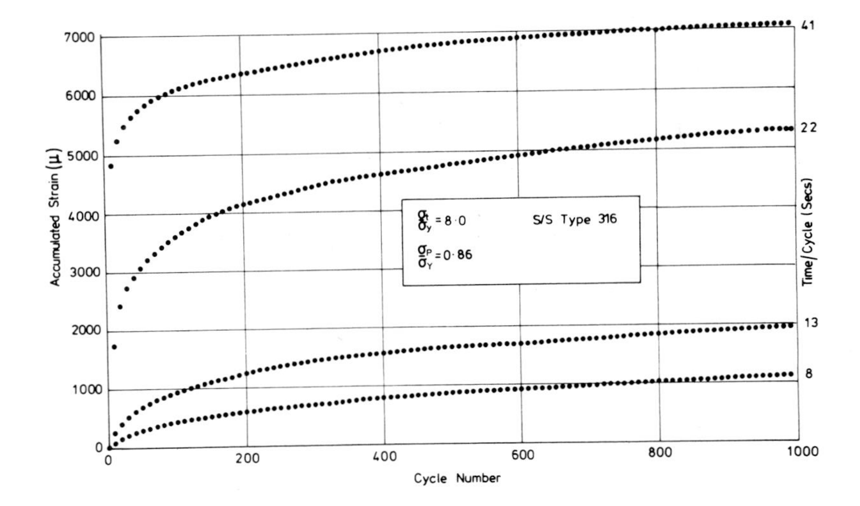

FIG. 17 The Effect of Cycle Time on Ratchet Strain Accumulation at the Same Loading Condition ($\frac{\sigma_t}{\sigma_y} = 8.0$ & $\frac{\sigma_p}{\sigma_y} = 0.86$).

It is known from classical constitutive relations of hardening that for a strain-hardening material ratchetting will cease within a small number of cycles. Mulcahy's results for linear kinematic hardening and the authors' analysis of non-linear kinematic hardening showed that ratchetting will progressively decrease with cycles and eventually stop, giving a finite total accumulation of strain. Experimental observations of ratchetting in the authors' work at room temperature however, showed that there is a tendency for ratchetting to continue at a low but constant rate for a large number of cycles. This continuous ratchetting was also observed by Yamamoto et al. [50] when thermal shock loadings were imposed on Type 304 stainless steel pipes for some 140 cycles. The tests of Goodall and Cook [47] and Anderson [49] did not show this aspect because of the fewer number of cycles tested.

This observation of continuous ratchetting has prompted the authors to use less conventional material models in their computer programs to attempt to represent this behaviour. The first step was to develop basic programs to model the Bree beam and the second was to select various properties from observed uniaxial material behaviour to implement in the programs.

Several programs have been developed for calculating the ratchetting strains of elastic/perfectly-plastic, linear and non-linear kinematic hardening materials. The results of the computations have been verified against the results of Bree's analysis for elastic/perfectly-plasticity and Mulcahy's results of linear kinematic hardening. The non-linear kinematic model initially utilised the Mroz Model [51] for piecewise linear approximation to a strain-hardening curve. This model was found to require a large computer core store and because of the inevitability that far greater core storage would be required when modelling cyclic properties it was replaced by an Osgood-Ramberg power law [35] to represent the strain-hardening curve with a least squares-fit optimisation to Type 316 stainless steel data. For computation, the Bree beam was divided into fifty elements across the thickness. The loading was applied in the form of a linear strain distribution (caused by the curvature and tensile load) on the first half cycle and a uniform strain distribution (due to the tensile load) on the second half cycle. For each element, the stresses were calculated using the chosen material model. Equilibrium and compatibility conditions were then applied to ensure that the summation of forces on the elements was equal to the applied tensile force. Two models have been implemented using the program originally written for non-linear kinematic hardening.

The first approach used a procedure reported by Jhansale [36] in which the cyclic stress/strain curves are related in shape to the "skeleton" monotonic stress/strain curve by means of an "elastic stress increment". It was shown by Jhansale that for a range of materials, it is possible to predict the transient cycle stress/strain curves at an intermediate cyclic between the first and the last at which the material cyclically stabilised. The technique required the experimental monotonic and stabilised cyclic stress strain curves and this latter requirement was obtained from the results of Jaske et al. [52] and Jaske and Frey [53]. These two references report an extensive study of fatigue and cyclic stress behaviour of Type 316 stainless steel at temperatures from 21 to 649℃. Taking only the 21℃ results into consideration, it was found that at a strain range of < .5%, there is a slight amount of cyclic softening followed by stable response. A strain range of 1.0% showed cyclic hardening. At higher temperatures the material consistently cyclically hardened. From the 21℃ data, a cyclic softening model was assumed and implemented in the program. The computations showed the total accumulated strains were greater than that predicted by the non-linear kinematic model. However, the total strain accumulation had a finite maximum value which was expected since the cyclic stress strain response of the material stabilised after a certain number of cycles. Based on this result, cyclic hardening will be expected to reduce the total strain accumulation at the same load condition.

The second approach used a model suggested by the yield

surface investigations of Moreton et al. [54] on a range of pressure vessel steels which included Type 316 stainless steel. This reference presented the results of a series of experimental investigations on the yield surface movement under a range of prestress directions and magnitudes. An empirical relationship was derived from these results to represent the yield locus movement using only three parameters, the yield stress, the prestress magnitude and the prestress direction.

Previous theoretically based models predict that the yield surface is constrained to pass through the point of prestress.

The empirical model of Moreton et al. is not thus constrained and suggests that the current yield surface will lie between the origin of stress and the prestress point. Briefly, this model gives the distance c of the centre of the yield locus from the origin of stress as a function of the effective prestress σ_{pE} and the prestress direction γ. The model states that

$$\frac{c}{\sigma_y} = g(\gamma).\frac{\sigma_{pE}}{\sigma_y} - g(\gamma)$$

where $g(\gamma) = \frac{1}{1.115} .f(\gamma)$ and $f(\gamma)$ is defined as

$$f(\gamma) = \{\frac{1}{1-\frac{1}{4}\sin 2\gamma}\}^{\frac{1}{2}}$$

For a uniaxial system of stresses $\gamma = \pi/2$ and hence $f(\gamma) = 1$. The model then reduces to

$$c = \frac{1}{1.115}(\sigma_{pE} - \sigma_y)$$

The yield locus is then assumed to move through the distance c without rotation or change of size. This model may be compared with the kinematic model for which the co-efficient 1/1.115 above would be unity. The work of Moreton et al. showed that the direction of translation was that of the prestress vector. This uniaxial representation gives predictions of the prestress and reloading stress difference, the value of which decreases with the number of reloadings after a single prestress and increase with the magnitude of the prestress. This model has been implemented in the non-linear kinematic hardening program. Fig. 18 shows the computer prediction of ratchetting and the experimental results plotted as total strain accumulation against the cycle number. The prediction of the model tended to underestimate the ratchet strain accumulation during the initial few tens of cycles but overestimated after this. The model is able to predict the steady "indefinite" ratchetting seen in the experiments.

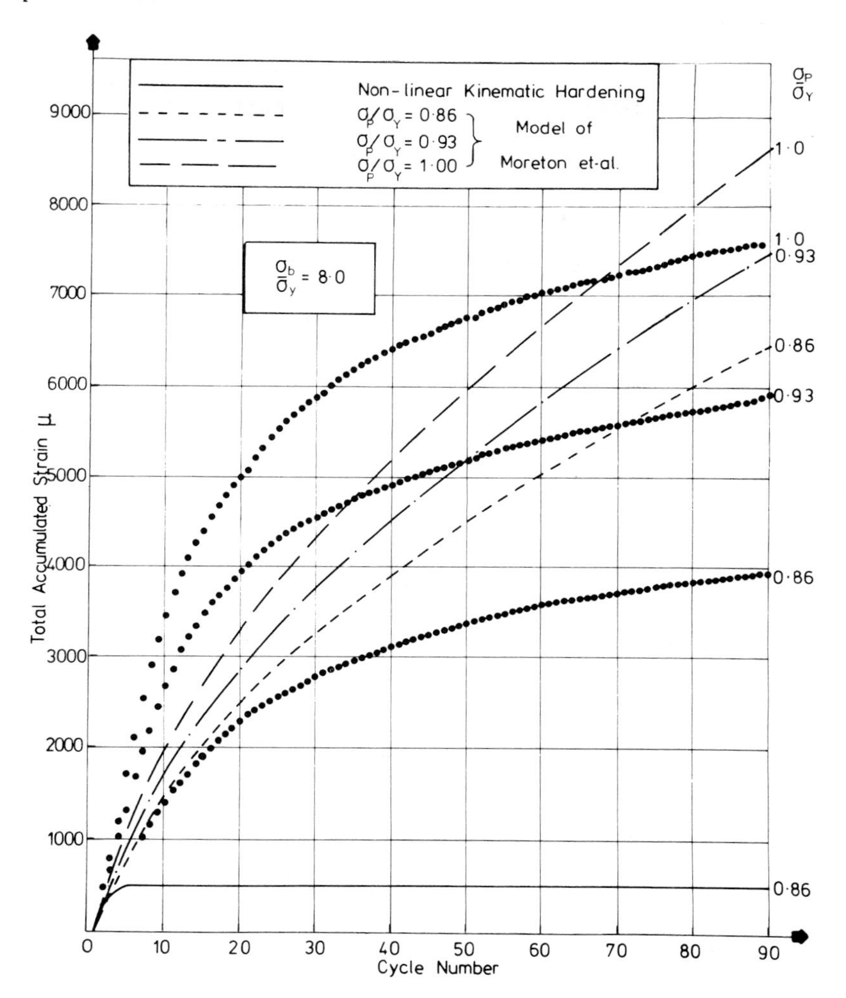

FIG. 18 A Comparison of Predicted and Experimental Ratchet Strain Accumulation for S/S 316.

The mechanism by which the "indefinite" ratchetting occurs is not clearly understood. It is almost certainly not due to the material itself ratchetting. Such a phenomena would only occur at much higher strains than the beam element experienced in this series of tests. Material models which include cyclic hardening or softening do not give satisfactory results when compared with the experimental data. The only material model which has so far permitted the authors to compute all the observed ratchetting characteristics is that of Moreton et al. [54][II]. The reasons why this should be so

II Some qualitative experimental support for this model is seen in the work of Goodman [55].

are not understood. The model was derived from monotonic tests on a range of steels of widely differing stress/strain characteristics. It should therefore be expected to give information which is not cycle dependent. However, when applied in the context cyclic loading between fixed strain limits it can be shown that the model gives a softening effect.

Finally, from the computing viewpoint, the difference between uniaxial tests and structural tests should be noted. In a uniaxial cyclic test, the cross section of the specimen is assumed to be subjected to a uniform stress and strain. A structural cyclic test such as on the Bree beam subject each element in the beam to a unique stress and strain history and the behaviour of such a beam is the sum of the responses of the individual elements. Calculations performed by the program are complicated and involve the repeated computation of the stresses in every element according to the rules of the models described until equilibrium and compatibility are attained. The above models are described in necessarily quantitative terms because they are not easily amendable to mathematical description. However some alternative analytical models are described by Goodman and Goodall [56] which bear some similarities.

8. DISCUSSION

The work conducted to experimentally verify the ratchetting boundary of the Bree diagram for elastic/ perfectly-plastic materials has proved satisfactory. The ratchetting boundary given by Bree is seen to agree with that of the experimentation with sufficient accuracy. The rate of strain accumulation obtained experimentally is in poor agreement with Bree. It is thought that this disagreement is due to rate effects and the presence of cold creep in real materials behaviour. Alternative diagrams have been presented for loading conditions other than that prescribed by Bree. It is hoped that these will prove useful to designers.

Experiments have also been conducted on the wheel rig using commercially available type 316 stainless steel strip. Heat treatment procedures for this material are not clearly prescribed by the various pressure vessel codes. However, it was felt that the most representative procedure which the authors could follow was to subject the strip to a solution heat treatment at 1050°C. This treatment was conducted in a vacuum furnace with strip cut to length and laid flat in the furnace. The uniaxial stress/strain characteristic obtained from the strip after this treatment has been given. The authors are well aware of the variability of this particular material and so care was taken to ensure a sufficient supply of strip for all the planned experiments.

After gaining confidence from the results of experimentation on mild steel strip the authors have experimentally obtained a "Bree diagram" for one particular strain-hardening material. The choice of stainless steel type 316 for this purpose was taken in the light of the recent designation of stainless steel type 316 for some proposed nuclear reactor components. This "Bree diagram" shows that, the onset of ratchetting lies to the right of the Bree line (designated 1st yield on Fig. 16). The line has been drawn assuming the material to be elastic/perfectly-plastic yielding at the limit of proportionality. Present design codes suggest that austenistic steels should be treated as elastic/perfectly-plastic yielding at some designated proof stress. The ASME recommendation for example has been given in Section 6. This results in a "design line" which lies well into the ratchetting regime. This line is shown on Fig. 16 and it is estimated that approximately 3% strain would be accumulated in 1000 cycles if a Bree cylinder were operated on this design line at $2.5 < \sigma_t/\sigma_y < 2.0$ on the ASME III scale.

The computer studies reported here have been undertaken with the intention of formulating an understanding of the mechanism by which the observed "indefinite" ratchetting can occur. After establishing computer programs which model the loading and which ensure equilibrium and compatability conditions, various models which describe the material's characteristics have been added. Such models tried included kinematic, Mroz and mechanical sub-layer - all of which showed ratchetting would cease after relatively few cycles. By introducing cyclic softening to such models the duration of ratchetting was extended but it was found that this then gave erroneous predictions in the shakedown and plasticity regimes.

The model of Moreton et al. [54], which is fundamentally a kinematic translation of the yield locus, was employed. Moreton and Moffat [57] had previously used this model to predict the shakedown pressures of three pressure vessel components. These were two torispherical drumheads and a nozzle/sphere intersection manufacture from stainless steel type 321. Predictions of the shakedown pressure were in agreement with experimentation by better than 10% for the drumheads and 19% for the nozzle. When employed on the Bree problem the model was found to give the required "indefinite" ratchetting observed in the experiments. The rate of ratchetting was found to be in poor agreement with experimentation but it should be noted that this model was derived from monotonic tests on a wide range of pressure vessel steels. It is probable that only a small change to the coefficient 1.115 would produce a better agreement since the kinematic model written in this form has a coefficient of unity.

The comparison made in Fig. 18 between the computed

ratchet strains and those found experimentally is somewhat misleading. The experimental data is known to be rate dependent during the first 200 or so cycles whereas the computed data is not. This rate dependency may be due to the development of "cold creep" strains during these early cycles of loading. This effect is seen, in Fig. 17, to decay after about 200 cycles and so comparison with experimental data after this point would then be more meaningful. Computations made up to 5000 cycles of loading however, showed ratchet strains of the order 6μ/cycle persisting in the range 4500 to 5000 cycles. Limited experimental work to this number of cycles shows that ratchet strains for the same σ_p/σ_y and σ_t/σ_y combinations drop to less than 1μ/cycle. Clearly, the model of Moreton et al. [54] used in these computations could be modified to correct this error and such a modification would not be unreasonable since this is a very general model. Even if this were done, poor agreement during the early cycles of loading would persist because of rate effects.

An apparent discrepancy in the experimental data of Figs. 17 and 18 should be highlighted. Fig. 18 shows experimental data obtained at 41s/cycle. The data for $\sigma_p/\sigma_y = 0.86$ in Fig. 18 shows an accumulation of about 4000μ after 90 cycles. The corresponding data of Fig. 17 shows about 6000μ accumulated after 90 cycles. These tests were conducted on different strips although from the same original stock and the discrepancy shown in these sets of data is typical of the scatter of data which has been seen in this series of experiments. In general it is the early cycles which have been prone to such scatter.

All of the work reported here has been conducted at room temperature. The primary use for this type of data is for structures operating at elevated temperatures. The authors are therefore currently engaged on extending the capability of the wheel rig so that tests at 400°C may be conducted. It is hoped that results of this work can be reported soon.

9. CONCLUSIONS

(a) The origins and subsequent literature concerning the Bree diagram have been reported.
(b) Three alternative diagrams for elastic/perfectly-plastic materials have been presented.
(c) The experimental verification of the Bree diagram has shown the ratchetting boundary in good agreement with Bree [1] although the rate of ratchetting is not so.
(d) A similar diagram has been produced experimentally for stainless steel type 316 which is strain-hardening.
(e) Computer studies have shown that the yield surface model of Moreton et al. [54] gives the correct form of strain behaviour when compared with the experiments of (d) above.

ACKNOWLEDGEMENTS

The work reported here is presently sponsored by the National Nuclear Corporation Limited. Their permission to publish these results is gratefully acknowledged.

REFERENCES

1. BREE, J. - 'Elastic-Plastic Behaviour of Thin Tubes Subjected to Internal Pressure and Intermittent High-Heat Fluxes with Application to Fast-Nuclear Reactor Fuel Elements'. J. of Strain Analysis, 1967, Vol. 2, 3, pp. 226-238.

2. A.S.M.E. Boiler and Pressure Vessel Code, Section III, Nuclear Power Plant Components.

3. TOWNLEY, C. H. A. - 'Designing Pressure Vessel Details'. The Engineer, March 1969, p. 438.

4. FINDLAY, G. E., MOFFAT, D. G. and STANLEY, P. - 'Elastic Stresses in Torispherical Drumheads: Experimental Verification'. J. of Strain Analysis, 1968, 3, 214. 'Torispherical Drumheads: A Limit Pressure and Shakedown Investigation'. Ibid., 1971, 3, p. 147.

5. SAMPAYO, V. M. and TURNER, C. E. - 'Computed Elastic-Plastic Behaviour and Shakedown of Some Radial Nozzle-on-Sphere Geometries'. 2nd Int. Conf. on P.V. Tech., San Antonio, 1973, paper 1-24.

6. MORETON, D. N. and MOFFAT, D. G. - 'Shakedown of Three Stainless Steel Pressure Vessel Components'. 3rd Int. Conf. on P.V. Tech., Tokyo, 1971, pp. 233-245.

7. LECKIE, F. A. and PAYNE, D. J. - 'Some Observations on Spherical Pressure Vessels with Flush Cylindrical Nozzles'. Proc. Instn. Mech. Engrs., 1965-66, 180, pt. 1, p. 497.

8. LECKIE, F. A. - 'Shakedown Pressures for Flush Cylinder-Sphere Shell Intersections'. J. Mech. Engrg. Sci., 1965, 7, p. 367.

9. PROCTOR, E. and FLINDERS, R. F. - 'Experimental Investigation into the Elastic/Plastic Behaviour of Isolated Nozzles in Spherical Shells. Part 2 - Shakedown and Plastic Analysis'. Central Electricity Generating Board, Berkeley Nuclear Laboratories, Report No. RD/B/N 881.

10. MORETON, D. N. and MOFFAT, D. G. - 'Effect of Proof Testing on the Shakedown Behaviour of Two Stainless Steel Pressure Vessel Components'. University of Liverpool, Dept. of Mech. Eng., Report No. A/047/80, May 1980.

11. MULCAHY, T. M. - 'An Assessment of Kinematic Hardening Thermal Ratchetting'. J. of Engineering Materials and Technology, Trans. ASME, Series H, Vol. 96, 1974, pp. 214-221.

12. MULCAHY, T. M. - 'Thermal Ratchetting of a Beam Element Having an Idealised Bauschinger Effect'. J. of Engineering Materials and Technology, Trans. ASME, Vol. 98, 1976, pp. 264-271.

13. ROARK, R. J. and YOUNG, W. C. - Formulas for Stress and Strain, McGraw-Hill, New York.

14. WEIL, N. A. and RAPASKY, F. S. - 'Experience with Vessels of Delayed Coking Units'. Preprint of paper submitted to a session on pressure vessels, American Petroleum Institute, Division of Refining, Los Angeles, California, May 13, 1958.

15. COFFIN, L. F. - 'The Resistance of Materials to Cyclic Thermal Strains'. Paper No. 57, A-286.

16. PARKES, E. W. - 'Wings Under Repeated Thermal Stress'. Aircraft Engineering, Vol. 26, 1954, pp. 402-406.

17. PRAGER, W. - An Introduction to Plasticity, Addison Wesley Publ. Ltd., 1959.

18. GILL, S. S. - 'A Simple Illustration of Incremental Collapse'. Bull. Mechanical Engineering Education, Vol. 6, No. 1, January 1967.

19. RUIZ, C. - 'The Inclusion of Incremental Collapse Problem in Undergraduate Courses'. Bull. Mechanical Engineering Education, Vol. 6, No. 1, January 1967.

20. HILL, R. - The Mathematical Theory of Plasticity, Oxford University Press, London, England, 1950, pp. 287-294.

21. MILLER, D. R. - 'Thermal Stress Ratchet Mechanism in Pressure Vessels'. Trans. ASME, Series D, Vol. 81, No. 2, June 1959, pp. 190-196.

22. EDMONDS, H. G. and BEER, F. J. - 'Notes on Incremental Collapse in Pressure Vessels'. J. of Mechanical Engineering Science, Vol. 3, 1961.

23. BURGREEN, D. - 'The Thermal Ratchet Mechanism'. J. of Basic Engineering, Trans. ASME, Series D, Vol. 90, No. 3, 1968, pp. 319-324.

24. A.S.M.E. Boiler and Pressure Vessel Code, Section III, Nuclear Power Plant Components, Division 1, Subsection NB-3222.5, 1974.

25. BURGREEN, D. - 'Structural Growth Induced by Thermal Cycling'. J. of Basic Engineering, Trans. ASME, Series D, Vol. 90, No. 3, 196, pp. 469-485.

26. PARKES, E. W. - Structural Effects of Repeated Thermal Loading, Thermal Stress, (Ed. Benham et al.), Pitman Ltd., London, 1964.

27. MEGAHED, M. M. - 'Influence of Hardening Rule on the Elasto-Plastic Behaviour of a Simple Structure under Cyclic Loading'. Int. J. Mechanical Sciences, Vol. 23, 1981, pp. 169-182.

28. AINSWORTH, R. A. - 'An Experimental Study of a Three Bar Structure Subjected to Variable Temperature', Int. J. Mechanical Sciences, Vol. 19, 1977, pp. 247-256.

29. MEGAHED, M. M., PONTER, A. R. S. and MORRISON, C. J. - 'A Study of Time Independent Material Ratchetting', Department of Engineering Report 80-14, University of Leicester, December 1980.

30. UGA, T. - 'An Experimental Study on Thermal-Stress Ratchetting of Austenitic Stainless Steel by Three Bars Specimen'. Nuclear Engineering and Design, Vol. 26, 1974, pp. 326-335.

31. COFFIN, L. F. - 'A Study of the Effect of Cyclic Thermal Stresses on a Ductile Metal'. Trans. ASME, 1954, Vol. 76, pp. 931.

32. BREE, J. - 'Incremental Growth due to Creep and Plastic Yielding of Thin Tubes Subjected to Internal Pressure and Cyclic Thermal Stresses'. J. of Strain Analysis, Vol. 3, No. 2, 1968, pp. 122-127.

33. FREDERICK, C. O. and ARMSTRONG, P. J. - 'Convergent Internal Stresses and Steady Cyclic States of Stress', J. Strain Analysis, Vol. 1, No. 2, 1966, pp. 154.

34. MORETON, D. N. and NG, H. W. - 'The Extension and Verification of the Bree Diagram'. Paper L10/2, Vol. L, Trans. Sixth Int. Conf. Struct. Mech. in Reactor Techn., Paris, 1981.

35. RAMBERG, W. and OSGOOD, W. R. - 'Description of Stress-Strain Curves by Three Parameters'. NACA Report No. 902, Washington, July 1943.

36. JHANSALE, H. R. - 'A New Parameter for the Hysteretic Stress-Strain Behaviour of Metals'. J. of Engineering Material and Technology, Trans. ASME, Vol. 97, January 1975, pp. 33-38.

37. O'DONNELL, W. J. and POROWSKI, J. - 'Upper Bounds for Accumulated Strains due to Creep Ratchetting'. J. of Pressure Vessel Technology, Trans. ASME, Vol. 96, 1974, pp. 150-154.

38. LECKIE, F. A. - 'A Review of Bounding Techniques in Shakedown and Ratchetting at Elevated Temperatures'. W.R.C. Bulletin No 195, June 1974, pp. 1-32.

39. PICKEL, T. W., YAHR, G. T., SARTORY, W. and GREENSTREET, W. L. - 'Study of Thermal Ratchetting Behaviour and Elastic Design Requirements'. High-Temperature Structural Design Methods for LMFBR Components, Quarterly Progress Report for period ending September 30, 1972, ORNL-TM-4058, p. 190, January 1973.

40. A.S.M.E. Boiler and Pressure Vessel Code, Code Case 1592, 1974, ASME, New York.

41. A.S.M.E. Boiler and Pressure Vessel Code, Code Case N-47-12, 1977, ASME, New York.

42. O'DONNELL, W. J., POROWSKI, J. S., CORUM, J. M., YAHR, G. T. and SARTORY, W. K. - 'Biaxial Model for Bounding Creep Ratchetting in Shells'. Paper L11/15, Trans. Sixth Int. Conf. Struct. Mech. Reactor Techn., Vol. L, Paris, 1981.

43. GOODMAN, A. M. - 'The Influence of Rapid Thermal Transients on Elastic-Plastic Ratchetting'. C.E.G.B. Berkeley Nuclear Laboratories, Report No. RD/B/N4492, February 1979.

44. PHILLIPS, J. - 'The Accumulation of Damage in Plates Subjected to Mechanical and Thermal Loading'. C.E.G.B. Berkeley Nuclear Laboratories U.K., Report No. RD/B/N4314.

45. ROBINSON, E. L. - 'Effect of Temperature Variations on the Long Time Rupture Strength of Steels'. Trans. ASME, Vol. 74, 1952, pp. 777-784.

46. MINER, M. A. - 'Cumulative Damage in Fatigue'. J. of Applied Mechanics, Trans. ASME, Vol. 12, 1945, pp. A159-164.

47. GOODALL, I. W. and COOK, T. - 'Behaviour of Beams Subjected to Combined Direct Load and Varying Curvature'. Int. J. Mechanical Sciences, Vol. 14, 1972, pp. 137-143.

48. ANDERSON, W. F. - 'Ratchetting Deformation as Affected by Relative Variation of the Loading Sequence'. Part I - Design and Analysis, 2nd Int. Conf. Pressure Vessel Techn., San Antonio, Texas, USA, ASME, 1973, pp. 277-289.

49. ANDERSON, W. F. - 'Creep Ratchetting Deformation and Rupture Damage Induced by a Thermal Transient Stress Cycle'. Design for Elevated Temperature Environment, ASME, New York, 1971, pp. 1-11.

50. YAMAMOTO, S., KANO, J. and YOSHITOSHI, A. - 'Thermal Ratchetting Experiment of Type 304 Stainless Steel Pipes under Alternating Cold and Hot Thermal Shocks with Varying Axial Loads'. Elevated Temperature Design Symposium, ASME, New York, 1976, pp. 25-32.

51. MROZ, Z. - 'On the Description of Anisotropic Hardening'. J. Mechs. and Physics of Solids, 15, 1967, pp. 163.

52. JASKE, C. E., MINDLIN, H. and PERRIN, J. S. - 'Development of Elevated Temperature Fatigue Design Information for Type 316 Stainless Steel'. Paper C163/73, Proc. of the Inst. Mech. Engrs., Conference Publication 13, 1973.

53. JASKE, C. E. and FREY, N. D. - 'Long Life Fatigue of Type 316 Stainless Steel of Temperatures up to 593°C'. J. of Materials, Trans. ASME, 1981, February.

54. MORETON, D. N., MOFFAT, D. G. and PARKINSON, D. P. - 'The Yield Surface Behaviour of Pressure Vessel Steels'. J. of Strain Analysis, Vol. 16, No. 2, 1981, pp. 127-135.

55. GOODMAN, A. M. - Private Communication.

56. GOODMAN, A. M. and GOODALL, I. W. - 'Constitutive Relations for Stainless Steels'. Berkeley Nuclear Laboratories, Report RD/B/5040N81, 1981.

57. MORETON, D. N. and MOFFAT, D. G. - 'The Use of an Empirical Yield Surface Model in Predicting Shakedown Loads'. IUTAM Symposium/France 1980. Eds. Hult and Lemaitre. Pub. Springer-Verlag, Berlin, p. 181.

Chapter 5

REPRESENTATION OF INELASTIC BEHAVIOR IN THE PRESENCE OF ANISOTROPY AND OF FINITE DEFORMATIONS

E. T. Onat

Section of Applied Mechanics, Yale University,

New Haven, Connecticut 06520, U.S.A.

SUMMARY

Deformation of a solid is, in general, accompanied by the development of anisotropy and internal texture. When this happens, the orientation of a deforming element becomes an issue, especially in the range of finite deformations. Section 2 of the paper is devoted to the study of the notion of internal state and orientation and the related role of superimposed rigid body rotations. We show that the state and orientation of a material element can be represented by a number of irreducible even rank tensors. These tensors not only define the orientation of the element, but also its internal symmetries and internal state. In Section 3 we discuss the laws of evolution for the internal state and we emphasize the very special role played by the rate of rotation Ω. Section 4 shows that the use of irreducible tensors as state variables leads to a convenient measure of strength of anisotropy. Representation of internal damage that accompanies tertiary creep is briefly considered in Section 5. It is seen that state variables in the form of irreducible tensors arise quite naturally from the statistics of internal happenings. In Section 6 a rigid-plastic body that hardens kinematically provides a second example for the illustration of the ideas used in the paper. Shear flow of such a material is considered in detail. Recent work on this subject is reviewed and clarified.

1. INTRODUCTION. REPRESENTATIONS BASED ON THE NOTION OF INTERNAL STATE AND ORIENTATION

Often it is of vital importance that the engineer be able to understand and predict the behavior of a structure (say, a component of a nuclear reactor) under various and sometimes extreme operating conditions. With this aim in mind, a vast amount of data are produced and collected on the mechanical behavior of the materials used in the structure of interest. The data come from phenomenological experiments and more recently from various kinds of microscopy that shed welcome light on the physical processes that accompany deformation.

Next the data must be combined and studied to develop mathematical models that reproduce, with various levels of accuracy, the observed relationship between histories of stress and deformation for a given material.*

The engineer will then choose from this set, a model that is appropriate to the task at hand and will use the model, together with the usual conservation laws of mechanics and appropriate numerical techniques to solve the structural problems of interest. We must emphasize that the choice of model depends on the nature of the problem. If, for instance, dynamic effects are expected to be important, the representation chosen must be capable of describing the mechanical behavior of material in the presence of high rates of strain with sufficient accuracy.

It is our opinion that the task of developing reasonably truthful representations of mechanical behavior with the help of experimental evidence and appropriate physics is still poorly understood. The reason for this is that most materials of interest exhibit nonlinear memory dependent behavior and the problem of identification and representation of such memory dependent behavior is a difficult one.

Nevertheless, it became clear during the past decade that representations based on systems of differential equations have definite advantages. Indeed most of the recent work on the representation of mechanical behavior, for simple nonaging solids and in the presence of isothermal deformations, can be put in the following "canonical" form based on the notion of internal state**:

$$\sigma(t) = f(S(t)) \qquad f: \Sigma \to T_2{}^s, \ \Sigma \subset R^N$$
$$\frac{dS(t)}{dt} = g(S(t), D + \Omega), \ g: \Sigma \times T_2 \to R^N \tag{1}$$

---

* For simplicity, we are considering here only isothermal deformations.

** The representation (1) is not appropriate for Voigt Solids unless one wishes to consider a Voigt like behavior as a limiting case of the one studied here.

where $\sigma(t)$ is the (Cauchy) stress tensor that the material element carries at time t; D and Ω are respectively, the tensors of rate of deformation and rate of rotation at time t: Thus, in components,

$$(D + \Omega)_{ij} = \frac{\partial v_i}{\partial x_j} = \text{velocity gradients.} \qquad i,j = 1,2,3.$$

$T_2(T_2^s)$ denotes the space of second rank (symmetric) tensors. And $S(t) \varepsilon \Sigma \subset R^N$ represents the internal state* and orientation of the element of interest. It is useful to think of S(t) as N parameters that measure those aspects of the arrangement of atoms or molecules within the material element at time t that are relevant to an approximate representation of mechanical behavior. It is assumed that the future behavior of the material will depend (to within a given approximation) only on these N parameters at time t and the stimulus applied to it in the future, i.e. for times $\tau > t$.

It is hoped that N will not be a large number. This expectation comes from the knowledge that many fine details of internal structure cannot influence strongly global aspects of mechanical behavior such as stress, displacement, and velocity. These are average quantities and they do not measure the details of the internal force distribution nor the details of deformation within grains. Thus we may say that S represents the internal state and orientation of the material as far as an approximate description of mechanical behavior is concerned. We refer to Σ as the state space. Clearly Σ is a subset of R^N.

The first equation in (1) simply states that the current stress is a function of the internal state and orientation. The second system of equations in (1) constitutes the laws of evolution or growth for S. It states that the rate of change of S is a function only of the current state and orientation and the velocity gradients applied to the element. Thus if $D + \Omega$ are given on the time interval (0,T) and the initial state and orientation S_o is known, 1(b) will enable us to obtain S(t) by integration. We can then read the stress $\sigma(t)$ from 1(a).

We will have much more to say about the structure of the representation in the remainder of the paper. But first we dispose of some preliminaries:

If S_o denotes the state and orientation of identically oriented and identical virgin material elements, then we insist that

---

* For the formal definition of state, see [3] and [4].

$$
\begin{aligned}
f(S_o) &= 0 \\
g(S_o, 0) &= 0
\end{aligned}
\tag{2}
$$

This means that the virgin state is stress free and that in the absence of deformations (no shape changes, no rotations) the state and orientation of the virgin material does not change over the course of time, i.e. the material is nonaging.

Now we let x_i and X_i denote rectangular coordinates of the current and initial positions, respectively, of a generic material point. The deformation of a body which occupies the domain D_o at $t = 0$ is then defined by the

$$x_i = x_i(X_j, t): \quad D_o \times R \to R^3 \tag{3}$$

The deformation gradient tensor F has the following components

$$F_{ij} = \frac{\partial x_i}{\partial X_j} \qquad (i,j = 1,2,3) \tag{4}$$

It is well known that

$$F^{\cdot} F^{-1} = D + \Omega \tag{5}$$

where $F^{\cdot}$ denotes material time derivative of F and D and Ω are, as before, the tensors of rate of deformation and rate of rotation respectively.

It must be noted that the map (3) used to define the deformation of the body is a "smoothed" map. We do not really expect $x(X,t)$ to give the position of the atom that occupies the position X at $t = 0$. (There may be no atom at all at that location initially. Moreover even if there were a particle there at $t = 0$, $x(X,t)$ will not be an exact location of it at time t.) What $x(X,t)$ measures is the average position of the material points that lie initially in a neighborhood of X. If we focus the attention to particles that lie initially within a small sphere centered at X, then we will find that these points will occupy after deformation a domain that has a "wiggly " boundary as a rule. But in ordinary experiments, devices used to measure strains are not capable of seeing atoms, and not even the grains of a polycrystalline solid. Hence the deformed shape of the sphere will look like an ellipsoid to these instruments and to us. Thus the act of measuring smooths the deformation map and in our calculations we work with this smoothed map. It is well-known that the deformation gradient tensor $F(X,t)$ completely determines the shape, size and orientation of the above (infinitesimal) ellipsoid at time t.

Finally we emphasize that (1) is so constructed that if

$S_1 \neq S_2$ at time t, then the elements in S_1 and S_2 will behave differently under some identical future deformation applied to them.

2. SUPERIMPOSED RIGID BODY ROTATIONS. TENSORIAL NATURE OF STATE VARIABLES.

We now discuss the special role played by superimposed rigid body rotations in the representation of mechanical behavior. The importance of superimposed rigid body rotations was stressed in the pioneering work of Green and Rivlin [1] and Noll [1]. This discussion is also closely related to the main subject of the paper: anisotropy and its representation.

Consider two identical and identically oriented virgin material elements. Apply to these elements, respectively, the homogeneous deformations

$$\text{(i) } F(\tau) \text{ and (ii) } Q(\tau)F(\tau) \quad \text{on } \tau \in [0,t] \tag{6}$$

where $Q(\tau)$ is a proper orthogonal transformation ($QQ^T = I$, $\det Q = +1$) that measures a rigid body rotation in R^3. Thus at time τ the deformed configuration of the element subjected to (ii) (or simply the element (ii)) differs from that of (i) by a superimposed rigid body rotation $Q(\tau)$.

What can we say about the internal state of these elements and the surface traction acting on them?

The reasonable and traditional answer to this query is that at time τ the arrangement of constituent particles in element (ii) will differ from that in (i) only by the applied rigid body rotation* $Q(\tau)$. Hence the surface tractions acting on (ii) will be those of (i) rotated by Q. The above assumption implies that the stress response of the two elements in tests (i) and (ii) would be, respectively,

$$\sigma(\tau) \text{ and } Q(\tau)\sigma(\tau)Q^T(\tau) \tag{7}$$

where Q^T denotes the transpose of Q.

We denote by $S(\tau)$ and $S'(\tau)$ the state and orientation of the elements (i) and (ii) respectively. In view of the assumptions just discussed, we may say that at any time the internal state of the rotated element (ii) differs from that of (i) only in orientation. More precisely this means that $S'(t) \in \Sigma$ will depend only on $S(\tau)$ and $Q(\tau)$ but not on the

---

* These assumptions that go under the name of principle of objectivity would not hold if these elements had gyroscopes in them or if there existed oriented external fields that affected mechanical behavior.

superimposed rotations applied previously, nor on the path that took the state point from S_o to S in the test. Thus

$$S'(\tau) = P(Q(\tau), S(\tau)) \quad P: \; O^+(3) \times \Sigma \rightarrow \Sigma \tag{8}$$

where $O^+(3)$ denotes the group of rigid body rotations in R^3. It is more instructive to write (8) as follows

$$S' = P_Q S \qquad P_Q: \Sigma \rightarrow \Sigma \tag{9}$$

and to think of P_Q as a mapping on Σ created by the superimposed rigid body rotation Q in all tests of the type (6) and in the manner described above.

Intuitively one could think of the action of P_Q on S as being induced by the application of an instantaneous rigid body rotation Q to a specimen in S.

The map P_Q is of basic importance: It will enable us to study the implications of the above assumptions. For example, we will find that f(S) in (1a) cannot be an arbitrary function but must satisfy the invariance requirement:

$$f(P_Q S) = Qf(S)Q^T \tag{10}$$

There are similar invariance requirements on the evolution laws that we shall discuss later.

On the other hand, P_Q also plays a central role in the discussion of questions of material symmetry (cf. Section 4):

For instance, if the initial internal state is isotropic then we must have

$$P_Q S_o = S_o \text{ for all } Q \in O^+(3). \tag{11}$$

It should be clear from these examples that we need to specify the precise nature of the transformation P_Q if we wish to arrive at explicit representations of mechanical behavior.

With this purpose in mind we make the following observations: The set of all P_Q as Q ranges over $O^+(3)$

$$G = \{P_Q: \; Q \in O^+(3)\}$$

defines a group of transformations on Σ. Indeed it is easily seen from the definition of P_Q that the following composition laws are obeyed:

$$P_{QR} = P_Q P_R$$

and

$$P_{Q^T} P_Q = I \text{ for all } Q,\ R \in O^+(3). \tag{12}$$

where I is the identify map on Σ. Moreover as noted by Geary and Onat [3] for solid materials G is isomorphic to $O^+(3)$ and hence is a Lie group. (These authors construct a theory of representation that starts with a definition of state based on observed behavior. They then establish a notion of distance in the state space Σ and introduce the transformations P_Q in Σ induced by the superimposed rigid body rotations Q. The metric for Σ is such that P_Q preserves distances.)

When G is a Lie group that acts on a finite dimensional metric space as is the case here ($\Sigma \subset R^N$) a theorem due to Mostow [5] concerning equivalent embeddings enables one (cf. [3] for details) to arrive at the following surprising and far-reaching result:

Without loss of generality one can choose Σ and N in representation (1) in such a way that P_Q can be considered to be a linear orthogonal transformation on R^N. Thus with this choice a rigid body rotation Q applied to a material element in $S \varepsilon \Sigma \subset R^N$ will move the state point to $P_Q S \varepsilon \Sigma$ where P_Q is a linear orthogonal transformation on R^N.

Thanks to this extended and new meaning of P_Q, the group $G = \{P_Q\}$ becomes a representation in R^N of $O^+(3)$. The theory of representation of $O^+(3)$ (cf. [6]) can now be used to specify the action of P_Q in R^N as follows:

S can be thought of composed of a number of irreducible tensors* $q_1, \ldots, q_n$:

$$S = (q_1, \ldots, q_n) \tag{13}$$

where each q_i defines through its components a point in an invariant subspace (under G) of R^N. Moreover

$$P_Q S = (P_Q q_1, \ldots, P_Q q_n) \tag{14}$$

where $P_Q q_i$ has the meaning of ordinary tensor transformation appropriate to its rank.

An example may be useful to illustrate the meaning of (13) and (14). Let us consider a second rank tensor $\underset{\sim}{a}$. Then $\underset{\sim}{a}$ has components a_{ij} (i, j = 1,2,3) in the laboratory frame. It is well-known that $\underset{\sim}{a}$ decomposes in the following way into three types of irreducible tensors:

---

* We shall soon see that in the present case these tensors can be taken to be of even rank without loss of generality.

$$a_{ij} = (\frac{1}{3}\mathrm{tr}\,\underset{\sim}{a})\delta_{ij} + \frac{1}{2}(a_{ij}-a_{ji}) + [\frac{1}{2}(a_{ij}+a_{ji}) - \frac{1}{3}(\mathrm{tr}\,\underset{\sim}{a})\delta_{ij}] \qquad (15)$$

corresponding to a scalar $(\frac{1}{3}\,\mathrm{tr}\,\underset{\sim}{a})$, an antisymmetric second rank tensor (3 independent components) and a symmetric traceless tensor (5 independent components). A second rank irreducible tensor is either antisymmetric or symmetric and traceless. As to the meaning of $P_Q\underset{\sim}{a}$, we have of course the familiar tensor transformation. In components

$$(P_Q\underset{\sim}{a})_{ij} = Q_{ip}Q_{jq}a_{pq} \; (= \underset{\sim}{Q}\,\underset{\sim}{a}\,\underset{\sim}{Q}^T). \qquad (16)$$

A fourth rank tensor $b_{ijk\ell}$ accepts a decomposition similar to (15). It may be useful to note that a traceless completely symmetrical fourth rank tensor, i.e., a tensor with the following properties

$$b_{ijk\ell} = b_{jik\ell} = b_{jki\ell} = \ldots;\; b_{iik\ell} = 0, \qquad i,j,k,\ell = 1,2,3$$

is irreducible and has nine independent components.

Irreducible tensors live in odd dimensional spaces. An antisymmetric second rank tensor has three independent components. A vector has also three components and is an irreducible tensor. What distinguishes these two irreducible tensors which live in R^3 is their behavior under coordinate inversions. The vectors are not invariant under coordinate inversions, whereas antisymmetric second rank tensors are. Now observe that the tensors σ, D and Ω that are the fundamental elements of the representation (1) are second rank and hence they are invariant under coordinate inversions. It follows then that one can take $q_1,\ldots,q_n$ to be even rank tensors without loss of generality. (For more details on this observation see [4]). Thus henceforward we consider the components q_i of S as even rank tensors.

3. INVARIANCE REQUIREMENTS ON THE LAWS OF EVOLUTION

The special role played by superimposed rigid body rotations places certain invariance requirements on the representation (1). We now discuss briefly these requirements by considering again the tests (i) and (ii) of the previous Section. In these tests the deformations applied and state and orientations and stresses produced are as follows:

	Deformation	Rate of Deformation	State and Orientation	Stress	
(i)	$F(\tau)$	$D + \Omega$	$S(\tau)$	$\sigma(\tau)$	
(ii)	$Q(\tau)F(\tau)$	$\dot{Q}Q^T + Q(D+\Omega)Q^T$	$P_Q(\tau)S(\tau)$	$Q(\tau)\sigma(\tau)Q^T(\tau)$	(17)

where $Q(\tau) \in O^+(3)$ and $Q^\cdot = \frac{dQ}{dt}$ and $Q^{-1} = Q^T$

The first equation in (1) and (17) yield

$$\sigma(\tau) = f(S(\tau)); \qquad Q(\tau)\sigma(\tau)Q^T(\tau) = f(P_{Q(\tau)}S(\tau)).$$

Comparing the above statements we arrive at the invariance requirement for f:

$$f(P_Q S) = Qf(S)Q^T. \tag{18}$$

The discussion of the invariance requirement on the growth law (the second equation of (1)) is more involved and rests heavily on the fact that P_Q is a linear transformation on R^N.

Applying (1b) to tests (i) and (ii) of (17) we find

$$\frac{dS}{st} = g(S(t),\ D + \Omega), \quad \frac{d(P_Q S)}{dt} = g(P_{Q(t)}S(t), Q^\cdot Q^T + Q(D+\Omega)Q^T) \tag{19}$$

Now we consider a deformation history (ii) such that

$$Q(t) = I \text{ and hence } \frac{dQ}{dt} = Q^\cdot = -\ Q^{\cdot T} \quad \text{at time t} \tag{20}$$

It can be shown [3] using the linearity of P_Q that when (20) holds

$$\frac{d(P_Q S)}{dt} = S^\cdot + T_{Q^\cdot} S \tag{21}$$

where $T_{Q^\cdot}$ is a linear map on R^N that depends linearly on $Q^\cdot$.

One way of deriving (21) is to consider the action of P_Q in each invariant subspace of R^N. For instance, if q_i is a second rank tensor denoted by $\underset{\sim}{a}$, then it follows from (16) and (20) that

$$\begin{aligned} \frac{d}{dt}(P_Q \underset{\sim}{a}) &= \underset{\sim}{a}^\cdot + Q^\cdot \underset{\sim}{a} + \underset{\sim}{a} Q^{T^\cdot}, \qquad Q = I \\ &= \underset{\sim}{a}^\cdot + Q^\cdot \underset{\sim}{a} - \underset{\sim}{a} Q^\cdot \end{aligned} \tag{22}$$

It should be clear that similar results will hold if $\underset{\sim}{a}$ were a tensor of higher rank and these results can be summarized by (21).

Now when (20) holds (19) and (21) yield

$$g(S, D + \Omega) + T_{Q^\cdot} S = g(S, Q^\cdot + D + \Omega) \tag{23}$$

(23) is valid for any antisymmetric $Q^\cdot$. In particular for $Q^\cdot = -\ \Omega$. This gives, since $T_\Omega S = -\ T_{-\Omega} S$,

$$g(S, D + \Omega) = g(S, D) + T_\Omega S \tag{24}$$

The above equation shows that the dependence of $S^{\cdot}$ on Ω must be of very special form, T_{Ω} being a linear map on R^N that depends linearly on Ω. One might say that the rate of rotation of the element spins the state of the element.

By going back to (19) we discover also that g(S,D) satisfies the invariance requirement

$$g(P_Q S, QDQ^T) = P_Q g(S,D). \tag{25}$$

It may be useful to rewrite (1) by using the above findings:

$$\begin{array}{|llr|}
\hline
\sigma(t) = f(S(t)), & & f(P_Q S) = Qf(S)Q^T \\
\dfrac{dS(t)}{dt} = g(S,D) + T_{\Omega} S, & & g(P_Q S, QDQ^T) = P_Q g(S,D) \\
\text{where} \quad f: & \Sigma \to T_2^{s} \quad , & Q \in O^+(3) \\
\quad\quad\quad g: & \Sigma \times T_2^{s} \to R^N & \\
\quad\quad\quad T: & R^N \times T_2^{as} \to R^N & \text{(multilinear).} \\
\hline
\end{array} \tag{26}$$

We consider now a special class of materials that include metals. What distinguishes these materials is that they behave elastically in the presence of (infinitely) fast deformations. For this class of materials the function g(S,D) in (26) takes the form:

$$g(S,d) = g(S) + h(S)D, \quad g: \ \Sigma \to R^N; \ g(P_Q S) = P_Q g(S) \tag{27}$$

where h is a linear map ($h \in L(T_2^{s}, R^N)$) that depends on S. h is easy to calculate in principle because it measures a well understood aspect of elastic behavior. We see that for this class of materials it is the function g(S) and the decomposition (13) that are the key elements of the representation.

In [7] and [8] (cf. also [15]) we have studied in the case of metals the basic attributes of the vector field and the trajectories defined by g(S).

Most materials including metals exhibit rate independence beyond their elastic limit. It is therefore natural to ask questions such as these: Under what circumstances (and why) solids such as metals can be considered as rate independent substances? Can studies of the rate dependence increase our understanding of plastic behavior? These questions have been discussed in [7], [8] and [15].

Finally we note that an application of the above ideas

to finite deformations of elastic-plastic materials is given in [9]. References [8] and [9] also contain discussions of "thermodynamic" requirements that must be placed on f(S) and g(S) of (26) and (27).

Another class of requirements to be imposed on a representation such as (26) arises from the behavior of "structures" composed of the material of interest. Here is an example. Suppose necking is observed to occur in tensile tests; then one would demand that the proposed constitutive laws together with conservation laws of mechanics must give rise to the observed necking behavior. Similar observations could be made on other forms of localization.

In the next section we consider the development of anisotropy and texture in a deforming material. We show that the decomposition of the internal state and orientation into even rank irreducible tensors provides a natural tool for the study of this phenomenon.

4. MEASURES OF ANISOTROPY.

We consider an element which is in the state and orientation $S \in \Sigma$. We assume, for simplicity, that the element was isotropic in its virgin state S_o. We wish to know whether the deformation history that took the internal state from S_o to S caused any loss of isotropy. If that is the case then we would like to know the nature and the strength of the deformation induced anisotropy. We wish now to show that the theoretical framework developed in previous sections provides answers to these questions.

The main phenomenological tool for studying anisotropy is a (suddenly applied) superimposed rigid body rotation and the comparison of the future behaviors of the rotated and unrotated elements. In the language of Section 2 the state and orientation $P_Q S$ of the rotated element is compared with the state and orientation S of the unrotated one. If

$$P_Q S = S$$

for some $Q \in O^+(3)$, then the rigid body rotation Q leaves S invariant, indicating that the material element in S possesses internal symmetry with respect to Q. These considerations lead one to introduce the isotropy group g_S of the state S. g_S is defined as follows

$$g_S = \{Q \in O^+(3): \quad P_Q S = S\} \qquad (28)$$

Thus g_S is composed of the rigid body rotations that leave S unchanged. If the material in S is isotropic then

$$g_S = O^+(3).$$

If, on the other hand, the material is totally anisotropic, then

$$g_S = I = \text{identity}.$$

We now show that g_S can be calculated in terms of the known symmetry groups of the irreducible tensors $(q_1,\ldots,q_n)$ that S comprises.

It is known (cf. [6]) that each type of irreducible tensor defines a different symmetry group: A scalar is invariant under $O^+(3)$. An antisymmetric second rank tensor is invariant only under rigid body rotations about an axis. A traceless second rank symmetric tensor with three distinct principal values is invariant only under 180° rotations about its principal directions. A traceless third rank fully symmetric tensor (2 x 3 + 1 = 7 independent components) may be invariant under one of the following rotations (and their repetitions) with respect to an axis: none; only 180°; only $120° = \frac{2\pi}{3}$; all. Which one of these cases will occur will depend on the orientation of the axis and the particular tensor under consideration. A traceless fourth rank fully symmetric tensor (2 x 4 + 1 = 9) independent components) may be invariant under one of the following rotations with respect to an axis: none; only 180°; only 120°; only $90° = \frac{2\pi}{4}$; all. For higher rank tensors such lists which also contain the angles $\frac{2\pi}{5}$, $\frac{2\pi}{6}$, ... can easily be written down. It follows from the above remarks that once $(q_1,\ldots,q_n)$ are known g_S can be calculated.

The above observations would also allow one to make decisions about the tensorial nature of state variables, based on the results of phenomenological experiments. Here is an example: Suppose that the experiments show that a material element possesses symmetry about an axis but only with respect to 90° rotations. Then we conclude that one of the internal state variables for this material must be an irreducible tensor of fourth rank or higher. Another example is concerned with the symmetries of the yield surface and their relationship to tensorial character of the internal state variables (Cf. [9] and Section 6).

The following considerations are also relevant to the study of internal symmetry of a deforming material. Consider a material which is isotropic at time t = 0. Thus we can take $q_1 = \ldots = q_n = 0$ at $t = 0$ without loss of generality. Assume further that the evolution law for this material is linear in D (cf. (27)). One can then show that $\left.\frac{dq_i}{dt}\right|_{t=0}$ will be nonzero only if q_i is a scalar or traceless second rank symmetric tensor. In other words, in the very initial deformation of a isotropic material with a growth law linear in D, the only state variables that can be created are scalars and second rank traceless symmetric tensors. This observation might explain the tendency to use these types of tensors as state variables.

Reflection will show that for many deformation histories of interest the material may lose all of its symmetries at the earlier stages of deformation. The representation (26) would enable one to discover, by calculating S and g_S, whether such a state of affairs is reached. If anisotropy has developed then we would like to know its strength. To be more precise about the notion of strength of anisotropy, consider the orbit O(S) of S under superimposed rigid body rotations. The orbit is defined as follows:

$$O(S) = \{P_Q S: \quad Q \in O^+(3)\} \tag{29}$$

The question that we pose is this: How different are the future behaviors of the material elements whose state and orientations are given by the points of the orbit O(S). This query leads to the following one. Consider two elements in S_1 and S_2. How different are their future behaviors? This question was studied at some length in [4]. The study showed that one can introduce a metric ρ in Σ such that

$$\rho(S_1,S_2): \Sigma \times \Sigma \to R^+. \tag{30}$$

measures in a precise and natural way the "distance" between future behaviors of the elements in S_1 and S_2. If such a metric is available then one can calculate the diameter d_ρ of the orbit O(S). d_ρ can then be used as the strength of anisotropy. If one starts with a representation of the type (26), with f and g and the decomposition (13) already fixed, then the metric (30) can be calculated, in principle, by integrating (26) under all deformations of interest. Of course such a calculation is practically impossible to carry out.

On the other hand, for a given representation of the type (26) there exists a natural metric ρ_E associated with the Euclidean structure on R^N. The hope is that if $(q_1,\ldots,q_n)$ are well chosen and they represent the physics of the situation accurately, then ρ_E might in some sense be close to ρ of (30), so that large distances according to ρ_E may correspond to large distances according to ρ. If such is the case then we could say that we have a fully reduced representation. Without this additional attribute, a representation of the type (26) would be only partially reduced: It would enable us to compute g_S for each S; hence we would know symmetries of the corresponding elements, but we would not be sure that the diameter of the orbit O(S) according to the metric σ_E would have a physical significance.

This may be a good place to discuss briefly other types of representations where the orientation of a deforming element is measured by the deformation gradient tensor F or by means of a triad of vectors (cf. [10]). It will easily be seen that in such representations the symmetry group of a deforming element is not available at the instant of interest unless a structure similar to the one discussed here is added to the state space. It is worth noting that the representations based on the current value of F suffer from other drawbacks. These are: (i) F cannot be determined by experiments and microscopy conducted here and now. (ii) Moreover in some materials, such as nearly ideally plastic materials, the two material elements with widely differing F's may be very close to each other as far as their future behavior is concerned.

We end this section by expressing the hope that the questions of reducibility discussed above will attract the attention of the workers in the field so that confusing claims regarding the merits of various types of representations may be settled.

5. REPRESENTATION OF INTERNAL DAMAGE

In the previous sections we emphasized the importance of the tensorial nature of state variables. Our considerations were based on general principles and we avoided any lengthy discussion of special cases. In the present section we discuss an example where the tensorial state variables of the type (13) arise naturally from the statistics of the internal events that accompany deformation. The example is concerned with the tertiary creep of certain polycrystalline solids. Detailed quantitative observations have shown (cf. [11] and [12]) that in these materials preferential cavity growth occurs on the grain boundaries during creep. The mathematical description of the distribution of cavities among the grain boundaries is one of the key steps in the construction of constitutive equations for such materials.

Here we discuss an aspect of this mathematical problem. We consider homogeneous deformations of the material of interest. (Our comments concerning the nature of the function x(X) must, nevertheless, be kept in mind! Thus we do not claim that deformation is homogeneous in the scale of grains). At a given time t we focus the attention on the material points that occupy a spherical region S_1. (Fig. 1). This spherical material element contains a very large number of grains. We assume that the grain boundaries are planar. We are interested in the mathematical description of the distribution of cavities among the grain boundaries that lie within this element. With this purpose in mind we consider another sphere. This one has unit radius and it defines space directions by the position vectors of points of its surface (which we denote by S_2). An infinitesimal area dA on S_2 about the end point of $\vec{n}$ will represent a bundle of directions about $\vec{n}$.

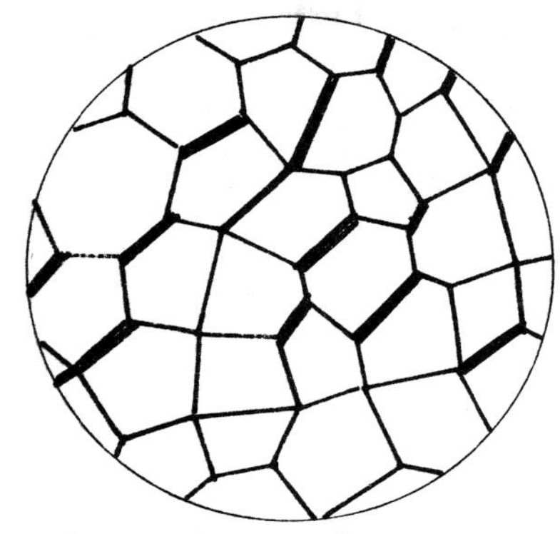

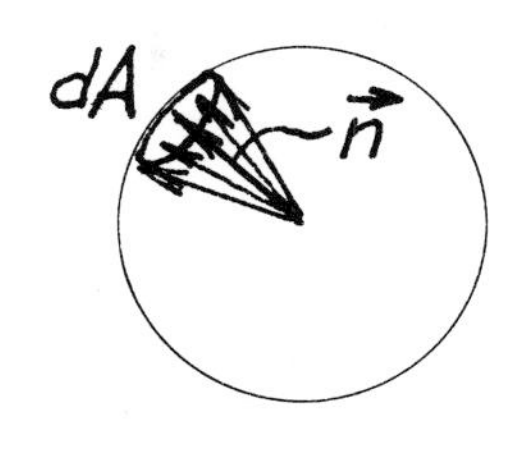

S_1 : Material Element S_2 : Unit sphere of Directions

(Dark boundaries are ⊥ to vector bundle)

Fig.1. Statistics of Damage.

Next consider those grain boundaries within S_1 that are orthogonal to the directions present in the vector bundle just described. Thus we concentrate the attention on grain boundaries that are nearly perpendicular to $\vec{n}$. We can now ask all sorts of relevant questions: What is the total area of these grain boundaries? What is the total volume of cavities within these grain boundaries. A typical answer to these questions, say to the last one, will be of the form.

$$V(\vec{n})dA$$

where $V(\vec{n})dA$ denotes the total volume of voids found on the grain boundaries perpendicular to the vector bundle associated with $\vec{n}$ and dA. Thus we have here a real valued function V defined on the unit sphere S_2

$$V: \quad S_2 \to R \tag{31}$$

that describes the directional dependence of the void volume distribution. We hasten to add that one encounters functions defined on the sphere in other studies dealing with the texture development in deforming materials (cf. [13] and references therein).

Representation of the functions defined on the sphere is a well studied problem (cf. [6], p. 46), a Fourier series based on spherical functions being the standard tool of representation. Here we offer a modified version of this representation. Our version has satisfactory properties of transformation under the rigid body rotations of the material element and it provides a simple example in which irreducible even rank tensors appear as state variables.

We observe first that, by its definition, V has the property

$$V(\vec{n}) = V(-\vec{n}) \tag{32}$$

Furthermore, we assume that V is square integrable. In view of (32), the standard Fourier representation of V has the form (cf. [6]):

$$V = \phi_o + \phi_2 + \phi_4 + \ldots + \phi_{2\ell} + \ldots \tag{33}$$

where $\phi_{2\ell} = \sum_{m=-2\ell}^{m=+2\ell} C_{2\ell}{}^m Y_{2\ell}{}^m$ and $Y_{2\ell}{}^m$ are the spherical functions that constitute an orthogonal basis for the subspace $M^{4\ell+1}$ in which $\phi_{2\ell}$ lives. $C_{2\ell}{}^m$ are scalars. These coeffitransform awkwardly under the rigid body rotation of the material element. To obviate this difficulty we modify (33) as follows

$$V = V_o \cdot 1 + V_{ij} f_{ij} + V_{ijk\ell} f_{ijk\ell} + \ldots, \qquad (i,j,k,\ell,\ldots 1,2,3) \tag{34}$$

where the usual summation convention is used. Here 1 (the unit valued constant function), $f_{ij}, f_{ijk\ell}, \ldots$ are the basis functions and V_o, $V_{ij}, V_{ijk\ell}, \ldots$ are real valued coefficients.

The above basis functions are linear combinations of spherical functions. Here we define them independently and do not pause to show that this is indeed the case (cf. [16]). We let (n_1, n_2, n_3) denote the coordinates of the unit vector $\vec{n}$ in a fixed rectangular frame. Thus

$$n_i n_i = 1. \tag{35}$$

The basis functions are defined as follows:

$$f_{ij}: S_2 \to R,$$

$$f_{ij}(\vec{n}) = n_i n_j - \frac{1}{3}\delta_{ij}, \qquad (i,j,k = 1,2,3)$$

and

$$f_{ijk\ell}(\vec{n}) = n_i n_j n_k n_\ell - \frac{1}{7}(\delta_{ij}n_k n_\ell + \delta_{ik}n_j n_\ell + \delta_{i\ell}n_j n_k + \delta_{jk}n_i n_\ell$$

$$+\delta_{j\ell}n_i n_k + \delta_{k\ell}n_i n_j) + \frac{1}{5\times 7}(\delta_{ij}\delta_{k\ell} + \delta_{ik}\delta_{j\ell} + \delta_{i\ell}\delta_{jk}), \quad (i,j,k,\ell=1,2,3)$$

$$\ldots \tag{36}$$

The definitions are based on the even symmetric products $n_i n_j$, $n_i n_j n_k n_\ell, \ldots$ of n_i. Additional terms in (36) are so chosen that f_{ij}, $f_{ijk\ell}, \ldots$ are completely symmetrical with respect to their indices and their traces are zero. Thus, by definition,

$$f_{ij} = f_{ji}, \; f_{ii} = 0 \tag{37}$$

and $f_{ijk\ell}$ have the symmetry properties of the fourth rank tensor $b_{ijk\ell}$ of the Section 2 with respect to its indices.

It follows then that there are only <u>five</u> independent functions f_{ij}, <u>nine</u> independent $f_{ijk\ell}$, etc.

In view of the properties that the functions f_{ij}, $f_{ijk\ell}, \ldots$ exhibit with respect to their indices (cf. (37), for instance) it is natural to assign the same properties to the coefficients V_{ij}, $V_{ijk\ell}, \ldots$. Thus we specify that the coefficients V_{ij}, $V_{ijk\ell}$, ... be completely symmetrical and traceless. For instance

$$V_{ij} = V_{ji}; \; V_{ii} = 0 \tag{38}$$

One can show using the above results and agreements that the Fourier coefficients V_o, V_{ij}, $V_{ijk\ell}$, ... in (34) can be expressed as inner products of $V(n)$ with the basis functions $1, f_{ij}$, $f_{ijk\ell}, \ldots$ respectively. For instance we have

$$V_o = \frac{1}{4\pi}\int_{S_2} V(\vec{n})\,dA,$$

$$V_{ij} = \frac{1}{4\pi} C_{ij} \int_{S_2} V(\vec{n}) f_{ij}\,dA \text{ with } C_{ij} = \begin{cases} 5 \text{ when } i = j \\ \frac{3\times 5}{2} \text{ when } i \neq j \end{cases} \tag{39}$$

(no summation)

The advantage of the representation (34) is that the coefficients V_o, V_{ij}, etc. transform as tensors under the rigid body rotations of the material element.

Since (34) is convergent we can say that the void volume distribution V: $S_2 \to R$ can be represented, to within a desired accuracy, by a finite number of coefficient tensors:

$$(V_o, V_{ij}, \ldots)$$

where each irreducible tensor represents in view of (36) and (39) a "moment" of the void distribution function V with respect to the origin.

In order to complete the task of constructing constitutive equations for tertiary creep of a given material, one needs (in addition to making the above choice) to write the laws of evolution for V_o, V_{ij}, ... Examples of such laws can be found in [14].

6. FINITE DEFORMATIONS OF RIGID-PLASTIC MATERIALS

It is well known that strain-hardening metals develop anisotropy during the course of plastic deformation. The importance of (deformation induced) anisotropy in certain technical problems (e.g. metal forming; cyclic loading in nuclear reactors) has led to the development of constitutive equation that attempt to represent, in an explicit and detailed way, the evolution of anisotropy and its effects on mechanical behavior. The earliest of such work is by Prager [17] who also introduced the term "kinematic hardening". In later work of Mroz [18], Dafalias and Popov [19] and Krieg [20] the yield surface or surfaces are allowed to move and change shape in the stress-space. All the above work is concerned with small deformations and rotations.

Lee [21] has clarified several basic issues that arise when attempts are made to generalize the classical laws of plasticity to the care of finite deformations. However the material considered by Lee remains isotropic in its unstressed state even after the occurrence of plastic flow. The case where the unstressed material becomes anisotropic was considered in [9]. In this work the very special occurrence of the rate of deformation tensor Ω to the constitutive equations (cf. (26)) was emphasized.

Recent work of Nagtegaal and de Jong is concerned with finite deformations of a kinematically hardening solid. These authors show that the joint presence of anisotropy and finite deformations could have unusual consequences in plastic behavior. In the present paper we reconsider the problem of finite deformations of a kinematically hardening material as an illustration of the ideas introduced in Sections 2-4 of the paper. We confirm the results of [22] by simple calculations and we clarify certain issues that arise in [22] and in the related work of Lee, Malett and Wertheimer [23].

In order to define the class of rigid-plastic materials of

interest we first let $\underset{\sim}{s}$ denote the deviatoric stress tensor:

$$\underset{\sim}{s} = \underset{\sim}{\sigma} - \frac{1}{3}\underset{\sim}{I}\ \mathrm{tr}\ \underset{\sim}{\sigma} \quad . \tag{41}$$

In this section we underline the tensorial quantities to avoid confusion in detailed calculations.

Next we introduce the symmetric traceless tensor $\underset{\sim}{\alpha}$. $\underset{\sim}{\alpha}$ defines the state and orientation of the plastic state of the element. The state S of the stressed element is given by the pair $(\underset{\sim}{s},\underset{\sim}{\alpha})$. The state variable $\underset{\sim}{\alpha}$ defines the yield condition for the stress free element in the state S $= (\underset{\sim}{0},\underset{\sim}{\alpha})$ by the modified v. Mises law:

$$Y(\underset{\sim}{s},\underset{\sim}{\alpha}) = \frac{1}{2}\ (\underset{\sim}{s}-\underset{\sim}{\alpha})\cdot(\underset{\sim}{s}-\underset{\sim}{\alpha}) - k^2 \leq 0 \tag{42}$$

where k is the yield strength of the virgin material ($\underset{\sim}{\alpha} = 0$) in pure shear. Here the dot indicates the usual inner product in the space T_2 of second rank tensors. Thus with $\underset{\sim}{a}$ and $\underset{\sim}{b}\varepsilon T_2$

$$\underset{\sim}{a} \cdot \underset{\sim}{b} = \mathrm{tr}\ (\underset{\sim}{a}\underset{\sim}{b}) = a_{ij}b_{ij} \quad (i,j=1,2,3).$$

For a material in the plastic state $\underset{\sim}{\alpha}$, a deviatoric stress $\underset{\sim}{s}$ satisfying (42) with the strict inequality sign cannot cause plastic flow. The set of such $\underset{\sim}{s}$ occupies a "flat" sphere centered at $\underset{\sim}{\alpha}$ in T_2. The boundary of this domain, $\{\underset{\sim}{s}: Y(\underset{\sim}{s},\underset{\sim}{\alpha}) = 0\}$, is denoted by $Y_{\underset{\sim}{\alpha}}$ and it is called the yield surface of the plastic state $\underset{\sim}{\alpha}$. (Fig. (2)).

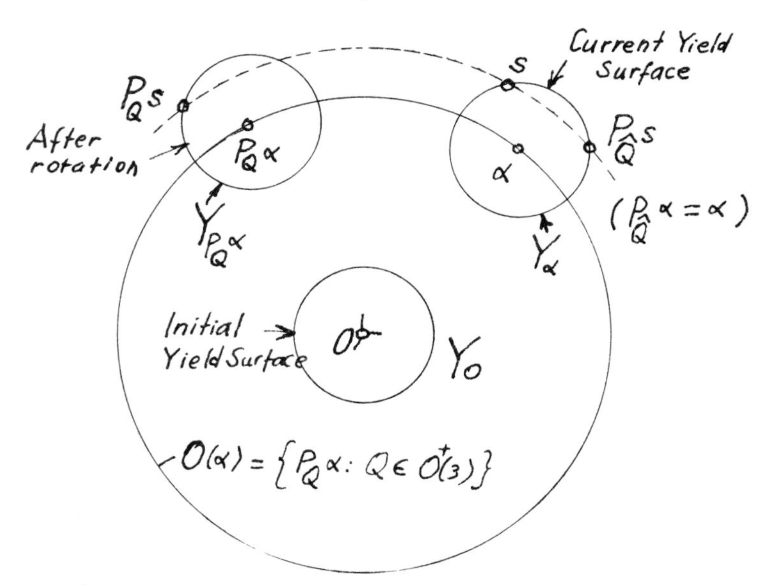

Fig.2 Rigid rotation of the element rotates the yield surface.

Note that if we rotate the unstressed element in the state $\underset{\sim}{\alpha}$ by an amount $\underset{\sim}{Q} \in O^+(3)$, then according to (9), the state $S = (0,\underset{\sim}{\alpha})$ becomes $P_{\underset{\sim}{Q}}S = (0, P_{\underset{\sim}{Q}}\underset{\sim}{\alpha})$. Thus the center of the yield domain moves to $P_{\underset{\sim}{Q}}\underset{\sim}{\alpha} = \underset{\sim}{Q}\,\underset{\sim}{\alpha}\,\underset{\sim}{Q}^T$. (Fig. (2)).

On the other hand it is clear from (42) that the yield function $Y(\underset{\sim}{s},\underset{\sim}{\alpha})$ has the invariance property

$$Y(\underset{\sim}{s},\underset{\sim}{\alpha}) = Y(P_{\underset{\sim}{Q}}\underset{\sim}{s},\ P_{\underset{\sim}{Q}}\underset{\sim}{\alpha}) \tag{6.3}$$

Thus if $\underset{\sim}{s} \varepsilon\ Y_{\underset{\sim}{\alpha}}$ then $P_{\underset{\sim}{Q}}s = \underset{\sim}{Q}\ \underset{\sim}{s}\ Q^T$ will be a point or the yield surface $Y_{P_{\underset{\sim}{Q}}\underset{\sim}{\alpha}}$.(Fig. (2)).

We know that $\underset{\sim}{\alpha}$ is invariant with respect to 180° rotations about its principal directions. Thus $\hat{Q}$ denoting such a rotation we have $P_{\hat{\underset{\sim}{Q}}}\underset{\sim}{\alpha} = \underset{\sim}{\alpha}$. Therefore in view of (6.3) if $\underset{\sim}{s}\ \varepsilon\ Y_{\underset{\sim}{\alpha}}$ then $P_{\hat{\underset{\sim}{Q}}}\underset{\sim}{s}$ will also be on $Y_{\underset{\sim}{\alpha}}$. This remark implies that the yield surface $Y_{\underset{\sim}{\alpha}}$ has symmetries inherited from the symmetries of $\underset{\sim}{\alpha}$. This fact can be used to verify experimentally, by studying the shape of the yield surface, the validity of representing the plastic state by a single second rank tensor $\underset{\sim}{\alpha}$. (cf [9], also Storåker's work).

We note from (6.2) and (6.3) that

$$\frac{\partial Y}{\partial s_{ij}} = \frac{\partial Y}{\partial \underset{\sim}{s}} = \underset{\sim}{s}-\underset{\sim}{\alpha} \quad \text{and} \quad \frac{\partial Y}{\partial \underset{\sim}{\alpha}} = -(\underset{\sim}{s}-\underset{\sim}{\alpha}) \tag{44}$$

and

$$\frac{\partial Y}{\partial \underset{\sim}{s}} \cdot (\underset{\sim}{\Omega}\underset{\sim}{s}-\underset{\sim}{s}\underset{\sim}{\Omega}) + \frac{\partial Y}{\partial \underset{\sim}{\alpha}} \cdot (\underset{\sim}{\Omega}\underset{\sim}{\alpha}-\underset{\sim}{\alpha}\underset{\sim}{\Omega}) = 0 \tag{45}$$

where $\underset{\sim}{\Omega}$ is any antisymmetric second rank tensor.

For a stress free virgin element we have

$$\underset{\sim}{s} = \underset{\sim}{\alpha} = 0 \tag{46}$$

What is needed now is the laws of evolution for the state $(\underset{\sim}{s},\underset{\sim}{\alpha})$ in terms of the rate of deformation $\underset{\sim}{D}$ and the rate of rotation $\underset{\sim}{\Omega}$ applied to the material element. These laws will have the general form (26), but now we can be specific. We must also be willing to live with a certain lack of analiticity that is inherent in plastic behavior. Thus separate laws are needed for the two cases of plastic and rigid behavior.

Plastic behavior. When a material element, which is in the current plastic state $\underset{\sim}{\alpha}$, deforms the deviatoric stress $\underset{\sim}{s}$ carried by the element and the current rate of deformation $\underset{\sim}{D}$ satisfy the yield condition: $Y(\underset{\sim}{s},\underset{\sim}{\alpha}) = \frac{1}{2}(\underset{\sim}{s}-\underset{\sim}{\alpha})\cdot(\underset{\sim}{s}-\underset{\sim}{\alpha})-k^2=0$

and (47)

the flow rule: $\underset{\sim}{D} = \lambda \frac{\partial Y}{\partial \underset{\sim}{s}} = \lambda(\underset{\sim}{s}-\underset{\sim}{\alpha})$ with $\lambda > 0$

It follows from (47) that

$$\text{tr } \underset{\sim}{D} = 0 \quad \text{(plastic flow preserves volume)} \tag{48}$$

and

$$\lambda = \frac{1}{\sqrt{2}\,k} \sqrt{\underset{\sim}{D}\cdot\underset{\sim}{D}} \quad . \tag{49}$$

Thus during plastic deformation we have

$$\underset{\sim}{s} = \underset{\sim}{\alpha} + \sqrt{2}\,k\,\underset{\sim}{D}/\sqrt{\underset{\sim}{D}\cdot\underset{\sim}{D}} \tag{50}$$

which shows that the material behaves as a non-Newtonian fluid during plastic flow. This equation shows also that once $\underset{\sim}{D}$ and $\underset{\sim}{\alpha}$ are known the deviatoric stress follows. Hence the unique importance of the law of evolution for $\underset{\sim}{\alpha}$ which we consider next.

We postulate, following the classical ideas, that $\dot{\underset{\sim}{\alpha}} = \frac{d\underset{\sim}{\alpha}}{dt}$ is linear in $\underset{\sim}{D}$; moreover we insist that it must have the proper dependence on $\underset{\sim}{\Omega}$ as dictated by (26) and, in particular, by (22). Thus

$$\dot{\alpha}_{ij} = K_{ijk\ell}(\underset{\sim}{s},\underset{\sim}{\alpha})D_{k\ell} + \Omega_{ip}\alpha_{pj} - \alpha_{ip}\Omega_{pj}.$$

But in view of the flow rule in (47) this law reduces to

$$\dot{\alpha}_{ij} = A_{ij}(\underset{\sim}{s},\underset{\sim}{\alpha})\lambda + \Omega_{ip}\alpha_{pj} - \alpha_{ip}\Omega_{pj}$$

or, more concisely and completely,

$$\dot{\underset{\sim}{\alpha}}=\underset{\sim}{A}(\underset{\sim}{s},\underset{\sim}{\alpha})\lambda+\underset{\sim}{\Omega}\underset{\sim}{\alpha}-\underset{\sim}{\alpha}\underset{\sim}{\Omega};\ Y(\underset{\sim}{s},\underset{\sim}{\alpha})=0;\ \underset{\sim}{D}=\lambda(\underset{\sim}{s}-\underset{\sim}{\alpha}) \qquad \lambda > 0 \tag{51}$$

where λ may be tought as the magnitude of the deformation $\lambda(\underset{\sim}{s}-\underset{\sim}{\alpha})$ associated with the yield state $(\underset{\sim}{s},\underset{\sim}{\alpha})$. We also recognize that $\underset{\sim}{\Omega}\underset{\sim}{\alpha}-\underset{\sim}{\alpha}\underset{\sim}{\Omega}=T_{\underset{\sim}{\Omega}}\alpha$ of (26).

$\underset{\sim}{A}$ is a tensor valued (symmetric and tracless) function of the yield state $(\underset{\sim}{s},\underset{\sim}{\alpha})$. We know from the considerations of Section 2 (cf (26) and (27)) that $\underset{\sim}{A}$ must obey the invariance requirement

$$P_{Q}A(s,\alpha) = QA(s,\alpha)Q^T = A(QsQ^T, Q\alpha Q^T) \tag{52}$$

In other words A must be a form invariant function of s and α. It is easily verified that the functions $s, \alpha, s^2, \alpha^2, \alpha s + s\alpha, \alpha s^2 + s^2\alpha, \alpha s\alpha, \ldots$ are symmetric and satisfy the requirement (52). Moreover it is known from the work of Rivlin et. al. (cf. [24]) that (52) implies that A can be expressed in the following reduced form:

$$A = C_1 I + C_2 s + C_3 \alpha + C_4 s^2 + C_5 \alpha^2 + C_6(s\alpha + \alpha s) +$$

$$C_7(\alpha^2 s + s\alpha^2) + C_8(s^2\alpha + \alpha s^2) + C_9(\alpha^2 s^2 + s^2\alpha^2) \; .$$

where $C_1, \ldots, C_9$ are functions of the joint invariants of s and α. These are $\mathrm{tr}s^2, \mathrm{tr}\alpha^2, \mathrm{tr}\alpha s, \ldots, \mathrm{tr}\alpha^3 s^3$. The functions $C_1, \ldots, C_9$ must be so chosen that $\mathrm{tr}A = 0$ as (51) requires.

We shall presently give examples of the often used forms of A.

We now come to the law of evolution for s. If deformation takes place during the time interval $(t, t+dt)$, then $Y(s,\alpha) = 0$ on this internal and hence

$$Y(s,\alpha) = 0 \quad \text{and} \quad \frac{d}{dt} Y(s,\alpha) = 0$$

which leads to

$$(\dot{s} - \dot{\alpha}) \cdot (s - \alpha) = 0 \quad \text{when } Y(s,\alpha) = 0 \text{ and } D = \lambda(s-\alpha)\lambda \quad > 0 \tag{53}$$

It is easily seen that the knowledge of the yield state (s,α) and the accompanying deformation and rotation rates (D,Ω) do not uniquely determine s. Indeed (53) implies that $\dot{s} = \frac{ds}{dt}$ can be written in the following form

$$\dot{s} = \frac{1}{2k^2}(A \cdot (s-\alpha))D + f + \Omega s - s\Omega \tag{54}$$

where f satisfies the requirement

$$f \cdot (s - \alpha) = 0$$

but otherwise arbitrary.

There is one more requirement that one places upon $A(s,\alpha)$ of (51). In order to motivate this requirement let us consider the geometrical interpretation of growth laws shown in Fig. 3. In this figure $\vec{\alpha\alpha'}$ and $\vec{ss'}$ denote the increments of α and s during $(t, t+dt)$ and when $\Omega = 0$. What distinguishes this case from the one of unloading is that $\vec{ss'}$ is directed

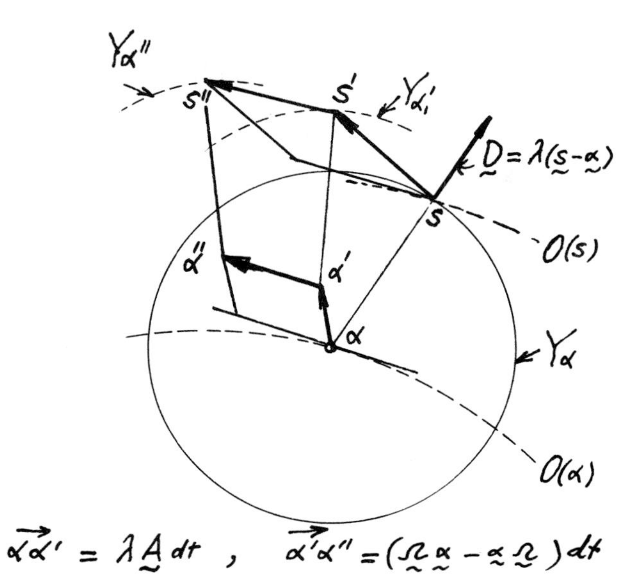

$\overrightarrow{\alpha\alpha'} = \lambda \underset{\sim}{A}\, dt \;, \quad \overrightarrow{\alpha'\alpha''} = (\underset{\sim}{\Omega}\underset{\sim}{\alpha} - \underset{\sim}{\alpha}\underset{\sim}{\Omega})\, dt$

Fig. 3. Plastic deformation when spin is present.

towards outside of the yield surface. In view of (54) this implies that

$$\underset{\sim}{A}(\underset{\sim}{s},\underset{\sim}{\alpha}) \cdot (\underset{\sim}{s}-\underset{\sim}{\alpha}) > 0. \tag{55}$$

It is easily seen that the requirement (55) implies that in uniaxial stressing the tanget modulus $\frac{d\sigma}{d\varepsilon} > 0$, but the converse of this statement may not be true.

For the sake of completeness we consider briefly the episodes of rigid behavior and the appropriate evolution laws for these episodes.

Rigid behavior. If at time t, $Y(\underset{\sim}{s},\underset{\sim}{\alpha}) < 0$, then no plastic deformation can take place, but the element can rotate as a rigid body with the spin $\underset{\sim}{\Omega}$. The accompanying growth laws are

$$\begin{aligned} &\underset{\sim}{D} = 0 \\ &\dot{\underset{\sim}{\alpha}} = \underset{\sim}{\Omega}\underset{\sim}{\alpha}-\underset{\sim}{\alpha}\underset{\sim}{\Omega} \qquad \text{when } Y(\underset{\sim}{s},\underset{\sim}{\alpha}) < 0 \\ &\underset{\sim}{s}^{\bullet} = \underset{\sim}{f} + \underset{\sim}{\Omega}\underset{\sim}{s}-\underset{\sim}{s}\underset{\sim}{\Omega} \end{aligned} \tag{56}$$

where $\underset{\sim}{f}$ is arbitrary.

If, on the other hand, the material is at the yield point but is not allowed to deform, i.e.

$$Y(\underset{\sim}{s},\underset{\sim}{\alpha}) = 0 \text{ and } \underset{\sim}{D} = 0$$

then (56) is still valid, but now f must be such that

$$\dot{Y} = \frac{\partial Y}{\partial \underset{\sim}{s}} \cdot s^{\cdot} + \frac{\partial Y}{\partial \underset{\sim}{\alpha}} \cdot (\underset{\sim}{\Omega}\underset{\sim}{\alpha} - \underset{\sim}{\alpha}\underset{\sim}{\Omega}) \leq 0 \tag{57}$$

In view of (45) and (56) this condition reduces to

$$\underset{\sim}{f} \cdot \frac{\partial Y}{\partial \underset{\sim}{s}} \leq 0$$

for the arbitrary tensor $\underset{\sim}{f}$ in (56).

We are now ready to consider some special cases and the work reported in [22] and [23].

The earliest choice for $A(s,\alpha)$ is found in Prager's work. In a slightly modified version Prager's choice has the form:

$$\underset{\sim}{A} = a(\underset{\sim}{s}-\underset{\sim}{\alpha}) \tag{58}$$

where a is a scalar which may be a constant or a function of the joint invariants of $\underset{\sim}{s}$ and $\underset{\sim}{\alpha}$ but often just a function of tr $\underset{\sim}{\alpha}^2$.

For the sake of later developments let us consider the response of this material to uniaxial tension. The rates of deformation and rotation associated with such a test performed with a virgin material are, in suitable units,

$$\underset{\sim}{D} = \begin{bmatrix} 1 & 0 & 0 \\ 0 & -1/2 & 0 \\ 0 & 0 & -1/2 \end{bmatrix}, \quad \underset{\sim}{\Omega} = 0 \tag{59}$$

If ε denotes the logarithmic strain in the direction of extension then according (59)

$$\varepsilon^{\cdot} = 1 \tag{60}$$

We find from (49) and (50) that

$$\lambda = \frac{\sqrt{3}}{2k} \quad \text{and} \quad \underset{\sim}{s}-\underset{\sim}{\alpha} = \frac{2}{\sqrt{3}}\, k\, \underset{\sim}{D} \tag{61}$$

and from (51)

$$\underset{\sim}{\alpha}^{\cdot} = a\, \underset{\sim}{D}$$

It follows from (60) and (61) that, σ denoting the only non-zero stress component, we have

$$\frac{d\sigma}{d\varepsilon} = \frac{3}{2} a$$

which shows that when a > 0, the increasing deformation in a uniaxial tensor test is accompanied by an increasing stress. This prediction is eminently reasonable for most materials of interest.

But when one considers the response of the material to shear flow then one meets with surprizes, as was first pointed out by Nagtegaal and de Jong [22].

Let us therefore look at the case of shear flow with some care.

Shear flow. The flow is defined by the time independent velocity field

$$v_1 = 2x_2, \; v_2 = 0, \; v_3 = 0 \tag{62}$$

where v_i are the velocity components in a rectangular frame with coördinates x_i. Since the material considered here is "rate independent" there is no need to specify the time units employed in (60). Under the action of this velocity field the unit material square OABC of t = 0 becomes the parallelogram OAB'C' at time t. Fig. 4. The rates of deformation and rotation for this flow are

$$\underset{\sim}{D} = \begin{bmatrix} 0 & 1 & 0 \\ 1 & 0 & 0 \\ 0 & 0 & 0 \end{bmatrix}, \qquad \underset{\sim}{\Omega} = \begin{bmatrix} 0 & 1 & 0 \\ -1 & 0 & 0 \\ 0 & 0 & 0 \end{bmatrix} \tag{63}$$

It is desired to find the stresses caused by this deformation in the material considered here.

We see right away from (50) and (62) that

$$\underset{\sim}{s} = \underset{\sim}{\alpha} + k \underset{\sim}{D} = \underset{\sim}{\alpha} + \underset{\sim}{k}$$

where

$$\underset{\sim}{k} = k \underset{\sim}{D} = \begin{bmatrix} 0 & k & 0 \\ k & 0 & 0 \\ 0 & 0 & 0 \end{bmatrix} \tag{64}$$

Thus the task of finding stresses reduces to finding $\underset{\sim}{\alpha}$. This requires the integration of the system of differential equation (51) governing the evolution of $\underset{\sim}{\alpha}$.

We write the first three equations without specifying the form of $\underset{\sim}{A}$. Observing that $\lambda = \frac{1}{k}$ for the present deformation (cf. (49)) we have from (51) and (63)

$$\alpha^{\bullet}_{12} = \frac{1}{k} A_{12} + (\alpha_{22} - \alpha_{11})$$

$$\alpha^{\bullet}_{11} = \frac{1}{k} A_{11} + 2\alpha_{21} \qquad (65)$$

$$\alpha^{\bullet}_{22} = \frac{1}{k} A_{22} - 2\alpha_{21}$$

where the right-hand side terms containing $\underset{\sim}{\alpha}$ are the relevant components of the spin tensor $\underset{\sim}{\Omega}\underset{\sim}{\alpha} - \underset{\sim}{\Omega}\underset{\sim}{\alpha}$.

We do not write the remaining equations. Because we can show that for the shear flow

$$\alpha_{13} = \alpha_{23} = 0 \qquad (66)$$

and, of course, $\alpha_{33} = -(\alpha_{11} + \alpha_{22})$ always.

The assertion (66) follows from these observations: A coordinate rotation of 180° about the 3-axis leaves $\underset{\sim}{D}$ and $\underset{\sim}{\Omega}$ in (63) invariant. This property is inherited by $(\underset{\sim}{s}, \underset{\sim}{\alpha})$ and hence (66).

Now we are ready to integrate (65) for various choices of $\underset{\sim}{A}$. In the case where A is given by (58), (63) becomes

$$\dot{\alpha}\, 12 = a + \alpha_{22} - \alpha_{11}$$

$$\alpha^{\bullet}_{11} = 2\alpha_{21} \qquad (67)$$

$$\alpha^{\bullet}_{22} = -2\alpha_{21}$$

It is convenient for the sake of later developments to introduce the following transformation

$$x = \alpha_{12}, \quad y = \alpha_{22} - \alpha_{11}, \quad z = \alpha_{22} + \alpha_{11} \qquad (68)$$

and to write (67) in the following form:

$$x^{\bullet} = a + y$$

$$y^{\bullet} = -4x \qquad (69)$$

$$z^{\bullet} = 0$$

and add the initial condition

$$x = y = z = 0 \quad \text{at } t = 0.$$

Here we consider the case of a = constant. The solution of (69) is then found easily:

$$x = \alpha_{12} = \frac{a}{2}\sin 2t$$
$$y = \alpha_{22} - \alpha_{11} = a(1-\cos 2t) \tag{70}$$

which in view of (62) leads to the following stress-shear displacement relations:

$$\sigma_{12} = k + \frac{a}{2}\sin\delta$$
$$\sigma_{22} - \sigma_{11} = a(1-\cos\delta) \tag{71}$$
$$\sigma_{22} + \sigma_{11} = -\,2\sigma_{33}$$

where δ is the displacement, say, of the point C in Fig. 3 and σ_{12}, σ_{11}, σ_{22} and σ_{33} are the stresses acting on the element.

We see that the shear stress remains finite and oscillates as a sine wave about the value k. Similar observations can be made on σ_{11} and σ_{22}.

These surprising results are very similar to those obtained in [22] by computer studies. We note that the stress varies periodically over the course of time with the period π (in time units chosen but not specified here). We compare this with the period 2π associated with the spin $\underset{\sim}{\Omega}$. Clearly these results follow from the overwhelming role played by the spin $\underset{\sim}{\Omega}$ in (67).

Two different aspects of these results may be cause for worry: (i) The shear stress does not go to infinity with increasing deformation, whereas it does so in uniaxial stressing (ii) The shear stress oscillates, giving rise to fears of instability.

The question arises (cf. [22] and [23]) as to whether these "undesirable" features can be avoided by a different choice of the function $\underset{\sim}{A}(\underset{\sim}{s},\underset{\sim}{\alpha})$.

In [22] the following modification to $\underset{\sim}{A}$ was suggested:

$$\underset{\sim}{A} = a(\underset{\sim}{s}-\underset{\sim}{\alpha}) - b(\bar{\alpha})k\underset{\sim}{\alpha} \tag{72}$$

where a is a constant but b is a dimensionless scalar which is a function of $\bar{\alpha}$ defined as

$$\bar{\alpha} = \sqrt{\underset{\sim}{\alpha}\cdot\underset{\sim}{\alpha}} \tag{73}$$

(72) satisfies the invariance property (52). It can be shown that for this material the requirement (55) takes the form

$$a > 0; \qquad \frac{\sqrt{2}\, a}{\bar{\alpha}} > b \tag{74}$$

Calculations based on (59)-(61) show that in the uniaxial stressing of this material the tangent modulus is given by

$$\frac{d\sigma}{d\varepsilon} = \frac{3}{2}\left(a - \frac{1}{\sqrt{2}}\, b(\bar{\alpha})\bar{\alpha}\right) \tag{75}$$

As noted in [22] by choosing

$$b(0) = \text{finite}$$

$$\lim_{\bar{\alpha}\to\infty} \bar{\alpha} b(\bar{\alpha}) = c < \sqrt{2}\, a \tag{76}$$

one can reproduce a stress-strain curve in uniaxial tension that has the initial slope 3/2 a and the asymptotic slope 3/2 (a-c).

We now study the shear flow of this material. By combining (6) and (72) and using the notation of (68) we find the following system

$$\begin{aligned} x^{\cdot} &= a - bx + y \\ y^{\cdot} &= -4x - by \\ z^{\cdot} &= -bz \end{aligned} \tag{77}$$

where $b = b(\bar{\alpha})$ and

$$\bar{\alpha}^2 = 2x^2 + \frac{1}{2}(y^2 + z^2) \; . \tag{78}$$

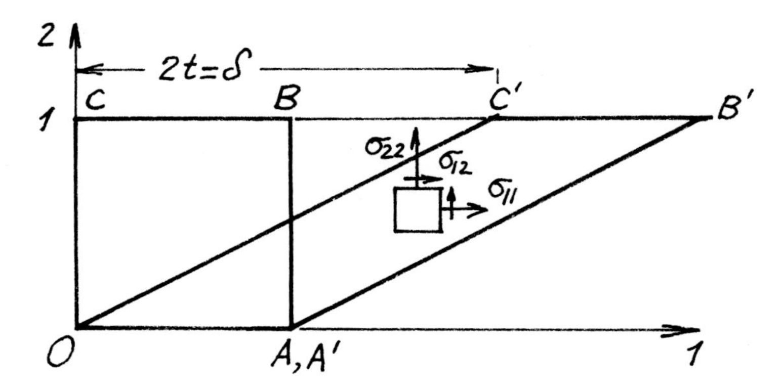

Fig.4. Shear Flow

One can make certain observations on the solutions of the non-linear system (78) without any numerical calculations. For this purpose we multiply these equations with 4x, y and z, respectively, and add to obtain

$$(\bar{\alpha}^2)^{\cdot} = 4\ ax - 2\ b\ \bar{\alpha}^2 \tag{77}$$

By using (77) we can show that the solution of (78) can never go to infinity. Indeed, since $b\bar{\alpha}$ is finite when $\bar{\alpha} = \infty$ (cf. (76)), the last term in (77) will dominate the right hand side when $\bar{\alpha}$ is large. This observation and (77) imply that $\dot{\bar{\alpha}} < 0$ whenever $\bar{\alpha}$ is very large ($\bar{\alpha} \geq 0$ always) and hence the desired result.

Thus we see that the modification suggested in [22] cannot give rise to shear stress that increases to infinity.

However numerical integration of (77) shows (cf. [22]) that for certain choices of $b(\bar{\alpha})$ the oscillations in shear stress (cf. (71)) can be avoided. A standard study of the state of affairs about the equilibria of (77) confirms this result.

It is seen, therefore, that the modified form (72) removes one undesirable feature, but as we noted (72) can never give rise to a steadily increasing shear stress.

Finally we discuss the modification suggested in [23]. The starting point in [23] is the following form for $\underset{\sim}{A}$:

$$\underset{\sim}{A}(\underset{\sim}{s},\underset{\sim}{\alpha}) = a(\underset{\sim}{s}-\underset{\sim}{\alpha}) + (\underset{\sim}{b}\underset{\sim}{\alpha} - \underset{\sim}{\alpha}\underset{\sim}{b}) \tag{78}$$

where a is a positive constant and $\underset{\sim}{b}(\underset{\sim}{s},\underset{\sim}{\alpha})$ is an <u>antisymmetric</u> tensor. In view of (51) this choice leads to the following expression for $\dot{\underset{\sim}{\alpha}}$:

$$\dot{\alpha} = a\underset{\sim}{D} + (\underset{\sim}{\Omega} + \lambda\underset{\sim}{b})\underset{\sim}{\alpha} - \underset{\sim}{\alpha}(\underset{\sim}{\Omega} + \lambda\underset{\sim}{b}) \tag{79}$$

We see tnat the choice of Lee et. al. has the potential of lessening the overwhelming influence that $\underset{\sim}{\Omega}$ had in shear flow of Prager material. For further progress one must choose a particular form for $\underset{\sim}{b}$. In [23] a definite expression is advocated for $\underset{\sim}{b}$. It seems to us however that many other choices for $\underset{\sim}{b}$ are available. Here is a simple one which is suggested by the work in [23]; but quite different from the one used in [23]:

$$\underset{\sim}{b} = c[(\underset{\sim}{s}-\underset{\sim}{\alpha})\underset{\sim}{\alpha} - \underset{\sim}{\alpha}(\underset{\sim}{s}-\underset{\sim}{\alpha})] \tag{80}$$

where $c > 0$.

Combining (79) with (80) and using $D = \lambda(\underset{\sim}{s}-\underset{\sim}{\alpha})$ we find

$$\dot{\underset{\sim}{\alpha}} = a\underset{\sim}{D} + (\underset{\sim}{\Omega} + c(\underset{\sim}{D}\underset{\sim}{\alpha} - \underset{\sim}{\alpha}\underset{\sim}{D}))\underset{\sim}{\alpha} - \underset{\sim}{\alpha}(\underset{\sim}{\Omega} + c(\underset{\sim}{D}\underset{\sim}{\alpha} - \underset{\sim}{\alpha}\underset{\sim}{D})) \tag{81}$$

$a > 0,\ c > 0$

We hasten to note that the above from satisfies the requirements (52) and (55). We also observe that we have in (81) a spin created by $\underset{\sim}{D}$ and $\underset{\sim}{\alpha}$.

We consider first the behavior predicted by (81) for uniaxial tension. It is easily seen that this behavior is identical with the one predicted for the Prager material; $\underset{\sim}{D}$ and $\underset{\sim}{\alpha}$ have the same principal directions and therefore there is no spin created by $\underset{\sim}{D}$ and $\underset{\sim}{\alpha}$.

Next shear flow of this material is considered. Combining (65) and (81) we find

$$\begin{aligned} \dot{\alpha}_{12} &= a + (1+c(\alpha_{22} - \alpha_{11}))(\alpha_{22} - \alpha_{11}) \\ \dot{\alpha}_{11} &= 2(c(\alpha_{22} - \alpha_{11})+1)\alpha_{12} \\ \dot{\alpha}_{22} &= -2(c(\alpha_{22} - \alpha_{11})+1)\alpha_{12} \end{aligned} \tag{82}$$

In the notation of (68), (82) becomes

$$\begin{aligned} x^{\bullet} &= a + (1+cy)y \\ y^{\bullet} &= -4(1+cy)x \\ z^{\bullet} &= 0 \end{aligned} \tag{83}$$

$a > 0,\ c > 0;\ x = y = z = 0$ at $t = 0$.

We observe immediately that when

$$c > \frac{1}{4a} \tag{84}$$

the right hand side of the first equation of (83) is always positive. This shows that the shear stress will increase monotonically to infinity when the constant c satisfies the above inequality.

(It can also be shown that when $c < \frac{1}{4a}$ a variety of behaviors are possible; when c is very small the behavior will be similar to that of Prager material. For larger values of c (but $c < \frac{1}{4a}$) the shear stress may reach a maximum followed by a minimum and then a climb to infinity.)

Thus we have shown that modification of $\underset{\sim}{A}$ aimed at producing a $\underset{\sim}{D}$ created spin can give rise to a reasonable behavior both in uniaxial tension and in shear flow.

We end this section with some notes of caution. The class of materials considered here is highly idealized. We know for instance, that they cannot represent cyclic behavior of many materials adequately. There is a clear need for introducing additional state variables. Nevertheless these materials can be said to have served as well by teaching the importance of joint presence of anisotropy and large rotations.

7. CONCLUDING REMARKS

The subject of constitutive equations received sustained attention during the past 30 years. Most workers in the field of Mechanics contributed to this subject and some of these contributions have been deep and enduring. Perhaps the time has come to write a comprehensive, comparative and critical review of these achievements. The present paper does not have such ambitious aims. Here the attention is focussed on a mode of representation that constitutes, we believe, a culmination of these efforts and provides a canonical form for the comparison and understanding of the previously or presently offered constitutive equations.

In this essay on constitutive equations the emphasis is placed on the current state and orientation $S(t)$ of the material. It is assumed that the parameters that represent $S(t)$ can, in principle, be determined from the relative statistics of the data provided by microscopy performed on the material element of interest at the time of interest. We do not claim that it is always necessary to perform microscopy. But we insist that $S(t)$ represents the relevant features of the arrangement of constituent particles at time t. The representation considered in this paper is for solid materials. (A fluid like behavior would result however if certain "relaxation times" associated with the evolution law in (26) are vanishingly small.)

Deformation of a solid is, in general, accompanied by the development of anisotropy and internal texture. When this happens the orientation of a deforming element becomes an issue, especially in the range of finite deformations. In Section 2 of the paper we considered the questions of orientation and the related role of superimposed rigid body rotations. We showed that the state and orientation S of a material element can be represented by a number of <u>irreducible</u> even rank tensors. These tensors not only define the orientation of a given element but also its internal symmetries and internal state. In Section 4 we explored the notion of state and orientation further and showed that the use of irreducible tensors as state variables leads to a convenient measure of the strength of anisotropy within the material element of interest. This section closes with a brief discussion of the notion of reducibility of representations.

In Section 5 we considered a problem of current interest (representation of internal damage in tertiary creep) that provides an example where the state variables in the form of irreducible tensors arise quite naturally from the statistics of internal happenings. This example should not be taken to mean that one must always use the statistics of the internal events to arrive at parameters that measure the internal state and orientation. The construction of reasonable and useful constitutive equations is a many sided enterprize and the answers may come from many directions. First there is the tradition to consider. Often a new set of constitutive equations are introduced as a modification or improvement of the older ones. Indeed, the constitutive equations used in Section 6 are based on well accepted notions, such as yield surface, kinematic hardening, etc., but they include modifications needed to cope with finite deformations. We hope that this Section of the paper clarifies certain issues that arise when anisotropy and large rotations are present.

Although we stress the importance of the traditional tools and concepts of representation, it would be safe to predict that the future work on constitutive equations will rest more on the physics and statistics of internal events.

It is also clear that there will always be an element of art and economics in this subject. We know that for a given material we really need a _set_ of constitutive equations that represent with _various levels of accuracy_, various _aspects_ of the observed behavior. The analyst will choose from this set a particular representation that is just right for the task at hand. This choice is difficult to make: it requires experience, a familarity with computing and its costs and above all a realistic understanding of the aims of analysis. These thoughts lead us to emphasize the following point. It is very important to know the limitations of a given representation. One would like to know: can it represent the behavior under cyclic loading accurately? Is it good for situations where high rates of straining are present? etc.

We hope that in future work the range of validity of a newly offered representation will be established with care and with scientific impartiality.

8. REFERENCES

1. GREEN, A. E. and RIVLIN, R. S. - "The Mechanics of Non-Linear Materials with Memory", Arch. Rat. Mech. Anal. 1, 1-34, 1957.

2. NOLL, W. - "A Mathematical Theory of Mechanical Behavior of Continuous Media", Arch. Rat. Mech. Anal. 2, 197-226 (1958).

3. GEARY, J. A. and ONAT, E. T. - "Representation of Non-linear Hereditary Mechanical Behavior", ORNL-TM-4225, 28p., 1974.

4. ONAT, E. T. - "The Notion of State and its Implications in Thermodynamics of Inelastic Solids", Proc. IUTAM Symp. Vienna, pp. 421-434, 1966.

5. MOSTOW, G. D. - "Equivalent Embeddings in Euclidean Space", Ann. Math. 65, (3), 432-446, 1957.

6. GEL'FAND, I. M., MINLOS, R. A. and SHAPIRO, Z. Ya. - "Representations of the Rotation and Lorentz Groups and Their Applications", Pergamon, Oxford, 1963.

7. ONAT, E. T. and FARDSHISHEH, F. - "Representation of Creep, Rate Sensitivity and Plasticity", SIAM, J. Appl. Math. 25, 522-538, 1973.

8. ONAT, E. T.- "Representation of Inelastic Behavior, Yale University Report for ORNL-Sub-3863-2, 1976.

9. ONAT, E. T. and FARDSHISHEH, F. - "On the State Variable Representation of Mechanical Behavior of Elastic-Plastic Solids", Proc. Symposium on Foundations of Plasticity, Warsaw, 89-115, Noordhoff, 1973.

10. MANDEL, J., "Relations de Comportement des milieux élastiques-plastiques et élastique-viscoplastiques. Notion de repère directeur, Problems of Plasticity, pp. 387-400, Noordhoff, 1973.

11. DYSON, B. F., LOVEDAY, M. S., and RODGERS, M. J. - Proc. Soc. London A, 349, 245, 1976.

12. LECKIE, F. A. and HAYHURST, D. R. - Proc. Roy. Soc. London A, 340, 323, 1974.

13. SOWERBY, R., DA C. VIANA, C. S. and DAVIES, G. J. - "The Influence of Texture on Mechanical Response of Commercial Purity Copper Sheet in Some Simple Forming Processes, Materials Science and Engineering 46, 23-51, 1980.

14. LECKIE, F. A. and ONAT, E. T. - "Tensorial Nature of Damage Measuring Internal Variables", IUTAM Symposium, Senlis, Frace, pp. 140-155, 1981.

15. ONAT, E. T. - Representation of Inelastic Behavior, Creep and Fracture of Engineering Materials and Structures , Ed. by B. Wilshire and D. R. J. Owen, pp. 587-602, Pineridge Press, U.K., 1981.

16. ONAT, E. T. - "Representation of Inelastic Behavior in the Presence of Anisotropy and Finite Deformations", to appear in Proc. Res. Workshop on Plasticity, Stanford University, 1981.

17. PRAGER, W. - "A New Method of Analyzing Stresses and Strains in Work-Hardening Solids", J. Appl. Mech. 23, 493-496, 1956.

18. MROZ, Z. - "On the Description of Anisotropic Work Hardening", J. Mech. Phys. Solids 15, 163-175, 1967.

19. DAFALIAS, Y. F. and POPOV, E.P. - "A Model of Nonlinearly Hardening Materials for Complex Loading", Acta Mech. 21, 173-192, 1975.

20. KRIEG, R. D. - "A Practical Two Surface Plasticity Theory", J. Appl. Mech. 641-646, 1975.

21. LEE, E. H. - "Elastic-Plastic Deformations at Finite Strains", J. Appl. Mech. 36, 1-6, 1969.

22. NAGTEGAAL, J. C. and de JONG, J. E. - "Some Aspects of Non-isotropic Workhardening in Finite Strain Plasticity", To appear in Proc. Res. Workshop on Plasticity , Stanford University, 1981.

23. LEE, E. H., MALLETT, R. L. and WERTHEIMER, T. B. - "Stress Analysis for Kinematic Hardening in Finite-Deformation Plasticity", SUDAM Report No. 81-11, Division of Applied Mechanics, Stanford University, 1981.

24. SPENCER, A. J. M. and RIVLIN, R. S. - "The Theory of Matrix Polynomials and Its Application to the Mechanics of Isotropic Continua", Arch. Ratl. Mech. Anal. 2, 309-336, 1959.

Chapter 6

THE INFLUENCE OF WELDING ON THE CREEP PROPERTIES OF STEELS

A.T.Price and J.A.Williams

Central Electricity Generating Board, Marchwood Engineering Laboratories, Marchwood, Southampton, SO4 4ZB, UK.

SUMMARY

Design criteria for welds in pressure parts are discussed and the most common fusion welding processes used in their fabrication are described. The factors affecting weld metal composition and weld metal microstructure are then introduced. The relationships between heat input into the weld, heat flow through the joint and the range of microstructures produced by single weld beads and in multipass weldments are next developed to complete the description of how weld metallurgy differs from that of the original material.

The development of residual welding stresses, their distributions and response to post weld heat treatment are reviewed. The mechanical and creep properties of the various microstructural constituents of welds are then compared with those of parent material concentrating on ferritic steels. The deformation of weldments at high temperatures is discussed taking into account the composite nature of weld structure and stress redistribution in real structures. Finally, the mechanisms by which welded components fail in service under creep conditions in CEGB plant are considered. These illustrate how the various metallurgical and residual stress factors arising from welding can result in failure modes that extend from relatively short times to the design life.

1. INTRODUCTION

In this paper, it is intended to examine how welding affects the structure and properties of materials, with particular emphasis on the high temperature creep behaviour of steels used in the manufacture of pressure parts. Fabrication by welding is fundamental to the construction of large scale plant for pressure containment as, for example, in chemical processing, the oil and gas industries and in power generation.

Welding produces both metallurgical and mechanical discontinuities and, not surprisingly, failures commonly occur at welds; the performance of welds often being the life limiting factor in many structures [1,2]. This is particularly true of creep applications and it is therefore remarkable to find that welds are virtually ignored in the design of components such as tubes and pipes. It is interesting to examine how this is justified.

Essentially, the design codes for pipework only take account of pressure and assume isothermal operation. The method used in the UK power generating industry, based on BS5500, employs the empirical mean diameter formula:

$$\sigma = \frac{PD_m}{2W} \quad \ldots \quad (1)$$

where σ is the design stress and is in the hoop direction
P is the pressure
D_m is the mean diameter
and W is the wall thickness.

This is probably the most common method for designing tubes and pipes, although variants are found in other industries, for example, the inside diameter formula, in which D_m is replaced by D_i, and the outside diameter formula, where D_m is replaced by D_o. Effectively, it is implied that under multiaxial stress conditions, the creep life will be determined by the maximum principal stress, in this case the hoop stress. Thus, any stress redistribution occurring through the wall is ignored.

In all codes, the design stress is then used in conjunction with uniaxial tensile data relevant to the temperature of operation and the material concerned. At temperatures where the material creeps, suitable data are required to define the stress to cause creep rupture in 10^5 h, the stress to produce 1% total plastic strain, or, alternatively, a creep rate of 10^{-7} per h. In practice, creep strain data for long times are expensive to obtain and are relatively scarce so that creep rupture data tend to be used. Having established the design stress by this route, it is then usual to apply a safety factor, normally $\sim$ 1.5, to define the actual stress at which the component will operate. Only in the ASME XI Boiler and Pressure Vessel Code, in section N47, which deals with design features for nuclear components, do welds merit special attention, being covered by requirements that limit creep strain to an average of 1% through the thickness or 5% locally.

The above codes are unsophisticated but quite adequate. They owe their success largely to the use of safety factors which provide blanket coverage to allow for inaccuracies due to lack of knowledge. Actually the difference in properties between the parent material and the weld is one of the most important factors limiting creep life. Whereas, the parent material composition, steelmaking process, production route and heat treatment are closely specified resulting in a near homogeneous product, welds, in contrast, have modified compositions and are subjected during welding to transient thermal conditions which produce non-equilibrium microstructures and residual welding stresses. These differences cause stress redistribution during service, modify creep rates and produce microstructural conditions that can favour local creep damage accumulation and lead to premature failure.

The causes, nature and consequences of the effects produced by welding will be described in the following sections. The first part introduces the factors affecting the metallurgy of welds, which arise as a function of the welding process. The discussion will concentrate on the welding of pressure parts and fusion welding processes, in particular, as these are usually employed in the fabrication of such plant. It is these pressure containing welds which are most critical. Although welding is also widely used in the construction of other high temperature structures, here the design can be cruder and, for example, sections can be made thicker to deal with uncertainties. In pressure parts, however, the necessity for good heat transfer and uniformity and conformity of weld shape to avoid erosion and corrosion etc., dictate that welds have to be the same thickness as the parent material.

The early sections will, therefore, discuss some of the welding and metallurgical issues that arise when materials are fusion welded, focussing on some of the steels used in the UK power generating industry. The literature is therefore reviewed selectively as an introduction to the subject to highlight those aspects which are relevant to the subsequent performance of components under creep conditions. Later sections describe the properties of welded materials relevant to high temperature performance and the various creep failure mechanisms that arise.

2. FUSION WELDING AND COMPOSITIONAL EFFECTS

The objective of welding is to achieve continuity in the load bearing structure. This is done by remelting some of the parent material and filling with additional compatible weld metal where necessary. The size and shape of the components will influence the choice of welding process, the detailed design of the weld geometry and the manner in which welding is carried out. Examination of some of the components in a power station steam raising circuit allows an appreciation of some of the issues.

In the boiler, preheated water enters and eventually emerges as superheated steam at temperatures of $\sim 600^{\circ}C$. In the low temperature regions C-Mn type steel is employed and to accommodate progressively higher temperatures 1Cr$\frac{1}{2}$Mo, 2Cr1Mo and finally austenitic stainless steels are necessary. As an example, superheater tubes in a modern coal-fired station might be in austenitic AISI Type 316 steel, 54 mm od and 8 mm wall. There would be typically $\sim$ 50,000 boiler tube butt welds, altogether, and welding would be done using the tungsten inert gas (TIG) Figure 1, or manual metal arc (MMA) processes.

Steam is conveyed from the top of the boiler via steam pipes to the turbo-generator. In the UK, the pipes are normalised and tempered $\frac{1}{2}$Cr$\frac{1}{2}$Mo$\frac{1}{4}$V typically 350 mm od and 60 mm wall. To date, these have been butt welded circumferentially with 2$\frac{1}{4}$Cr1Mo using the MMA process; a weld preparation being indicated in Figure 2. Each weld would be built up in $\sim$ 15 layers with $\sim$ 10 Kgm of metal being deposited. The arcing time, which is the time during which metal is being deposited, is about 20 h and the total working time to produce a single butt weld is 3 or 4 days.

Among the largest components in the pressure circuit is the boiler drum, which is a cylindrical vessel $\sim$ 60 m long, $\sim$ 1.8 m od and $\sim$ 150 mm wall. The manufacturing sequence is to fabricate cylindrical sections by forming two C-Mn type steel plates into semicircular shapes, which are butted together and welded. Several such cylinders are then butt welded to produce the required length. Large volumes of weld metal are required so that a high deposition process is necessary. In this case, the submerged arc (SA) process, which is mechanised, is commonly used. In this process, wire is fed into the liquid metal pool beneath a protective layer of flux.

Commercially, it is necessary to produce sound, fully fused welds, with suitable metallurgical properties and engineering strength while minimising handling and machining time, the quantities of consumable welding materials used and the total welding time. As a result, the TIG, MMA, and SA electric metal arc processes are in general use to meet these requirements. While these processes are essentially similar, there are significant differences between them which affect the weld metal chemistry and the metallurgical structures of both the weld metal and the adjacent heat affected zone (HAZ). In the following paragraphs TIG and MMA welding are discussed to familiarise the reader with the topic generally, while concentrating on those aspects which affect metallurgical and, in particular, creep behaviour. For those who are interested in pursuing the subject further, the basic principles and chief characteristics of welding processes are covered by Houldcroft [3] while Lancaster [4] deals with welding metallurgy.

Single - V butt

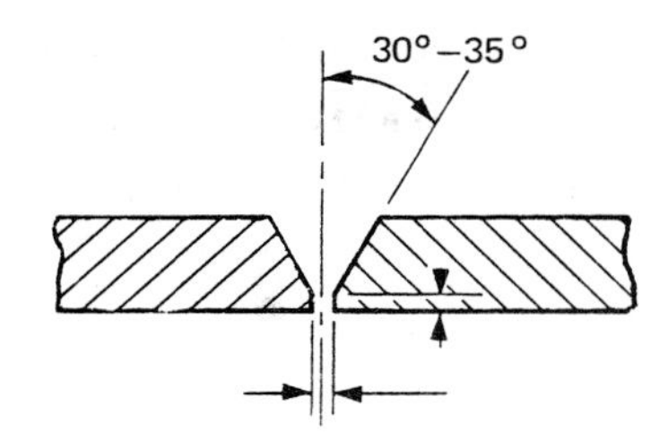

Root face: 0.75mm to 1.5mm

Root gap: 1.5mm to 3mm

U-type

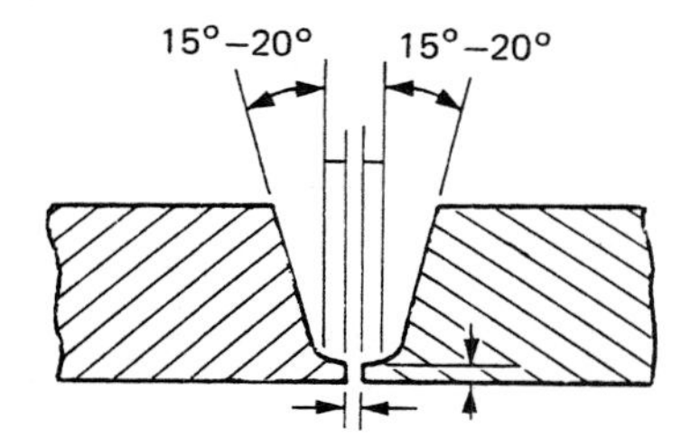

Autogenous Joint before and after welding.

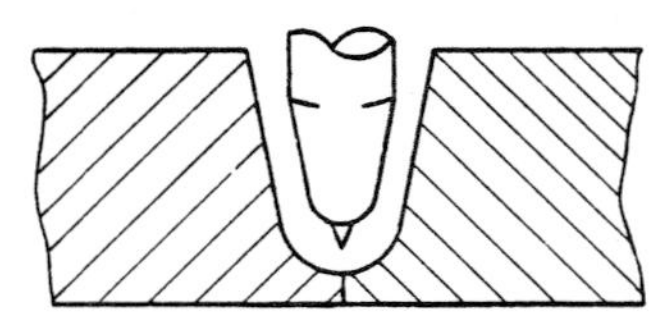

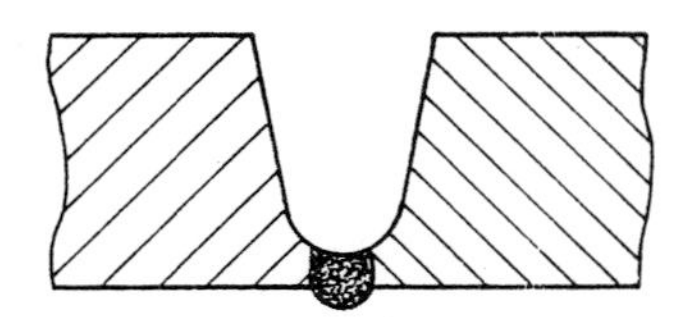

Fig. 1. Joint preparations and TIG welding of Open Butt Joints.

2.1 Tungsten Inert Gas Arc Welding

Welding of thin section material or the first fusion pass of thick sections is often carried out autogenously, that is, the material is simply remelted. Figure 1 illustrates the use of the TIG process for the butt welding of tube root runs. TIG welding is carried out using a non-consumable thoriated tungsten electrode, which simply acts as a source of electrical energy to melt the substrate. An inert shielding gas, commonly argon, is used at the back of the weld and around the electrode to prevent oxidation of the tungsten, at the very high temperatures reached at the root of the arc and at the weld pool. The requirement is for sufficient heat to be supplied to the joint to cause melting under conditions where arc forces, gas pressure, and surface tension effects allow the molten pool to be retained within the welded joint so that upon solidification a fully fused joint of uniform macroscopic appearance is produced. If excessive heat is applied, a large pool of liquid metal is produced which may sag, fall out of the joint under gravity, or be blown out by the presence of the electric arc. Welding can be carried out either manually or by using mechanical and electrical equipment designed so that the arc rotates around the job. In both cases, the travel speed around the joint must be neither too fast nor too slow, so that adequate but not excessive fusion is produced. This requires considerable skill and experience from manual operators or, alternatively, extensive development to produce well engineered electrical and mechanical systems to ensure consistent and reproducible welding conditions. TIG welding is particularly suited for low heat input applications where delicate control of the size of the fusion pool is important. Thus, the process is used to make the initial welding pass in many high integrity applications to ensure good root fusion and weld bead profile.

Despite the apparent simplicity of TIG welding, the metallurgical product can be very different from the original homogeneous parent material whose composition, structure, heat treatment and properties were carefully controlled during the production route. The differences arise from several causes. The inert gas used to prevent ingress of oxygen and nitrogen from the atmosphere may not always function adequately in practice. Apart from trivial, although not uncommon, reasons such as interruption to the gas supply, or draughts, turbulence can arise around the welding nozzle in the immediate vicinity of the weld pool, causing oxygen and nitrogen from the atmosphere to become entrained. Design of the welding nozzle is therefore very important, but is very much a pragmatic development exercise. However, in general, where inert gas protection is used there is little change in the chemistry of the fused metal. Nevertheless, solidification of the weld metal and the heating and cooling of the adjacent material can cause dramatic changes in structure and in properties as will

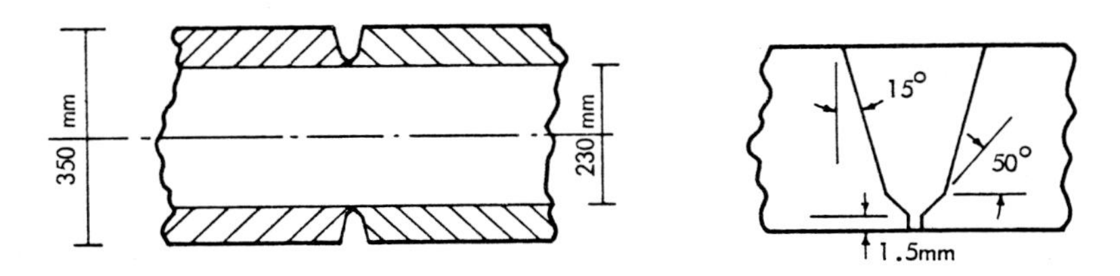

Fig. 2. Dimensions and weld preparation for a typical $\frac{1}{2}$Cr$\frac{1}{2}$Mo$\frac{1}{4}$V steam pipe.

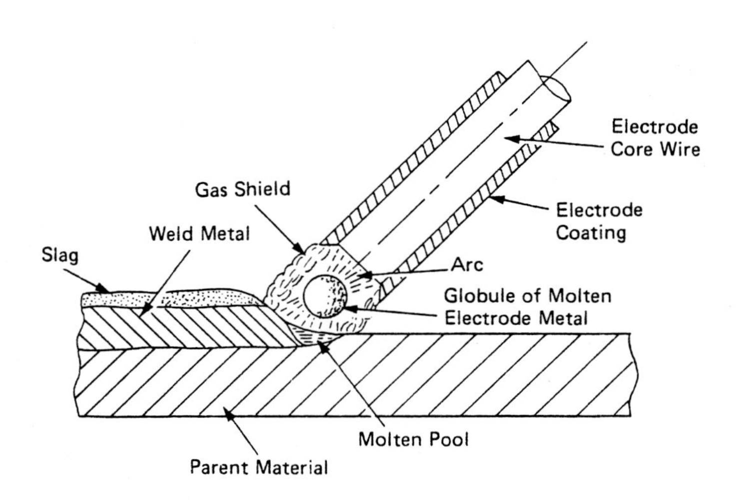

Fig. 3. MMA welding illustrating protection of the weld pool from atmospheric attack.

be discussed later.

For sections greater than ∿ 3 mm thickness, it is necessary to add material to fill up the joint and this is achieved by the addition of wire into the weld pool. The result is that the composition of the weld pool becomes an average value which reflects the composition and quantity of parent material melted and the amount of filler material added. Within the weld pool itself, mixing of the superheated liquid metal is quite efficient because electromagnetic forces produce a stirring action causing the weld pool to rotate, typically at speeds of the order of ∿ 10 mm per second. Thus, apart from a quasi-static stagnant layer that may exist very close to the fusion boundary, the liquid weld metal will be relatively homogeneous in composition. In practice, the proportions of added filler and remelted material can vary considerably but, typically, dilution of the filler by the substrate is of the order of 20 - 40% for TIG welding. During mixing and solidification, little chemical reaction occurs in the molten weld metal, so that the composition of the resulting weld is largely determined by the ratio of the two constituents.

It may thus appear that weld metal composition is easy to control if solid wire is used. However, although weld metal composition is usually required to match the parent material quite closely, there are many necessary exceptions to this. Also, it should be appreciated that the quantities of wire required to weld large boiler components are relatively small, often of the order of a tonne or less. These have to be produced as special casts and this may not be attractive commercially. Thus supply problems occur, particularly for welding development purposes, when wires of different compositions are needed to explore welding response and weld metal properties. Occasionally, this may result in available but non-optimum wires being employed.

2.2 Manual Metal Arc Welding

Historically, this is the most popular welding process, because the highly developed manipulative skills of welders allow very complicated geometries of poor accessibility to be welded to high standards. The process is illustrated in Figure 3. Essentially, the welding electrode consists of a metal wire core with a mineral coating. When an arc is struck between the electrode and the workpiece, a molten pool is formed on the parent material and at the same time the electrode begins to melt. Metal, in the form of drops or globules, transfers across the arc and is projected into the molten pool. Within the arc and weld pool complex gas-metal-slag reactions take place. The chemistry of such systems is extremely complex but, nevertheless, is sufficiently well understood to allow the basic principles to be discussed to develop a general picture of the chemistry of welding and how this influences the composition of weld metals.

The electrode wire is usually manufactured from rimming steel for low alloy ferritics and, for more highly alloyed material, from wire whose composition is close to that finally required. The function of the core wire is to conduct electricity to the arc and to supply additional metal filler. All electrodes for MMA welding have an extruded flux coating. The principal functions of the coating are to stabilise the arc, to flux away any impurities, to form a protective gas shield and slag over the weld and to add any metallic elements needed to adjust the final composition to that required. For high integrity applications, particularly in power plant construction, basic electrodes are usually employed, in which the coating contains calcium carbonate and fluorspar. These electrodes can be manufactured, using appropriate baking treatments, with very low moisture contents so that the hydrogen content of the weld metal as deposited is low enough to suppress hydrogen cracking [5]. The latter is a form of cracking that may easily arise in hardenable ferritic steels, especially above about 25 mm thickness.

2.3 Electrode Formulation and Slag-Metal Reactions

The major mineral constituents of basic electrode coatings are CaO-CaF_2-SiO_2-Al_2O_3-TiO_2. The functions of these constituents in slag formulation have been critically reviewed by Boniszewski [6]. Briefly, they form a slag in much the same way as in basic steel making processes and react chemically; the major difference being that the slag metal oxidation-deoxidation reactions occur at much higher temperatures and in much shorter times. CaO as $CaCO_3$, and CaF_2 are the chemically basic components of the welding slags; the CaF_2 also acting as a flux which dissolves the CaO and increases the fluidity of the slag. The other oxides combine with the CaO to form a basic slag. The quantities of the various oxides affect slag viscosity and thereby the shape and appearance of the solidifying weld metal pool. As in steel making, oxides are chemically active and take part in slag-metal reactions in ways determined by their reactivity. For low alloy steels, metallic elements and deoxidants are added to the coating in the form of ferro-alloy metal powders. Elements commonly concerned are Mn, Si, Cr and Mo. These elements are oxidisable and the amounts transferred to the metal depend on the oxidation potential of the slag and gas atmosphere.

Wolstenholme [7] has analysed the partition of elements between metal and slag for MMA welding with $2\frac{1}{4}$Cr1Mo electrodes, using a formal chemical thermodynamic approach and evaluating the relationships between coating additions and the weld metal composition. Using experimental and commercial electrodes, he showed that the composition of the weld metal and its slag are related through reactions of the type:

$$x/y[M]_{Fe} + [O]_{Fe} \rightleftharpoons 1/y[M_x O_y]_{slag} \qquad \ldots \quad (2)$$

where the partition of the metallic element, M, between the weld metal, Fe, and the slag depends on the weld metal oxygen, O, content. Experimental reaction constants for the oxidation of Fe, Cr, Mn, Si and Ti were examined and, by assuming an effective reaction temperature of about 1900°C, good agreement was obtained with equilibrium reaction constants calculated from published thermodynamic data. Thus, each element competes for the available oxygen and the partition of the metallic elements between the weld metal and slag approaches close to equilibrium.

In basic electrodes, the major source of oxygen arises from dissociation of carbonate:

$$CaCO_3 \rightarrow CaO + CO_2 \rightarrow CaO + CO + [O] \qquad \ldots \quad (3)$$

and some from dissociation of water vapour. The oxygen made available then takes part in reactions of the type:

$$x/y\ M + [O] \rightarrow 1/y\ M_x O_y \qquad \ldots \quad (4)$$

It was found, that for practical purposes, CO_2 and H_2O were reduced to CO and H_2, so that the extent of chemical reaction was largely determined by the amount of carbonate in the coating in relation to the weld metal and slag masses.

It is therefore possible to imagine a closed gas-liquid metal-slag system which approaches equilibrium, even at the very short times required for drop detachment, because of the high temperatures involved in arc welding. For basic coated electrodes, Wolstenholme pictured events occurring in the following sequence:

When the electrode coating is heated, water vapour is evolved together with CO_2 produced by the thermal decomposition of the carbonate in the coating. This occurs predominantly at the electrode tip, where a molten drop of metal forms on the core wire and is partially covered by molten slag, formed as the coating melts. Concurrently, gas is evolved, the metallic constituents pass into the metal droplet and the residual constituents of the coating pass into the slag. The droplet forms and transfers to the weld pool in about 0.5 s. In the formation of the metal droplet, hot spots develop locally at arc roots on the electrode and, under these conditions, the metal is oxidised and the CO_2 and H_2O are very rapidly reduced in high temperature reactions. Gas-metal reactions probably occur at effective temperatures of about 2300°C, while the slag-metal reactions occur at an effective temperature, in the bulk of the molten metal droplet, of about 1900°C.

In the 0.5 s for which the droplet exists before it is absorbed into the weld pool, there is violent agitation and the slag and metal are in close contact. During this period, the metal is rich in oxygen from the gas-metal reactions and reaction with metallic elements occurs to form oxides. The products of oxidation pass into the slag and the slag-metal reactions approach equilibrium. It is thought that the oxides formed largely escape from the metal into the slag during the period of the existence of the molten droplet. Thus when the droplet passes into the weld pool, little further reaction occurs.

The weld pool, which is formed as the individual metal droplets transfer across the arc, is less violently agitated than the molten droplet on the electrode tip and there is thought to be little opportunity for slag-metal reactions to take place in the weld pool outside the arc region. Thus the product delivered to the weld pool is metal at about 1900°C containing dissolved oxygen resulting from the slag-metal reaction equilibrium. This state is believed to be effectively frozen in as further reaction with the slag is inhibited by poor physical contact between the metal and slag and the relative quiescence of the weld pool. As the electrode moves forward, the temperature of the molten metal falls rapidly and solidification occurs. As the solid metal cools, the solubility of oxygen in the molten metal decreases and deoxidation products are precipitated. For practical purposes, the oxygen analysed in the weld deposit is that which was dissolved in the molten metal when the slag-metal reactions approached equilibrium. From this model, it can be seen that elements more noble than Fe, such as Cu are virtually unoxidised and transfer almost completely to the weld metal with negligible effect on the oxygen content. At the other extreme, elements such as Ca, which form very stable oxides, are not found in the metal phase. Elements of intermediate activity such as Mn and Si partition between the metal and the slag in a way that is dependent upon the concentration of oxygen in the metal droplet.

The basicity of the flux is important as a measure of silica activity which reduces in the direction:

$$SiO_2 \rightarrow CaO,\ SiO_2 \rightarrow 2CaO,\ SiO_2$$

with a progressive change from acid, through intermediate to basic fluxes.

The chemical nature of the flux therefore plays a large part in determining weld metal composition and much of the art of consumable design is to achieve the required composition by suitable metallic or alloy additions to compensate for flux chemistry. This is complex because of the other properties of fluxes. For example, highly basic fluxes have low oxygen activity, corresponding to a lower supersaturation of oxygen in the weld metal, but they produce more fluid slag. In practice, this means that they are less tolerant to all positional welding. They are also more susceptible to moisture pick-up and difficult to deslag. The less basic fluxes tend to produce high weld metal oxygen and silicon contents and form glassy slags.

The weld metal composition eventually obtained comprises:

Main alloying elements, including C, as a result of wire composition, additions to the flux and, where relevant, dilution from the parent material. These will determine general creep strength, for example, by their influence on solution strengthening and carbide stability.

Deoxidants such as Mn, Si and Ti, present both in solution and as oxides precipitated from supersaturation.

Noble elements, such as Cu which transfer across the arc unaffected, and impurity elements, such as As, Sb, P and Sn, arising from the consumable materials, which may affect grain boundary behaviour and produce, for example, low ductility in creep.

Thus, even with careful design of the electrodes, it is not possible to obtain complete identity of composition with the parent material. Commonly, the main alloying elements are controlled to within the same specification but elements such as oxygen, nitrogen and silicon may differ significantly. In addition, C is often lowered in the weld metal to obviate cracking associated with hardening mechanisms during cooling, and Mn is increased to ensure that cracking problems associated with the segregation of sulphur during solidification are controlled. These differences in chemical composition and any dilution effects can result in changes in the structure, transformation behaviour and properties of the weld metal compared with the parent material.

Finally, it is thought that similar metal-slag reactions apply in SA welding [8]. However, in this process heat input is usually higher and up to ∿ 60% dilution of the weld metal from the substrate is obtained, compared with ∿ 30-50% for MMA welding.

3. WELD METAL SOLIDIFICATION STRUCTURES

While the composition of weld metals differs from that of parent material for reasons such as those outlined above, differences in metallurgical structure arising during solid-

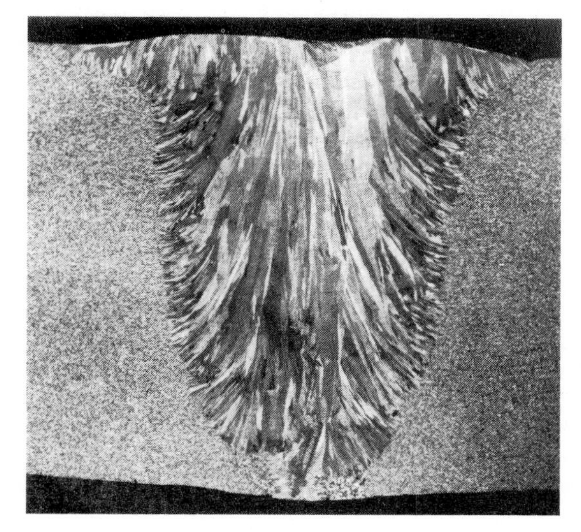

x 1.5

Fig. 4. Multipass MMA weld in Type 316 steel illustrating the solidification pattern. [11]

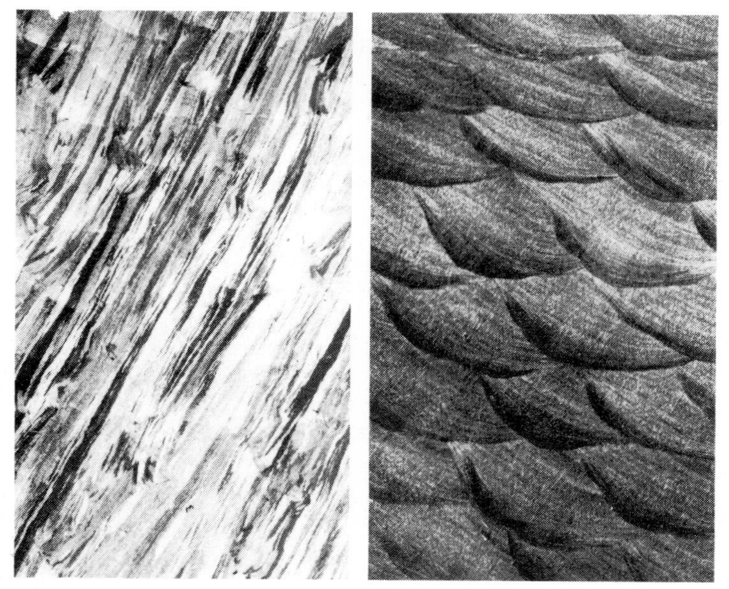

x 4

Fig. 5. Section through a Type 316 multipass MMA weld, etched to reveal the relationship between solidification structure and weld bead positions. [11]

ification and cooling are much more obvious. The features of solidification in weld metals have been reviewed by Davies and Garland [9] and by Savage [10].

The principles of solidification are similar for ferritic and austenitic steels, but transformations on cooling and reheating during deposition of subsequent beads complicate the picture in the case of ferritic steels. These effects are discussed later.

In the present context, it is therefore useful to refer to welds in austenitic steels, such as Type 316. This material is commonly used for applications where good corrosion resistance and high temperature creep strength are required. The essential features of solidification and grain growth are evident in Figure 4, which shows an austenitic MMA multipass weld in a plate. Solidification is generally cellular or dendritic in nature. Austenitic grain growth occurs as the result of cooling by heat conduction into the base material and into the already solidified weld metal. The columnar austenite grains grow parallel to the heat flow direction. Within each weld bead, grains grow continuously from the base upwards, converging inwards and towards the weld bead surface. Thus, a strongly orientated fibre structure is produced.

When further weld beads are deposited, epitaxial growth takes place from the previous weld bead, with the grains from the underlying bead propagating into the newly deposited bead. This results in continuity of grain structure even in multipass welds. This is illustrated in Figure 5 in which the same section has been etched to reveal both the individual weld beads, and the inherent grain structure which consists essentially of long columnar grains. Baikie and Yapp [11] in an elegant series of experiments, deposited welds in Type 316 weld metal in various welding positions and demonstrated that the direction of solidification was determined by the direction of heat flow which, in turn, could be predicted from a knowledge of the bead deposition sequence. In fact, they were able to produce a series of test blocks, in which different specific orientations were obtained based on strong(100) fibre textures produced by preferred grain growth in the <100> crystallographic direction.

Epitaxial growth is a common feature of fusion welds. It arises because the base metal is similar in composition to the weld pool, is readily wetted and acts as an ideal substrate upon which solid phase growth can occur. One of the features of this type of solidification is that the influences of welding parameters such as welding current, travel speed, etc., are second order. The orientation of the fibre structure is controlled largely by welding and the effects of surface tension and gravity, and also by the geometrical configuration in which the weld is being made, in that this determines the direction of heat flow.

The importance of preferred orientation in weld metals is that physical, mechanical and creep properties may be directional. For instance, Baikie and Yapp showed that good transmission of ultrasound in austenitic weld metals was possible only in certain orientations. They also found that the minimum creep rate in the direction parallel to the grain orientation was a factor of ~ 3 slower than in the direction perpendicular to the fibre structure.

A further consequence of the solidification processes in weld metal is that, as in conventional steelmaking, the highest melting point constituents solidify first on cooling. Segregation of lower melting point constituents and impurities, such as sulphur and phosphorus, may result in solidification cracking during the welding process [12,13]. This can be a serious problem in the welding of austenitic steels, which is usually overcome by adjusting the composition of the welding consumables so that the weld metal solidifies initially to δ-ferrite [14]. Austenite subsequently forms and grows into the δ-ferrite but a few percent of δ-ferrite is retained at room temperature. While the presence of this second phase allows sound welds to be manufactured, δ-ferrite is not stable at creep temperatures and transforms to intermetallic phases. Both constituents provide nucleation sites for cavitation by creep and can result in low ductility weld metals, particularly when the second phase forms a continuous network [15].

4. HEAT AFFECTED ZONES

The preceding section dealt with weld metal microstructure for materials which remain essentially single phase, so that weld metal deposition produces a continuation of the pre-existing microstructure by epitaxial growth. However, in materials such as ferritic steels which transform to a second phase on heating, the effects of thermal cycling are more complex. This section deals with this aspect of multipass welding. In principle, the reheated weld metal and the heat affected zone of the parent material are exactly equivalent, responding in much the same way to heating and cooling cycles. However, for simplicity, the effects produced in the HAZ will be considered first.

4.1 Heat Input

When electrical energy is supplied to melt the wire electrode, not all of the energy is transferred into the weld pool as latent heat and superheat. Losses arise from various sources such as conduction back along the electrode, gaseous convection losses, melting of the slag and radiation. Also, heat input, θ, to the job itself depends on the travel speed of the electrode, S, along the job. Thus, in practice, the heat input is given by:

$$\theta = \eta VI/S \qquad \ldots \qquad (5)$$

where η is the arc efficiency of the particular process [16]. V and I are the voltage and current, respectively, and depend on the physical dimensions of the electrode wire and its arcing characteristics. The physics of the welding arc is a complex subject [17,18,19] and for the present it is sufficient to consider only the more obvious measurable parameters. Typical values for $2\frac{1}{4}$Cr1Mo consumables are given below:

Process	Electrode or Wire Dia. mm	Voltage V Volts	Current I Amps	Arc Effcy. η	Travel Speed S $mm.s^{-1}$	Effective Heat Input ηVIS^{-1} $kJ.mm^{-1}$
TIG	1.2	12	200	∿0.6	∿2.0	0.7
MMA	2.5 3.2 4 5 6	22	50-75 85-125 110-175 150-240 195-340	∿0.70	∿2.5	0.4 0.65 0.9 1.2 1.6
SA	3.2	26	450	∿0.95	∿6	1.9

It can be seen that heat inputs can vary considerably. For example, for MMA the heat input can vary by a factor of about 4 depending on the choice of electrode core diameter. In addition, the speed at which the individual welder deposits the weld metal can vary by a factor of at least 2. Also, if side to side movement, called weaving, of the electrodes is permitted, large weld pools can be maintained under some conditions. Thus, the heat input can vary by up to about an order of magnitude even for a single wire diameter; although this is not usually the case. The values given for TIG welding are for fill-up applications and lie in the middle of the MMA range. For root run welding, low heat input conditions are employed with currents typically ∿ 80 amps. The SA values, which fall at the top end of the MMA range, are not typical of SA welding. For example, in C-Mn steels currents of the order of 1000 amps are employed. Finally, it should be noted that heat input in $kJ.mm^{-1}$ does not adequately characterise a process, since it does not take into account the mass of weld metal deposited. However, the latter type of information, $kJ.gm^{-1}$ deposited, is not readily obtainable.

In conclusion, it should be appreciated that heat input is very process and operator dependent and can be a major source of variance in the metallurgical structures in welds.

4.2 Heat Flow

In material ahead of the weld pool rapid heating occurs as the arc approaches, followed by cooling as the arc recedes. Regions immediately adjacent to the fusion boundary are heated very rapidly to temperatures which approach the melting temperature of the weld metal and then cool by conduction at rates determined by heat flow considerations. Positions further from the heat source experience lower peak temperatures and cool at slower rates. Thus, the general thermal field experienced by the material in the vicinity of the weld reflects the movement of the welding arc as it approaches the particular point in the structure, as a function of time and the rate of heat conduction in the particular geometry. It can be appreciated that the metallurgical changes that occur in the parent metal HAZ and in already deposited weld metal can be understood by combining heat flow theory to specify the thermal history of any region with knowledge of the basic physical metallurgy of the material.

It is easy to visualise that in the welding of say, a thin sheet or tube which is completed in a single pass, heat flow will be essentially in two dimensions (2-D). Depending on heat input and travel speed, the rate of heat conduction away from the fusion boundary may be rapid resulting in fast cooling rates, or relatively slow, if the job becomes hot. In heavy section plate containing deep weld penetrations, three-dimensional (3-D) heat flow is approached and cooling rates are fast. Rosenthal [20] derived equations to describe heat flow in these different situations:

$$\text{for 2-D, } T-T_o = \frac{\Phi}{2\pi kg} \exp(-\lambda v\xi) K_o(\lambda vR) \quad \ldots \quad (6)$$

$$\text{for 3-D, } T-T_o = \frac{\Phi}{2\pi k} \exp(-\lambda v\xi) \frac{\exp(-\lambda vR)}{R} \quad \ldots \quad (7)$$

These apply to a point source of heat moving with constant velocity v, along the x-axis of a fixed rectangular coordinate x, y, z system. T is the instantaneous temperature in K, T_o is the initial temperature in K, g is the plate thickness, k is the thermal conductivity, K_o is a Bessel function of the first kind, zero order, $0.5/\lambda$ is the thermal diffusivity. $\xi = x = vt$, where t is time, Φ is the arc power, ηVI, and $R = \sqrt{\xi^2+y^2+z^2}$. Christensen, et al. [16] and Hess, et al. [21] have shown that the equations accurately predict isotherm widths and cooling rates for 2-D and 3-D heat flow situations for a wide range of MMA welding conditions.

These equations demonstrate that the thermal history of the HAZ depends, inter alia, on the physical properties of the base material, the heat input of the welding process, the temperature of the base material, prior to each welding run, and, of course, the distance of the particular location from the fusion boundary. Figure 6 shows typical thermal cycles experienced at various distances from the fusion boundary when a single weld bead is deposited on a 75 mm ferritic steel plate using TIG, MMA and SA processes. Immediately adjacent to the boundary, the material experiences cycles characterised by relatively rapid heating to a maximum temperature that approaches its melting temperature followed by more or less rapid cooling, which is related to the magnitude of the heat input. Further from the boundary, peak temperatures are lower and cooling rates slower. The metallurgical changes that result from such temperature cycles can be very complex and can profoundly affect the properties of the base material. They will be considered next.

4.3 HAZ Microstructures in Ferritic Low Alloy Steels

4.3.1 Austenitic Grain Growth

When low alloy ferritic steels are heated, transformation to the stable high temperature austenite phase begins at the Ac_1 temperature, and is complete at the Ac_3 temperature. Subsequently, austenite grain growth occurs to an extent determined by temperature and time; the driving force being the reduction in grain boundary surface energy. The reverse process, the decomposition of austenite to ferritic transformation products on cooling is, inter alia, critically dependent on cooling rate and for welding where much of the cooling is complete in the order of 30 s, a variety of non-equilibrium structures can be produced. Thus, the metallurgical structure of any position in a HAZ depends on the detailed thermal cycle it experiences, its austenite nucleation and grain growth kinetics and continuous cooling transformation (CCT) behaviour.

During the heating cycle produced by welding, austenite is the stable phase thermodynamically at temperatures $> Ac_3$ and grain growth will occur. Initially, grain boundaries are pinned by carbides, nitrides and inclusions until a temperature is reached at which these become soluble in the austenite. This will depend on the initial size and distribution of the particles. In practice, in low alloy ferritic steels, carbides are the main obstacle to grain growth but dissolve relatively rapidly on heating to temperatures above $\sim 950^{\circ}C$. Grain growth is then related through an activation energy, dependent on diffusion criteria, to temperature and time. Thus the problem in considering a welding cycle is to sum grain growth over the times and temperatures of the heating and cooling cycle experienced by the particular region of the HAZ.

The factors affecting the kinetics of grain growth in metals are well understood and, indeed, an empirical relation-

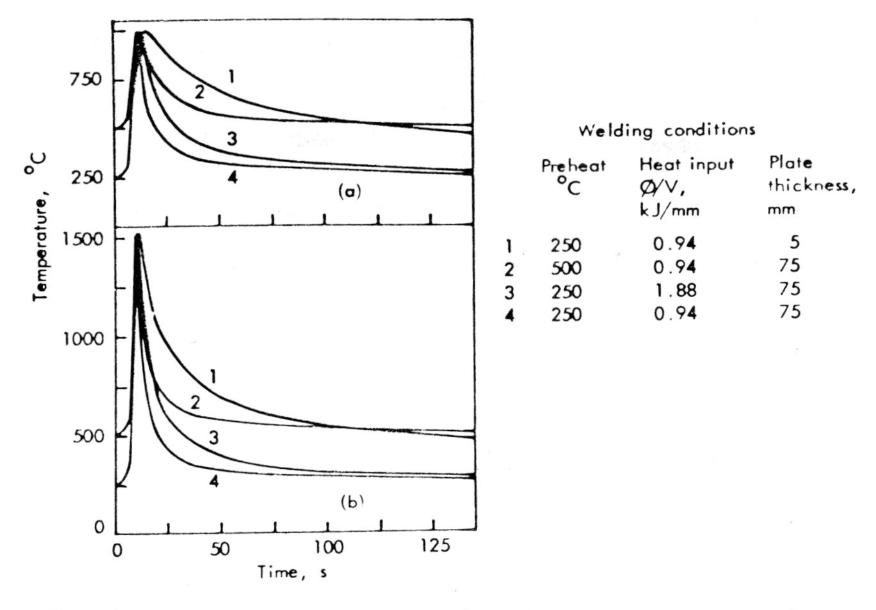

Fig. 6. Weld thermal cycles with peak temperatures of 1000°C and 1520°C for various welding conditions.

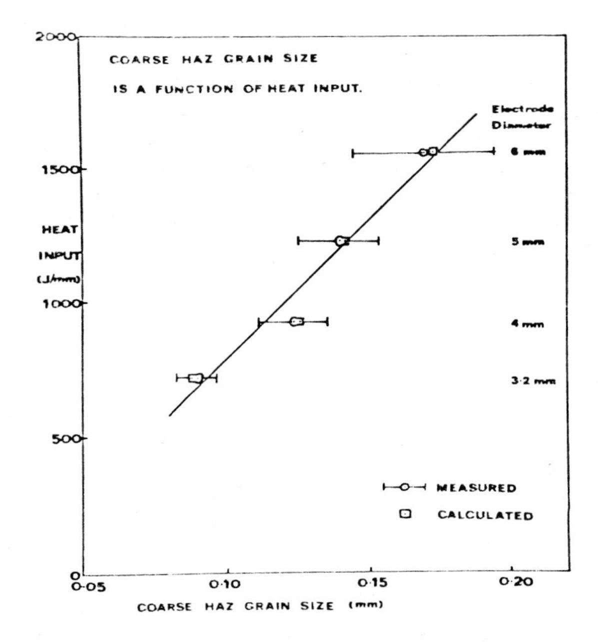

Fig. 7. Measured and calculated grain size as function of heat input. [24]

ship has been obtained [22] for the growth of austenite in low alloy steels. However, the analysis of grain growth under the rapid temperature-time cycle experienced by weld HAZs presents many difficulties. This has been investigated experimentally [23,24,25] using **weld HAZ simulators; equipment** capable of subjecting laboratory specimens to heating and cooling cycles close to those actually found in welding. Alberry and Jones [23] carried out a comprehensive series of measurements of grain size in ½CrMoV material as functions of temperature and time. They also demonstrated [24] that the final grain size achieved could be calculated from an equation of the form:

$$D_t^{2.7} - D_o^{2.7} = A[\exp(-Q^*/RT)]t \qquad \ldots \quad (8)$$

where D_t is the final average grain size after a hold at temperature of duration, t,
D_o is the original grain size,
A is a material dependent constant,
Q^* is related to an experimental activation energy for grain growth; other symbols have their usual meanings.

They then used this equation to estimate the grain growth by representing each cycle as a series of discontinuous isothermal steps of small constant duration. The grain growth was then summed numerically over the temperature interval in which austenite was stable; in their case from ∿ 950°C on heating, to the peak temperature and down to ∿ 550°C on cooling. Good agreement was obtained between grain sizes calculated in this manner, using the Rosenthal method to specify temperature history, and those measured in real welds made with different size welding electrodes chosen to cover a range of heat inputs, see Figure 7.

This approach has been extended [26] and appears to apply generally to low alloy ferritic steels. Thus, it can be appreciated that the basic principles of physical metallurgy that describe grain growth under isothermal equilibrium conditions apply in a modified form to welded materials. In practice, the mechanism of growth is rather more complex than indicated, in that the steep temperature gradients that exist with distance away from the fusion boundary exert a pinning force on grain growth. However, the effect is second order. Also, it is found that, because the activation energy is large most grain growth takes place at temperatures near the peak value and the contribution during cooling is small.

Figure 7 demonstrates a direct relationship between grain size and heat input for regions of a ½CrMoV HAZ immediately adjacent to the fusion boundary, illustrating how the selection of the welding process parameters can influence materials properties, through its effect on grain size.

Clearly, as distance from the fusion boundary increases, heating rates become less rapid, peak temperatures lower and cooling rates slower. This results in a gradual decrease in austenitic grain size with distance from values of the order of 150 μm to that of the original parent material, say ∿ 10 μm. Beyond this location, there will be a region where the parent material grains are subjected to relatively low peak temperatures such that austenite nuclei form on the existing grain boundaries, but do not grow to any extent. This occurs at temperatures of the order of 770°C for ½CrMoV corresponding to the intercritical temperature range between Ac_1 and Ac_3; the structures produced are said to be decorated. Finally, more remote from the weld in the regions where lower peak temperatures are found, some degradation of the structure occurs, mainly resulting in increased vanadium carbide precipitation. Optically, the microstructure appears very similar to that immediately adjacent to the parent ½CrMoV material which contains ferrite and about 5% bainite, produced by the original normalising heat treatment. The range of structures arising from the deposition of a weld and their distributions are illustrated schematically in Figure 8.

4.3.2 Transformation Products

The other main feature of structure produced by welding is the complexity of the transformation products formed within the austenite cooling. Essentially these are meta-stable reaction products whose formation depends critically on the composition of the original material and on cooling rate. The ranges of temperature and times over which they are produced may be found from continuous cooling transformation data. An example is given in Figure 9. It can be seen that, in the particular case of ½Cr½Mo¼V, transformation to ferrite and/or bainite are possible. However, in practice, cooling rates for welding from temperatures near the melting point to below about 700°C are so rapid for thick section welds that the ferrite field is avoided and bainitic structures are obtained. In general, as the heat input and preheat temperature are increased and the section size decreased, the slower the cooling rate becomes and the more likely is the formation of the higher temperature, softer, transformation product. This has again been described in detail by Alberry and Jones [27].

Other factors also affect transformation. For example, it is easier to initiate transformation products from fine grained austenite than from coarse; the general effect being to shift the phase field to higher temperatures and shorter times. Such CCT diagrams are therefore specific to a particular prior austenite grain size. Also, as can be seen from Figure 9, the transformation start and finish temperatures are cooling rate sensitive. Thus the variety of potential structures is large [28,29]. Finally, transformation behaviour is

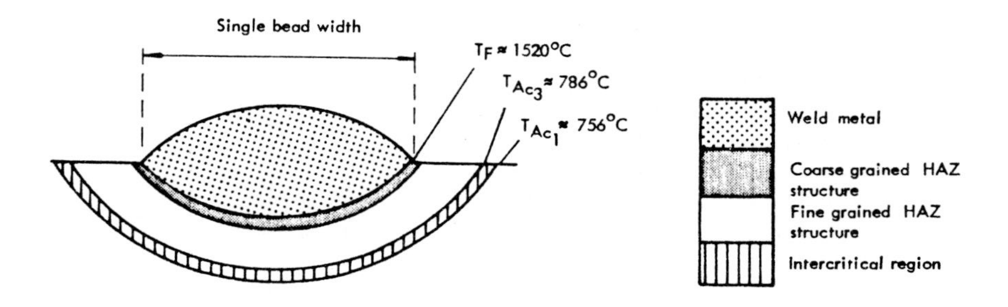

Fig. 8. Schematic representation of the distribution of structures of the HAZ of a weld bead.

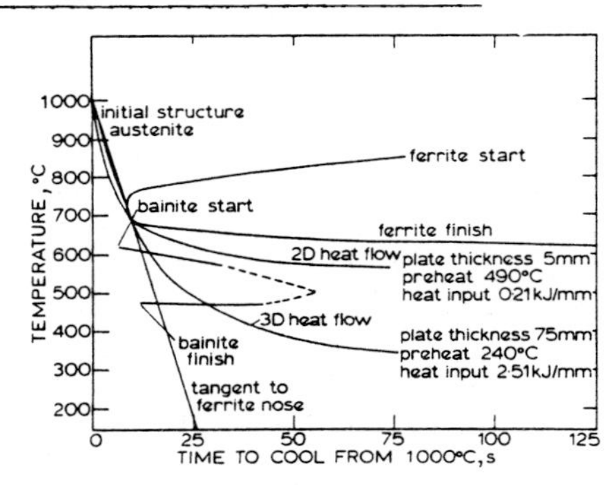

Fig. 9. Continuous cooling transformation diagram; fine-grained simulated 0.5CrMoV HAZ. [27]

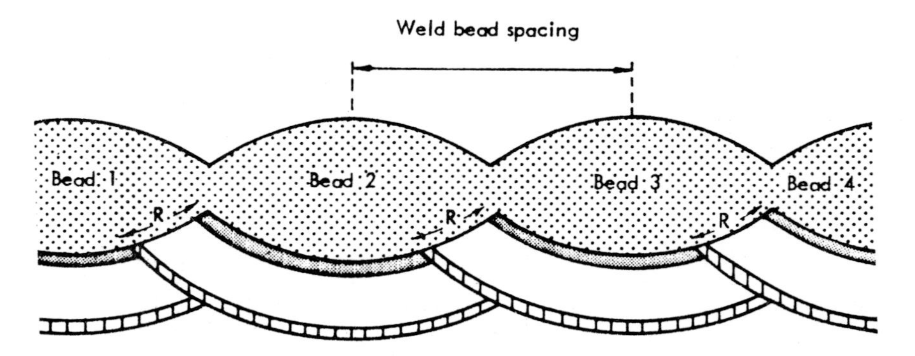

Fig. 10. Schematic representation of the distribution of structures resulting from the deposition of weld beads in sequence. [33]

strongly dependent on alloy composition. For example, the $2\frac{1}{4}$Cr1Mo composition is much more hardenable than $\frac{1}{2}$Cr$\frac{1}{2}$Mo$\frac{1}{4}$V so that bainite structures are invariably produced for all practical welding conditions.

It can therefore be appreciated that the range of structures present in the weld vicinity is a complex function of material composition and welding procedure. Nevertheless, experimental information of the kind indicated above, taken with calculations of heat flow, can be used to construct an accurate picture of the metallurgical constitution of the heat affected material. For present purposes, it is not necessary to discuss compositional effects. Main alloying elements such as Cr, Mo, Mn, Ni and Si, etc., affect precipitate types and solution strengthening and, therefore, the mechanical and creep properties of the various classes of low alloy steels. However, microstructurally, they have little effect on grain size but affect the type of transformation product in much the same way as in parent materials.

4.3.4 Multipass Considerations

It remains now to consider how the above processes combine in a multipass weld. From the above discussion it can be seen that a single weld bead deposited on ferritic material will produce a range of microstructures in the HAZ, whose character changes predictably with distance away from the fusion boundary. In $\frac{1}{2}$Cr$\frac{1}{2}$Mo$\frac{1}{4}$V steel, for example, these might be coarse grained bainite which merges into fine grain bainite, followed by an intercritically transformed region and finally tempered parent material. In a multipass weld, each weld bead effectively heat treats the underlying material in a manner exactly as described above. Thus a periodic spatial distribution of structures will be developed in the plane normal to the welding direction, with a wavelength whose dimensions will be determined by, inter alia, the dimensions of the weld beads and the heat input arising from their deposition. This is shown diagrammatically in Figure 10. Local to the fusion boundary, therefore, pockets, or more properly, bands of different microstructures are created which may be visualised, roughly, as separate fibres running parallel to the welding direction. Apart from weld stop-start positions and fluctuations caused by the variance of the welding process itself, these bands will be virtually continuous in the welding direction, albeit meandering slightly as differences in level of the weld metal surface are accommodated. Actually the spatial distributions of these regions and, in particular, the extent to which coarse grain bainite structures are replaced by fine grain and other more ductile structures can be critical in fabrication. This is because during post-weld heat treatment (PWHT) cracking can occur in low ductility, low strength coarse grain bainite under the action of residual welding stresses [30]. This HAZ cracking has been a serious industrial problem and, in practice, one of the ways of eliminating HAZ cracking is to control weld deposition sequence to obtain fully refined structures. There are also longer term creep implications, since incipient HAZ cracking may also impair long term integrity.

Further away from the fusion boundary, the thermal field becomes more diffuse and the relatively sharp distributions of structure identified with each individual weld bead are lost. Thus, the far field structure of the HAZ will approximate to a succession of narrow bands lying roughly parallel to the mean position of the fusion boundary, which are continuous through the material. This feature may have important consequences for component performance, if the properties of such regions are inadequate; an example of this is the so-called Type IV cracking which is discussed later, which is located typically ∿ 10 mm into the parent material in the intercritical temperature region.

4.4 Ferritic Weld Metal Microstructures

Within the body of a multipass weld, the effects of interacting thermal fields which heat treat adjacent and underlying weld beads is entirely analogous to that in the HAZ. However, microstructurally, whereas coarse and fine grained regions etc.,are produced immediately adjacent to each deposited bead, the original as-deposited columnar grain structure is only partially eliminated and often comprises the bulk of the weld metal volume. Again the structure is periodic but in this case the properties of the original weld metal are directional and, often inferior to those of refined material. The columnar and coarse prior austenite grain boundaries may act as sites for crack initiation by various mechanisms during cooling and later, such as solidification cracking [12,31] and hydrogen cracking [5,32]. They are also prone to temper embrittlement. Importantly, for high temperature applications, creep cavitation and cracking can occur during post weld heat treatment and early service life when tempering is incomplete and when residual stresses are present. Also, creep life in the very long term tends to be determined by the development of creep damage at these boundaries. This is discussed more fully later.

A section through an MMA multipass weld is given in Figure 11. The example shown is actually an experimental weld made in C-Mn steel and has been used because it can be etched readily to show refined and columnar regions; this difference is not very obvious macroscopically in 2CrMo weld metal. It can be seen that the coarse columnar regions comprise about half of the total volume. In general, more extensive, coarser weld metal areas are produced with large gauge electrodes so that for creep applications a balance has to be struck between the high rates of weld deposition required

x 2

Fig. 11. Experimental MMA ferritic steel weld showing columnar and reheated microstructural areas typical of multipass welding.

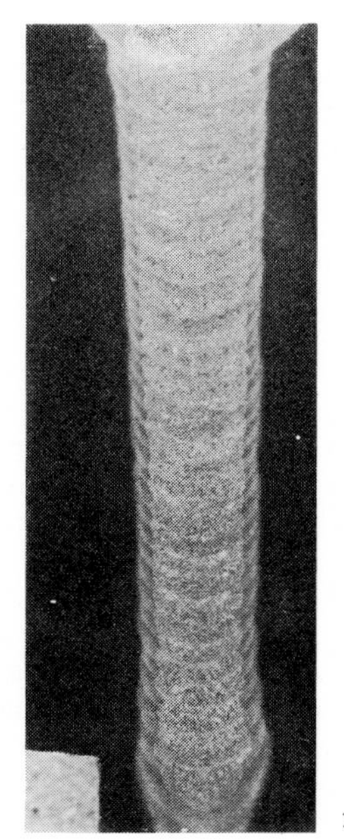

x 2

Fig. 12. Narrow gap TIG weld illustrating thin weld beads and low total volume of weld metal.

by economic considerations and the need to limit the extent of possible microstructural weaknesses. Recently, narrow gap processes have been developed and are being applied in the construction of power plant. Such a weld in $2\frac{1}{4}$Cr1Mo is shown in Figure 12, where it can be seen that this particular TIG variant has succeeded in producing rather thin weld beads which completely span the weld preparation. The result is that both original weld metal and HAZ are completely refined producing structures that are resistant to the various possible cracking mechanisms and ductile in creep. This is a good example of how welding procedure can produce markedly different structures in two, nominally, very similar weldments. Nevertheless, it should also be appreciated that the deformation patterns that arise during creep of these two welds could also be significantly different as a result of the different macroscopic shapes of the weld preparations and the relative volumes of weld metal deposited.

4.5 General Description of Weldment Microstructures

The preceding sections have sought to demonstrate that heat flow theory can be used to specify the thermal history of material during weld metal deposition, from a knowledge of the welding process parameters. The metallurgical structures that arise at any given position can then be related to the thermal history using the basic physical metallurgy to describe grain growth kinetics and phase transformation behaviour. The final step required to relate these two elements and so provide a general method for determining weldment microstructures is to characterise the weld bead shape. A theoretical treatment is not available to do this, but the information can be obtained empirically by depositing weld beads on plates under various conditions and relating bead dimensions to welding parameters. This has been done by Alberry and Jones [33] for MMA welding and has been extended to the TIG process by Alberry, Brunnstrom and Jones [34]. This development takes the form of a computer model that outputs the spatial distribution of single and multipass HAZ structures as a graphical display, thereby allowing the relationships between microstructure and welding process parameter to be explored systematically and rapidly. This is an extremely powerful technique which enables welding procedures to be optimised without the necessity of carrying out large numbers of expensive and time-consuming welding trials. Also, such detailed knowledge can be used to manufacture specimens for mechanical and other property determinations, so that the likely performance of the weldment in service can be evaluated with confidence. The method require further development and needs to be extended to other fusion welding applications but promises to provide a sound basis from which welding procedures, weldment structures and properties can be evaluated.

5. RESIDUAL STRESS

The previous sections have dealt with welding and how it can affect the metallurgical character of ferritic and austenitic steels. However, welding may also give rise to residual stresses which may necessitate PWHT. Subsequently, both factors may influence behaviour in service under creep conditions. This section deals with residual stresses in welds and later sections then consider how weld metallurgy and stress environment affect creep behaviour, deformation and cracking in service.

The build up of residual stresses in a weldment can be visualised very simply, although complete characterisation of the 3-dimensional distributions in welds is difficult experimentally. In addition, computer numerical analysis can be prohibitively complex, time consuming and expensive. For simplicity here, a circumferential weld fabricated in a cylindrical pipe geometry is considered. During welding, a band of heated, molten metal is applied around the cylinder. During cooling, this band shrinks relative to the material away from the weld and can be considered to act much as a tourniquet[35]. On further cooling, the weld metal tightens on to the pipe generating thermal stresses both in itself and in the surrounding pipe. The magnitude and distribution of the thermal stresses will be related primarily to the relative displacements of the weld and parent material during cooling. These will be a function of the temperature differences between the parent and weld materials, the coefficient of thermal expansion differences between the parent and weld materials, and the temperature sensitivity of the coefficients of thermal expansion. In addition, any metallurgical changes in the weldment material which result in volume changes during cooling will affect the residual stresses generated.

The build up of these stresses will now be considered. Firstly, the idealised system of a uniaxial sample, held at a fixed displacement at elevated temperature and subsequently cooled will be examined to identify the effects of the physical and metallurgical factors. Secondly, experimental measurement and analyses which have been carried out on welds of specific geometry will be examined to characterise magnitude and distribution of the stresses. Finally, the effect of PWHT on stress distribution and magnitude is considered. The emphasis throughout is on butt welds fabricated in cylindrical pipe sections as these are appropriate to the general theme of the paper. However, the basic principles are relevant to all welds. It is stressed that this review is not intended as a critical appraisal of residual stress calculational methods in welds but is intended to act as a basis from which realistic residual stress magnitudes and distributions may be visualised as one data input in assessing the creep behaviour of welded materials.

5.1 The Cooling of a Material Constrained to a Fixed Displacement at Temperature

A relatively simple picture of the factors affecting residual stresses in weldments is now developed before considering analytical solutions obtained for real weldments. This will allow the effect of materials' properties to be identified and understood. The model consists of a bar of material, locally heated at the centre so that a symmetric temperature gradient exists along the bar. Following Jones and Alberry [36], the displacement is fixed at some high temperature and the system is cooled. A stress free state exists at the starting temperature and the stress to maintain this fixed displacement can be monitored. On cooling, the bar attempts to contract and a tensile stress is generated. For a stable, transformation free material, the stress will be a function of:

$$E\alpha\Delta T$$

where E is the elastic modulus, α is an average coefficient of thermal expansion and ΔT is the cooling temperature range.

On further cooling, stress will increase until the material's flow stress is reached. This process is shown in Figure 13 for Type 316 material, which is essentially single phase [36].

On cooling, ferritic steels undergo phase transformation from austenite and the transformation products and the temperature range over which they occur are composition, peak temperature, cooling rate, and grain size dependent [36]. Such processes have been discussed earlier. As transformation, for example to bainite, occurs, a volume expansion is generated locally over the temperature range of the transformation. This volume expansion will mechanically relax the stresses built up during cooling. When the transformation is complete, further cooling will generate additional thermal stress, again due to coefficient of expansion effects. Stress generation will cease when the weldment and surrounding pipe have reached uniform temperature; in welding this will correspond to the preheat or interpass temperature.

This model was examined experimentally by Alberry and Jones [36] using a weld simulator coupled to a servohydraulic loading system which allowed accurate stress measurement during cooling. The steels tested ranged from C-Mn to 12CrMoV so that a variety of transformation start and finish temperatures was covered. Detailed analysis of the data which is presented in Figure 13 showed that predictions of stress reduction due to expansion during transformation were qualitatively correct. However, further calculations indicated that significant compressive stresses should have been developed but these were not

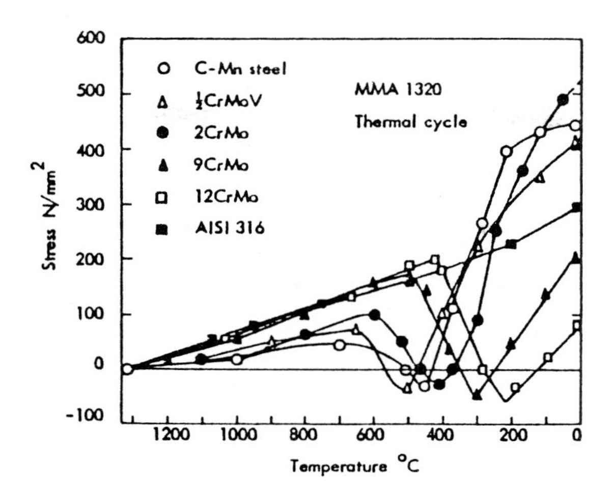

Fig. 13. Stress accumulated during cooling under restraint for several steels.[36]

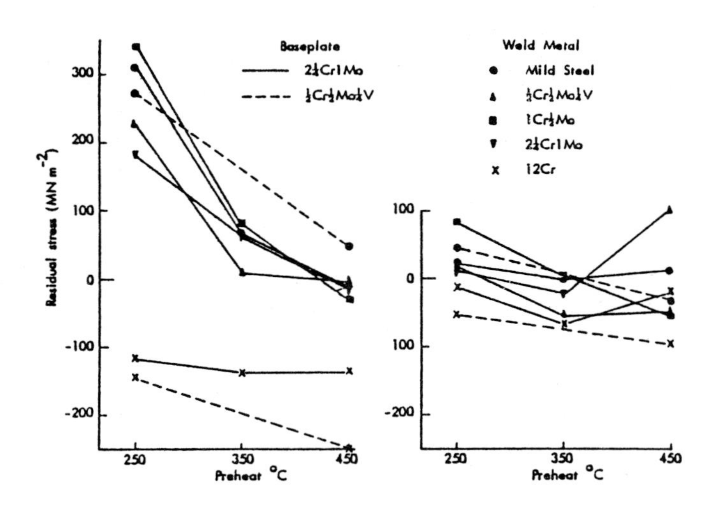

Fig. 14. Effect of weld metal type on residual stress in single pass welds.[38]

found. Indeed, only small compressive loads could be detected. Jones and Alberry [36,37] attributed this to the effect of transformation induced plasticity which is known to occur during phase transformations. Under these conditions extensive plastic flow can take place at low stress. With a fixed displacement applied to the sample, the stress build-up recommences only after transformation is complete.

Thus, the simple model suggests that in ferritic steels, the stresses built-up during cooling to the transformation start temperature are first reduced to near zero by the volume expansion associated with the transformation and then remain at this level because of transformation plasticity effects. After the transformation ceases, stress build-up recommences, due to α differences and occurs over the temperature interval from the transformation finish to the ambient temperature, which in welding is to the preheat or interpass temperature. The magnitude of stress is of the order of $E\alpha\Delta T$ and for low alloy ferritic steels corresponds to 0.1 - 0.2% strain. Residual stresses in the weld metal of the order of the ambient temperature flow stress are therefore expected. For high alloy ferritic steels, lower stresses are predicted.

These predictions were examined [38] for a series of single pass bead-on-plate welds. The weld metal compositions were C-Mn, $\frac{1}{2}$Cr$\frac{1}{2}$Mo$\frac{1}{4}$V, 1Cr$\frac{1}{2}$Mo, 2$\frac{1}{4}$Cr1Mo and 12Cr1MoV which give transformation finish temperatures in the range $\sim$ 500°C - 200°C; the measured stresses are given in Figure 14. A similar series of experiments [38] was carried out with 2$\frac{1}{4}$Cr1Mo weld metal using a series of preheat temperatures from $\sim$ 20°C to 400°C and these results are shown in Figure 15. As can be seen in Figures 14 and 15, high residual stresses are obtained for the low alloy steels and these stresses decrease with increasing preheat temperature. For high alloy steels, values are compressive. Thus, the lowest stresses are generated with the combination of weld metal composition and interpass temperature where there is a minimum difference between transformation finish and interpass temperatures. Based on this argument, the 12CrMoV weld metal and 200°C interpass or the 2$\frac{1}{4}$Cr1Mo weld metal and 500°C interpass combinations generate the lowest stresses as shown. For the non-transforming austenitic steel, later work by Fidler [39,40] has confirmed that the residual stress achieves the 0.2% proof stress value of the weld metal as there will be no compensating effect of either volume changes resulting from phase transformations or transformation induced plasticity.

Based on these simple but illuminating concepts, residual stresses present in full size, cylindrical pipe butt welds are now considered. Two aspects of residual stress distribution are examined; firstly, the macroscopic distribution across the complete weldment and secondly, the local stress fields which are influenced by the microstructural changes that occur

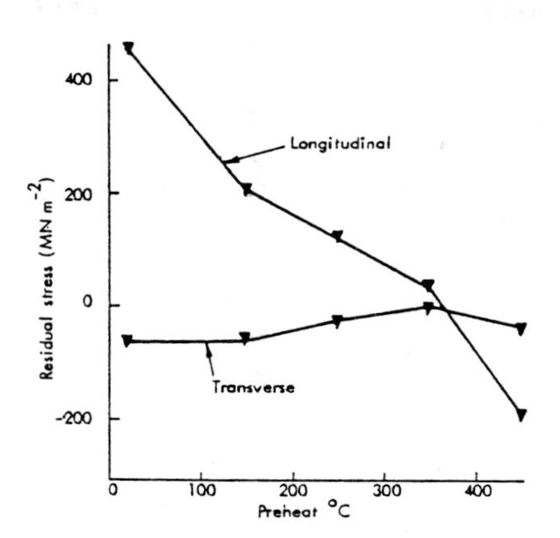

Fig. 15. Effect of preheat on residual stress in single pass 2CrMo welds on $\frac{1}{2}$Cr$\frac{1}{2}$Mo$\frac{1}{4}$V plate. [38]

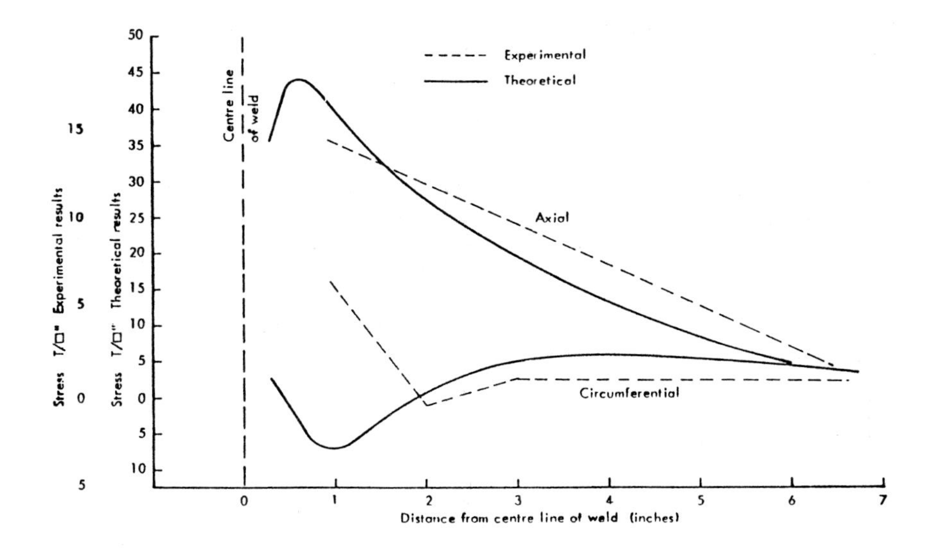

Fig. 16. Comparison of theoretical and experimental residual stresses in a thick section ferritic weld. [35]

in the reheated region of individual weld beads.

5.2 Stress Distributions in Cylindrical Butt Welds

In principle, the parameters considered in the previous section are important. However, the simple single bead phenomena are complicated for multipass welds by the constraints imposed by realistic geometries and successive interacting thermal fields generated by the welding process. In general, therefore, applying even the simplest approach to calculate stresses in real welds requires massive computational facilities generally using finite element or difference methods [41]. The following, selective, treatment attempts to outline the general principles and to illustrate some of the distributions that occur. These will act as a basis for their use later in discussions of creep behaviour.

5.2.1 Macroscopic Stress Distributions

Chubb, Fidler and Wallace [35] suggested the tourniquet model for representing the residual stress system in a welded pipe. They carried out a finite element analysis, in which each layer of weld metal was visualised as being instantaneously wrapped around the pipe at a high temperature. This weld metal then cooled building up thermal stresses both in itself and in the adjacent material. This process was repeated until the complete weld was modelled; the stresses in each layer being the numerical sum of the stresses arising from later layers. The results of the original analysis are shown in Figure 16. It should be noted that the general stress distributions agree reasonably well with experiment and surface stresses have the following characteristics:

1. High tensile hoop stresses are present in the weld metal and reduce with increasing distance from the weld into the parent material, reaching zero at some distance remote from the weld.

2. Tensile axial stresses arise from local bending and peak within the parent material adjacent to the weld.

This can be shown to be a sensible stress distribution by considering the simple elastic loading generated by shrinking a ring on to a cylinder.

In general, experimental measurement is limited to surface stress distributions and it is essential that through thickness distribution also be quantified. Fidler [42], using a combined measurement and analytical method, evaluated the complete through thickness residual stress distribution in a thick section as-welded ferritic pipe. In essence, the outer surface residual stress distribution was measured using a trepanning technique which produced average stresses over a radial thickness of ∿ 10 mm. The outer layer was then removed by machining and the stresses measured in the new surface. This sequence was repeated incrementally through the pipe wall. On completion, the section was reconstructed by a finite element technique, by applying surface forces to each ring of metal in turn to regenerate the measured loading. Using this method the complete through thickness macroscopic stress distribution was obtained. Typical results are shown in Figures 17 and 18, which give the stress distributions across the centre line of the weld and at selected depths. The main points to note are:

1. High tensile hoop stresses are generated in the weld metal outer surface.

2. The magnitude of the hoop stresses reduces towards the pipe bore, with zero stress approximately two thirds of the way through the thickness and compressive stresses at the bore. The stress distribution is approximately linear.

3. Tensile axial stresses of somewhat lower magnitude are found in the weld:parent interface region at or near the outer surface. They are compressive at the bore.

4. The radial stress component is zero or slightly compressive.

Such a distribution is compatible with a tourniquet model.

In addition, it highlights one other very important feature, namely that equi-triaxial stresses do not arise anywhere in the system. In this case, the stresses are essentially biaxial in nature.

5.2.2 Microscopic Stress Distributions

The above work examined macroscopic stress distributions and utilised the trepanning method [43], diameter ∿ 10 mm, for stress measurement. An alternative technique referred to as the centre hole drilling method [44,45] is capable of measuring the average stress over a diameter of ∿ 2 mm. Using this technique, finer detail associated with individual weld beads can be resolved. For example, work on as-deposited welds in heavy section ferritic pipe has shown that in the weld metal, the residual stresses are not uniform but vary significantly with position relative to the weld beads [46]. Investigations initially reported by Cheetham, et al., [2] and later by Hepworth [47] showed conclusively that these stress variations were systematic. The lowest stresses were found at the weld bead centres, whereas the highest stresses were found in the region of weld metal subjected to the thermal cycle imposed by deposition of the subsequent weld bead, i.e. in the HAZ of the

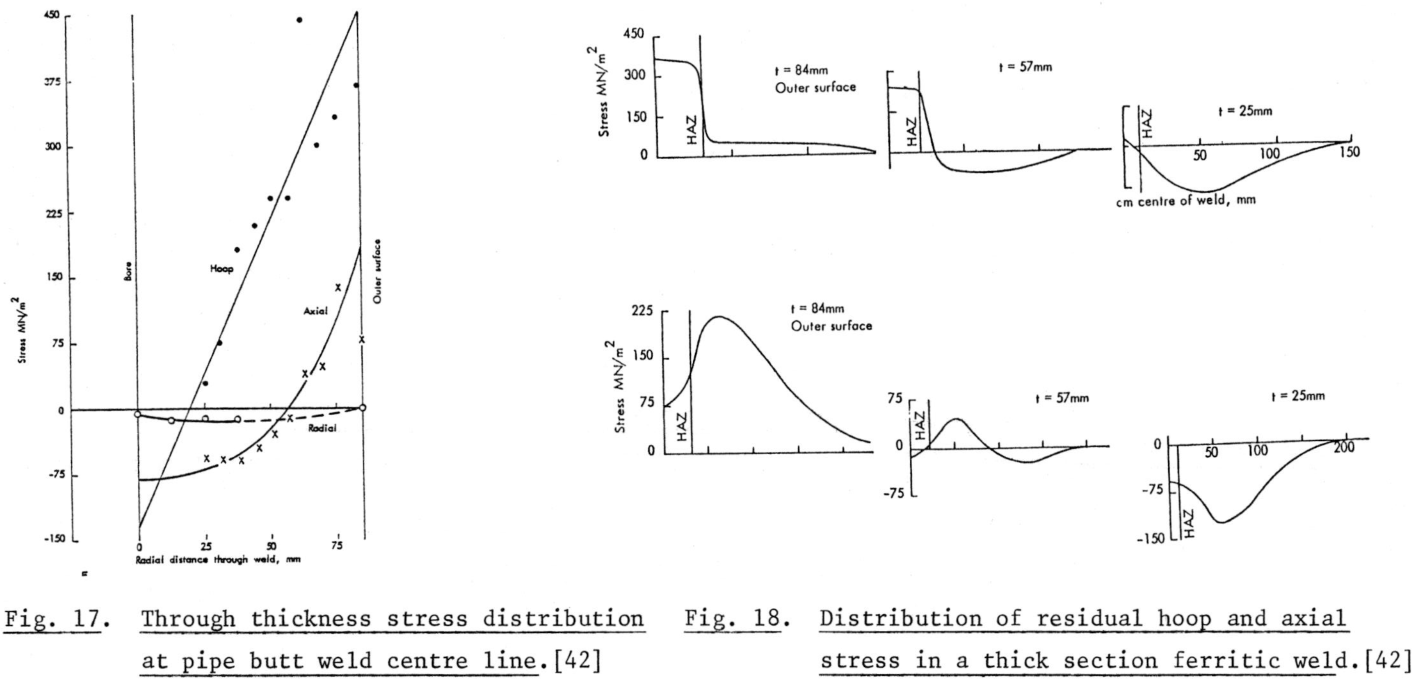

Fig. 17. Through thickness stress distribution at pipe butt weld centre line.[42]

Fig. 18. Distribution of residual hoop and axial stress in a thick section ferritic weld.[42]

bead. Figure 19 taken from the work of Hepworth [47] illustrates this for a series of overlapping SA bead-on-plate welds, but the effect is general and has been confirmed for multipass butt welds in ferritic pipe. The periodicity is readily seen. A second characteristic distribution [48] is found in the direction of weld bead deposition. Initially, the stress is high at the weld bead start position, becoming lower as welding proceeds and high stresses recur at every welding restart position.

Such investigations have confirmed that the characteristic features of the residual stress distribution in as-deposited butt welded ferritic joints are a macroscopic stress distribution, generally consistent with a simple tourniquet model, with a microscopic peak and trough distribution superimposed on this. It is apparent therefore that when residual stresses are measured, particularly in the weld metal, care must be taken to identify correctly the position of measurement relative to the metallurgical structure.

Similar residual stress measurements have been made in fully austenitic Type 316 weldments in thick plate [40]. These results while showing the macroscopic distributions expected from consideration of the distortion changes resulting from cooling after welding, do not show the microscopic peak and trough variation characteristic of ferritic welds. This is not unexpected because Type 316 weld metal is essentially single phase and will not show either the effects of volume expansion or the transformation plasticity associated with ferritic materials during cooling.

It is believed that the results and conclusions indicated above are reflected in through thickness distributions. However, currently, no methods exist to establish experimentally the subtle effects of variance in welding parameters such as preheat and cooling rate or any systematic changes which occur during the deposition of multipass welds.

5.3 Calculation of Residual Stresses in Weldments

In parallel with experimental work, many attempts have been made to calculate residual stress build-up in weldments by taking into account the welding sequence and materials properties. These are well represented in the literature by the work of Masubuchi and co-workers [49] and of Rybicki and co-workers in the USA [50], Ueda [41,51] and Hepworth [52] in the UK. The method used is as follows:

1. The effect of welding procedure and conditions on the thermal fields generated in the parent and weld metal during welding are determined. Correct description of the thermal transients, gradients and maximum temperatures as a function of time is essential. Good agreement has been reported on this

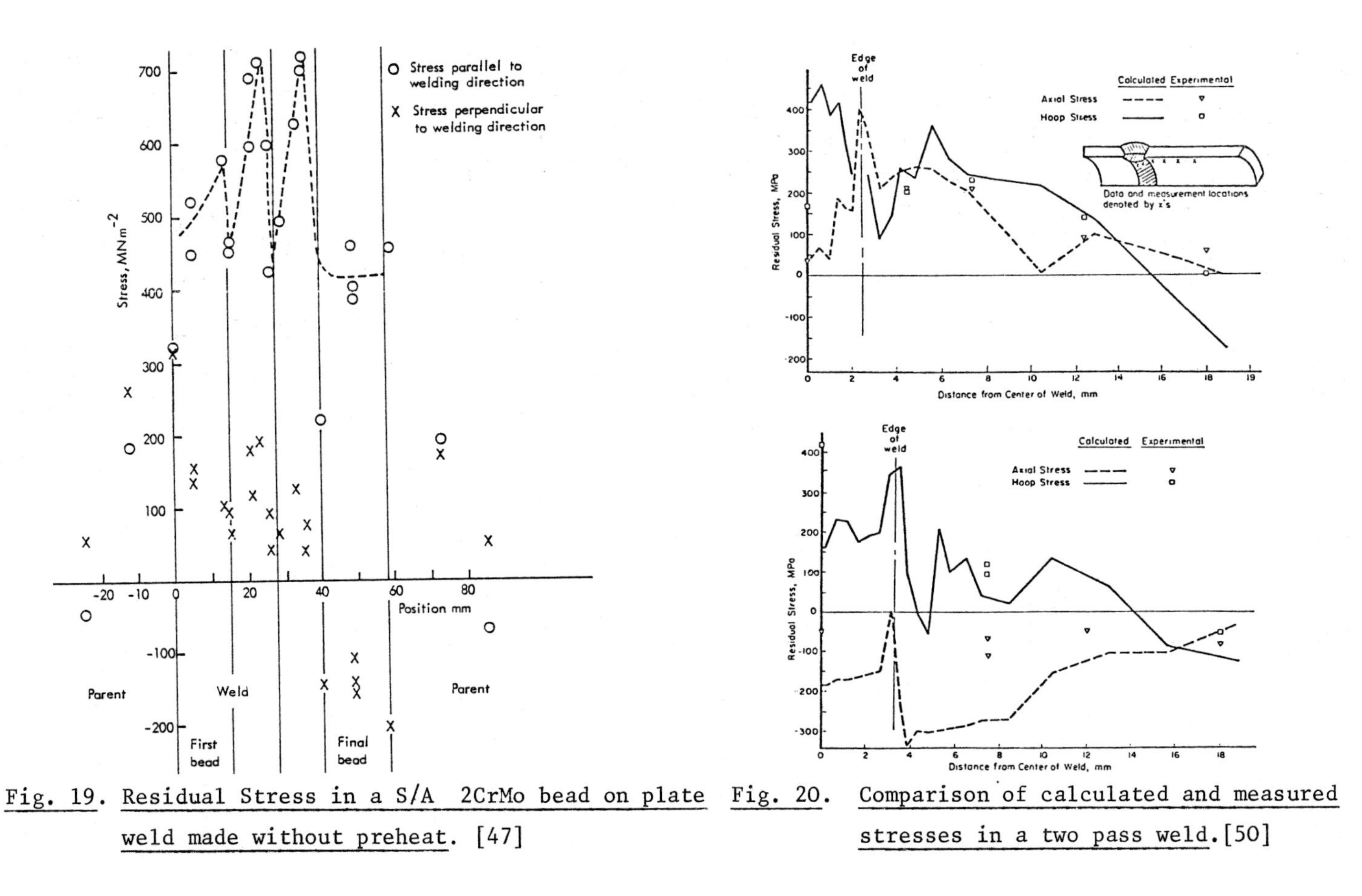

Fig. 19. Residual Stress in a S/A 2CrMo bead on plate weld made without preheat. [47]

Fig. 20. Comparison of calculated and measured stresses in a two pass weld.[50]

aspect between calculated and measured temperature distributions for controlled welding conditions [49,50,51,53]. The effects of changes in the welding parameters, in particular, the presence of heat sinks has also been studied [54].

2. The required uniaxial mechanical and physical property data, plus the creep constitutive relationships for the relevant materials in the weldment are established. This is no mean task and data are available for a limited number of systems only; primarily these are austenitic [50] although more data are becoming available for ferritic materials.

3. The temperature and physical, mechanical and creep property data are then used in an iterative manner, in finite difference or finite element programmes, to calculate the residual stresses.

The size of such a task, albeit set out very simply above, should not be underestimated. In practice approximations, such as the definition of a temperature, say 750 - 850°C, above which the materials cannot sustain load, are often incorporated to facilitate a solution.

The best documented example of this approach in the open literature is that produced by Rybicki and co-workers [50] for austenitic 304 type weldments. Figure 20 taken from their work, illustrates the comparison between calculated and measured stresses. Clearly, there is some room for improvement but such procedures show promise. Similar comparisons have been reported for ferritic welds [52]. However, they do have one inherent advantage over other methods, namely, the possibility of computer simulation allowing optimisation of a welding process to minimise or modify the residual stress distributions. This was an important issue in the Rybicki programme, which was undertaken to determine methods of reducing residual stresses locally so that stress corrosion cracking could be minimised. A proposed modification of the heat transfer process during welding was identified and allowed one practical solution to be suggested [54] although there were other optional routes.

5.4 The Effects of Post Weld Heat Treatment

For many low temperature and high temperature applications the design codes require that post weld heat treatment, is carried out. In practice, this entails heating a weld, in a controlled manner to minimise thermal stresses generated by temperature gradients, to a temperature at which the welding residual stresses can be relaxed. The weld is maintained at this temperature for a predetermined time before cooling to ambient, again in a controlled manner. Such procedures are commonly required for heavy section welds although they are not usually necessary for fabrications of less than 10 - 12 mm thickness. Historically, temperature and time are set from examination of the uniaxial stress relaxation behaviour of the materials. However, there are two effects of PWHT that must be considered. The first, and most obvious, is the stress relaxation requirement and the second is the effect of tempering on the constituent parts of the weldment. This latter factor is important as both beneficial and detrimental effects can be found under different combinations of time and temperature in the weld metals and HAZs. For example, heat treating a $\frac{1}{2}Cr\frac{1}{2}Mo\frac{1}{4}V$ steel at 690 - 710°C can improve the ductility [55] whereas heat treating $2\frac{1}{4}CrMo$ weld metal at $\leqslant$ 650°C can lead to creep embrittlement and to secondary hardening [56]. Similarly, for austenitic weldments, the PWHT temperature chosen can lead to different precipitation sequences and consequent changes in properties [57,58,59]. Commercially, the lower the PWHT temperature, the cheaper the procedure and this is the main driving force for minimising PWHT temperatures. For clarity here, both stress relaxation and tempering are treated independently initially although later, it will be seen that practically, this separation is not viable.

5.4.1 Relaxation of Residual Stresses

The flow stress of most materials decreases with increasing temperature and thus, during initial heating of a weldment, both macro- and micro-stresses must revert to the flow stress value at temperature. This procedure should be virtually instantaneous with subsequent relaxation occurring by creep processes. At sufficiently high temperatures, both processes will occur concurrently and will be indistinguishable. Thus, some stress relaxation will occur during the heating period.

It is important to recognise that residual stresses are the manifestation of local elastic displacements so that, in general, the strain to be relaxed is fixed. However, as welds are inhomogeneous some local stress and strain redistribution may occur during stress relaxation and should be taken into account in any exact analysis. More generally, such elastic follow-up processes effectively increase the local displacements available for relaxation and arise from longer range compatibility considerations.

Uniaxial stress relaxation, by creep, can be adequately represented by a time:temperature parameter, for example, the Larson-Miller parameter, $T[20 + \log_{10}t]$,

where T is the temperature in K
and t is the time at temperature

although other parametric options can be used. The original Larson-Miller parameter is derived from the Norton steady state creep rate law, $\dot{\varepsilon} = A\sigma^n$ and the Monkman-Grant relationship, $\dot{\varepsilon}t_f$ = constant,

where $\dot{\varepsilon}$ is the steady state creep rate
σ is the applied stress
A and n are material dependent constants.

In practice, the constant, 20, has been chosen so that, within experimental error, the relationship holds for a wide number of materials. However, it should be noted that a linear relationship is not required, *a priori* by the Larson-Miller parameter, merely that a complete family of stress, temperature and time curves can be collapsed on to one master line using this parameter. This approach may be applied practically to the initial stages of stress relaxation at high temperature provided the thermally activated component of flow stress data can be incorporated, for example, as a first order correction for strain rate effects via the time to attain a fixed strain. This is illustrated in Figure 21 for Type 316 materials using data from a number of sources. It can be seen that both, initial, flow stress dependent and long term creep dependent, stress relaxation can be described satisfactorily by this unifying approach.

Such relationships are based on uniaxial data whereas in real weldments, the stress fields are three dimensional and complex. Creep relaxation under multiaxial stress conditions has been examined by many workers, for example, Finnie and Heller [60] for biaxial fields and Mackenzie [61] for full triaxial fields. As shown by Mackenzie, stress relaxation by creep cannot occur in an equi-triaxial field. However, in a biaxial field the relaxation behaviour, although slower, is not markedly different to the uniaxial case, see Finnie and Heller. An early comparison produced by Chubb, et al., [35] supported the latter prediction. In addition, Fidler's work on the complete through thickness distribution of residual stress in a heavy section ferritic weld [42] has shown that the radial stress was zero or slightly compressive. Therefore, it appears that the extreme case of equi-triaxial tension is not encountered in welds so that uniaxial or near uniaxial approximations should be valid.

5.4.2 Stress Relaxation of As-welded Joints

Experimentally, the effect of PWHT has been studied extensively, [62,63] particularly for low alloy ferritic steel welds. However, following Fidler and co-workers, the results can be summarised as follows:

1. On heating, the local stresses revert to the flow stress values at the particular temperatures and the microscopic peak and trough distribution is rapidly smoothed out by both flow stress and creep relaxation effects. Thus, the absolute values of residual stress reduce and the distribution becomes more uniform [46].

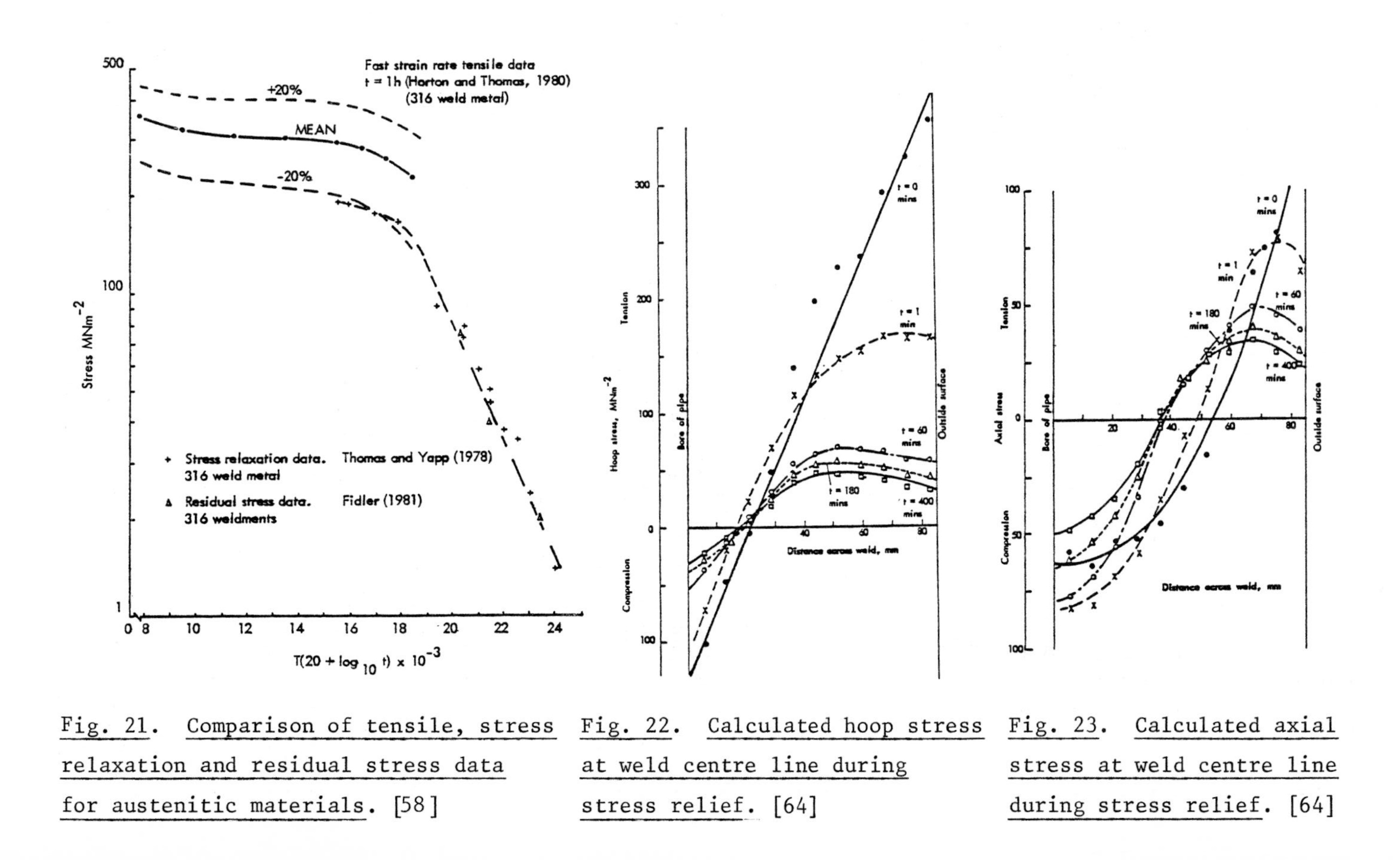

Fig. 21. Comparison of tensile, stress relaxation and residual stress data for austenitic materials. [58]

Fig. 22. Calculated hoop stress at weld centre line during stress relief. [64]

Fig. 23. Calculated axial stress at weld centre line during stress relief. [64]

2. With longer time at temperature, the stress relaxes further by creep with the stress corresponding, to a first approximation, with predicted parametric time-temperature behaviour.

3. The hoop to axial stress ratios, which were previously fairly random, converge to unity.

These results were obtained from surface measurements only and it is essential that the through thickness behaviour is also characterised. As a continuation of work described above on the as-welded stress distribution in a thick walled butt welded ferritic pipe, Fidler [64] carried out a finite element calculation of the effect of a conventional PWHT at 700°C on the stresses existing in the joint. The hoop and axial stresses across the centre line of the weld are shown in Figures 22 and 23 respectively, as a function of time at temperature. The data fully support his other results that initial fast relaxation is followed by a much slower stress reduction. In this case, the dominant deformation mechanism was creep. In addition, the maximum stresses existing in the weld were consistent with predictions based on Larson-Miller and during relaxation, the hoop to axial stress ratio tended to unity.

An equivalent series of surface measurements were made[40] as a function of time and PWHT temperature for Type 316 weldments, fabricated to a patch in plate geometry. The results are summarised [40] for two temperatures in Figures 24 and 25. Again, the characteristic pattern of a rapid decrease in stress followed by more limited stress reduction with time was observed. The distribution became more uniform with time and within experimental error, the peak stresses were $\sim$ 0.2% flow stress values and consistent with Larson-Miller parameter predictions. There are numerous other publications in the literature which support these general concepts, although few provide the near complete picture given by Fidler's work.

5.4.3 Microstructural Changes during Post Weld Heat Treatment

Concurrent with the stress relaxation process, the microstructures in welds are modified by time at temperature. Primarily this occurs because non-equilibrium structures are formed in both weld metal and HAZ materials during the welding process. The tempering reactions in most commercial ferritic and austenitic steels are complex. In the case of low alloy, ½CrMoV:2¼Cr1Mo weldments, the main reaction envisaged is carbide precipitation in the bainitic phases and modification of the carbide composition by transfer of transition elements. Full descriptions of the possible reactions are given in references 65 and 66 for conventional ferritic materials whereas those occurring in, for example, 2¼Cr1Mo weld metal have been examined by Wolstenholme [67] and others [68].

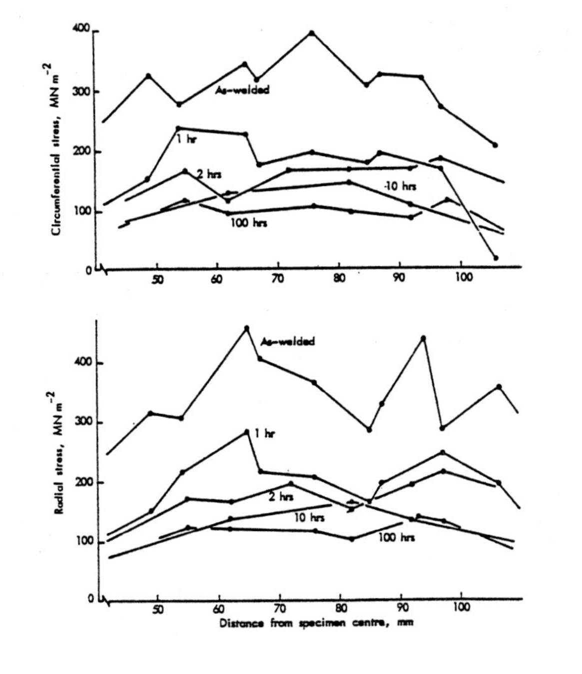

Fig. 24. Effect of heat treatment time at 650°C on residual stress in austenitic weld.[40]

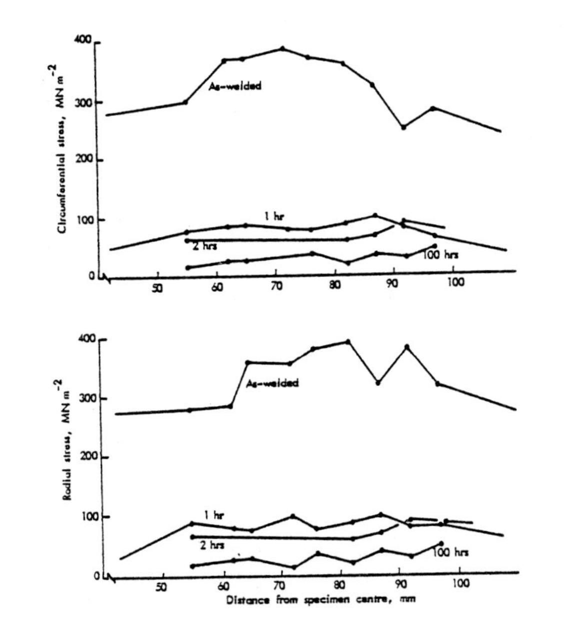

Fig. 25. Effect of heat treatment time at 750°C on residual stress in austenitic weld.[40]

In austenitic steels, the temperature of PWHT is not defined in the codes and, here, the reactions can produce $M_{23}C_6$, σ, χ and Lavé phases, particularly in austenitic weld metal which contains delta ferrite for weldability reasons, [57,58]. The incidence of these products is determined by the specific temperatures and times chosen and the kinetics of these reactions are now better understood for commercial weld metals.

It is not intended to describe the details of these reactions, merely to indicate their presence so that their effects on properties can be dealt with later.

Briefly, in ferritic materials generally, PWHT at recommended temperature results in softening, reductions in the flow and UTS values and improvements in ductility, of which the last is the most important. As both stress relaxation and tempering occur in parallel, creep strain can accumulate in regions of low ductility. It is therefore important that, during stress relaxation, the particular microstructural region accumulating the strain has sufficient ductility to sustain strain without cracking.

6. PROPERTIES OF CONSTITUENT PARTS OF A WELDMENT

The previous sections have shown that during welding, the weld metal and HAZ undergo local structural changes due to the heating and cooling cycles produced during weld deposition. These modifications are manifest as a range of microstructures across and through the weldment and form repetitive patterns dictated by the detail of the welding procedure. In principle, each structure may have different physical, mechanical and creep properties, when measured uniaxially in the laboratory.

Obviously, the scope for examining such properties is immense but, for the purposes of this paper, discussion will be focussed mainly on the behaviour of ferritic weldments. However, where available, data for other systems will be used to illustrate the generality of the effects.

It is convenient to consider the properties of the individual constituents of the weld in three groupings, namely weld metal, parent material and heat affected zone structures, particularly as not all weldments are fabricated with weld metal of identical composition.

Within these groupings, the more local structural variations will be discussed. Particular emphasis will be given to $\frac{1}{2}Cr\frac{1}{2}Mo\frac{1}{4}V$ ferritic steel welded with $2\frac{1}{4}Cr1Mo$ weld metal, which is used for conventional power plant within the UK for pipework operating in the range 540 - 565°C and at pressures which give a design life of up to 2 x 10^5 h. This weld metal is also used in the USA for fabrication of $2\frac{1}{4}Cr1Mo$ pipework.

6.1 Measurement Variability

Any measurement of physical, mechanical or creep property will be subject to variability and this has to be recognised and, where possible, incorporated into any analysis that attempts to assess long term behaviour. One of the more comprehensive discussions illustrating these effects in parent materials was held in 1966 [69] and is one of the few sources which identifies how the available data are incorporated into a design presentation. The causes of variation considered were laboratory to laboratory scatter [70], country to country scatter [71] and temperature measurement effects [72]. Obviously, in 1981, experimental methods have improved but it should be remembered that much of the long term creep data has been produced to earlier, possibly less exacting, standards. Examination of the 1966 data does show variation with test house location and with country. However, the deviation of creep rupture results for pedigree casts from the average value is adequately covered by a ± 10% stress factor for all data, particularly for > 3000 h life. Deviations on failure strain can, however, be much higher.

Another cause of variability, particularly with creep rupture data for the long times relevant to design, is the method of extrapolation used to derive rupture times at low stress levels. The study of extrapolation methods has been a prolific research field for many decades and several conferences and seminars have been held specially on this topic[73, 74]. There is no generally accepted solution available, at present, although certain methods have national support and are used to define design stress rupture relationships. However, it should be noted, that the different extrapolation methods give different life predictions and care should always be taken to ensure that a consistent data base is used when comparisons of different methods are made.

Finally, commercial steels are manufactured to conform to specified composition ranges. Thus, in addition to the mechanical causes of creep data variability discussed earlier, there is the inherent feature that the elements present can vary within an allowed range and can affect the measured low and high temperature properties. Variations in the component fabrication procedure, the amount of remaining cold or warm work and any subsequent heat treatment can also affect the measured values of properties.

Codes for plant design and construction recognise these factors and quite specifically identify the range of composition, heat treatment conditions and fabrication processes to which the data apply. The complete data base is then analysed and fitted to parametric relationships so that a mean line representing all accepted data can be produced. The effect of variability arising from the numerous possible causes is then

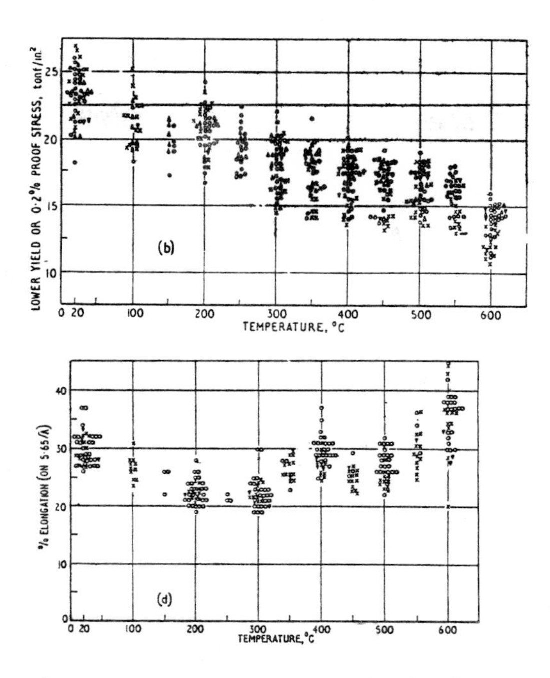

Fig. 26. Mechanical property data for $\frac{1}{2}$Cr$\frac{1}{2}$Mo$\frac{1}{4}$V material of different casts and products.[75]

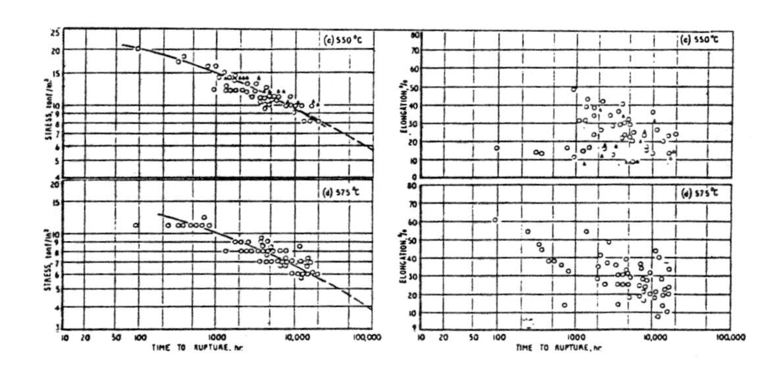

Fig. 27 Creep data for $\frac{1}{2}$Cr$\frac{1}{2}$Mo$\frac{1}{4}$V material.[75]

covered by the use of upper and lower bound limits, normally quoted as ± 20% of the stress level around the mean parametric value. An alternative approach, also commonly used, is to utilise mean data but covering the variability by incorporating a reduced level of allowed stress. This is the design factor approach referred to earlier.

For completeness, it is of interest to consider the actual data spread represented by a mean line for a variety of casts. The most comprehensive published data for ferritic alloys are given by Johnson, May Truman and Micklewraith [75] and are reproduced in Figures 26 and 27. These illustrate the variability for tensile yield strength and ductility and for stress rupture for ½CrMoV steel. It is emphasised however, that lower bound data are sufficient for design purposes.

In many other cases, such as remanent life estimation and for definitive benchmark experiments to validate analytical models, design approaches are inadequate because accurate placement within the error band is necessary. An example of the data spread from one cast of ½Cr½Mo¼V material [76], creep tested at a fixed stress is given in Figure 28a. Here, the results from creep tests carried out on 10 samples cut from a main steam pipe in the radial, circumferential and axial directions are shown. The results, albeit at high stresses, are presented in the form of a Weibull probability function, [77]:

$$\ell_n . \ell_n \left[\frac{1}{1\text{-Cumulative \% failures}} \right] = A . \ell_n . t_f + C \quad \ldots (9)$$

where t_f is the failure life and A and C are constants. The relationship is linear with a gradient of ∿ 3.5, corresponding to a Gaussian distribution function. The confidence limits are given by:

$$\pm L_{x\%} = \pm \frac{t_{x\%}}{\sqrt{n}} S$$

where S is the standard deviation, n the number of samples and t is obtained from tables for the required confidence level. In Figure 28b, the confidence limits are transposed on to the experimental data with some additional stress levels incorporated to define the stress rupture line. This demonstrates that even with single casts of material, some variability is still found. It further suggests that for multiple casts, the rupture data could be represented by a series of Gaussian distributions centred about the mean value for each individual case. However, such data are rare so that this argument cannot be extended at present.

Scatter is undoubtedly caused by compositional and fabricational factors, which implicitly control the microstructures present in the finished product. For ferritic steels, the problem is compounded by the complexity and kinetics of the

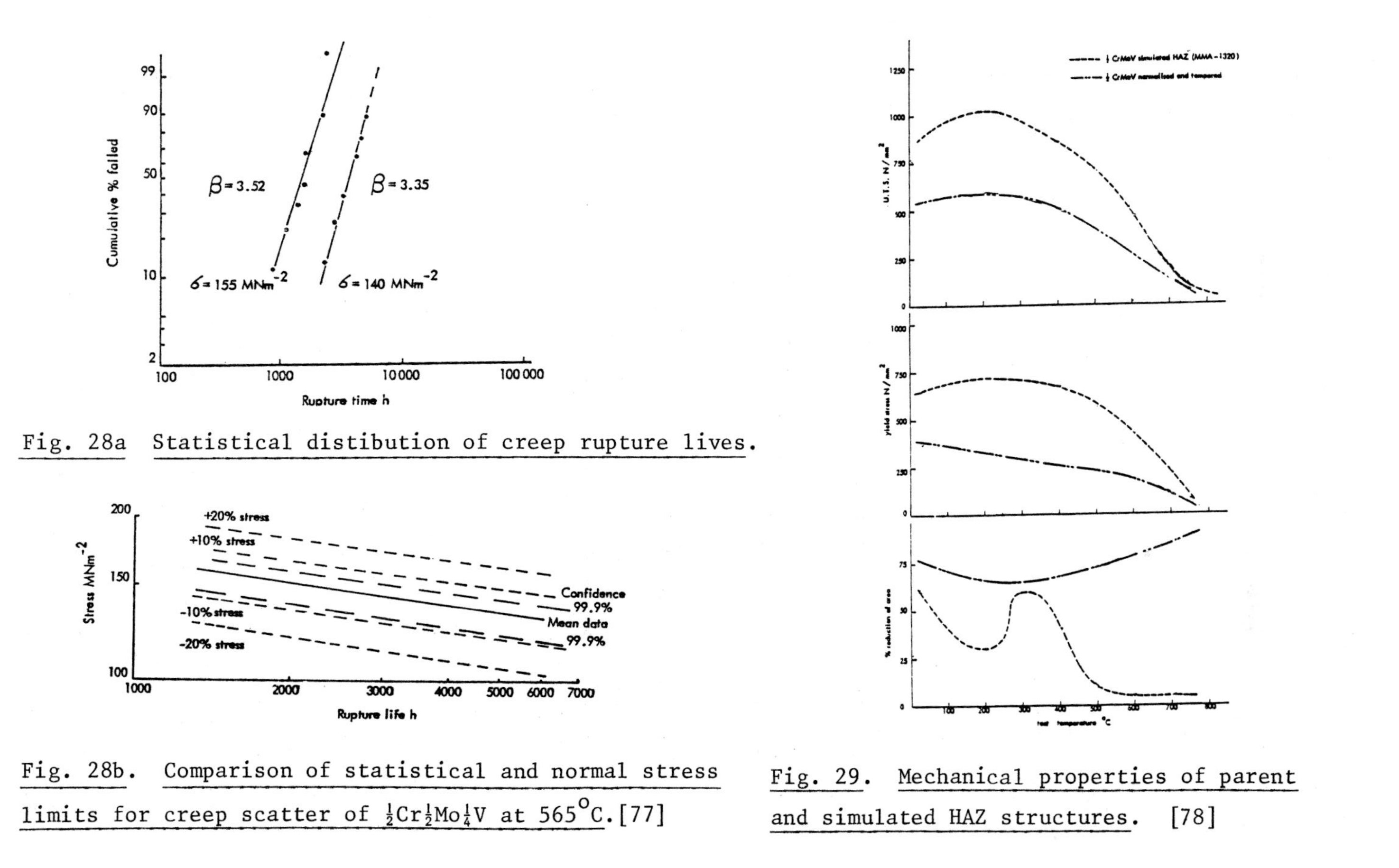

Fig. 28a Statistical distibution of creep rupture lives.

Fig. 28b. Comparison of statistical and normal stress limits for creep scatter of $\frac{1}{2}Cr\frac{1}{2}Mo\frac{1}{4}V$ at 565°C.[77]

Fig. 29. Mechanical properties of parent and simulated HAZ structures. [78]

phase transformation and tempering reactions. These are affected indirectly by many factors such as prior austenite grain size, cooling rate etc., and are difficult to define for a real component. However, it is just these structural differences that are important when welded joints are considered.

The procedure that will be used here for comparison of the properties of the individual microstructures associated with weldments is twofold. Firstly, where possible, the published data spread is identified as a base line for each property. Secondly, where more specific aspects of the microstructure are examined, additional results produced, if possible, from specific casts, are used to highlight the relative effects of the microstructural changes. This sequence is important as it is necessary to place the particular material within the more general scatter band before comparison is made for use in analysis.

6.2 Mechanical Properties

The data spread for ½Cr½Mo¼V materials has been shown in Figure 26. As the data includes samples taken from small and large component sizes and variations in heat treatment, it is difficult to use as a basis for comparison. However, Alberry and Jones [78] examined the effect of weld simulation cycles, and the consequent differences in metallurgical structure, on the mechanical properties of a single cast of ½Cr½Mo¼V. They used a thermal cycle to 1250°C to reproduce a coarse grained bainitic structure similar in grain size and hardness to that found in HAZ material. As shown in Figure 29, at all temperatures up to ∿ 600°C, the yield strength and UTS are increased over the parent material value by a factor of ∿ 2. Above ∿ 700°C, the differences are small. Comparable values for 2¼Cr1Mo weld metal, which is also fully bainitic, show a smaller increase but with the same basic characteristics. Of special note, however, is the ductility of coarse grained bainite, given here as reduction in area, which is very low when compared to the parent material, particularly above ∿ 500°C.

Near the HAZ:parent material interface, overaging and partial transformation of the parent material produces decorated ferritic or mixed ferrite, bainite structures. The properties at 550°C of specimens containing such structures are shown in Figure 30, [78] where they are compared with coarse grained and refined bainites. Refining the grain size or producing ferrite:bainite structures reduces the strength relative to the coarse grained variant but has a more significant effect on ductility. Similar effects are shown for 2¼Cr1Mo, and three casts of ½CrMoV, one with high residual element content.

The effect of compositional variation within the code specification has been investigated [79] for ½Cr½Mo¼V alloys with particular reference to the effect of trace elements,

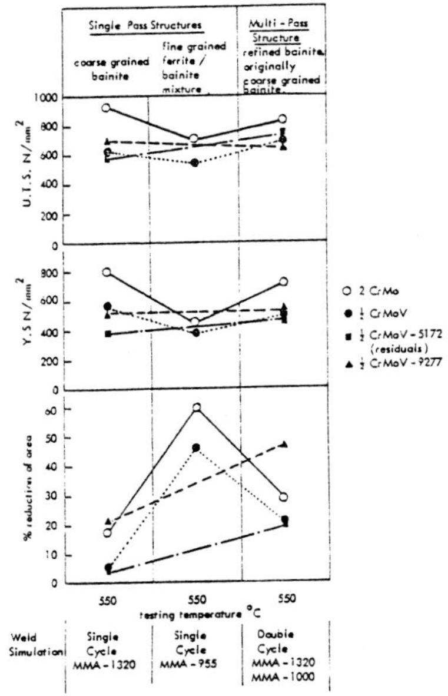

Fig. 30. Effect of refinement of coarse grain HAZ on mechanical properties of $\frac{1}{2}Cr\frac{1}{2}Mo\frac{1}{4}V$ [78]

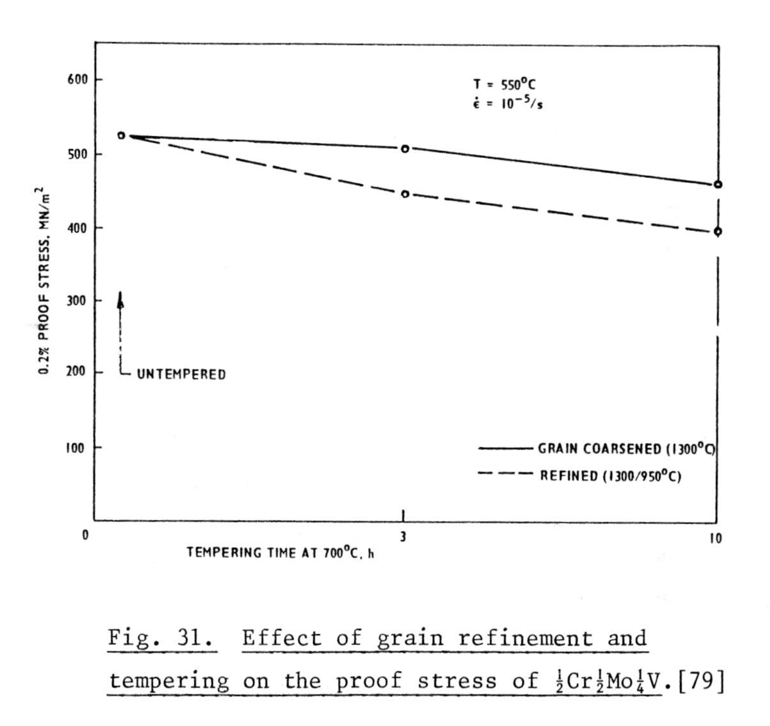

Fig. 31. Effect of grain refinement and tempering on the proof stress of $\frac{1}{2}Cr\frac{1}{2}Mo\frac{1}{4}V$.[79]

which are sometimes not even covered by the code. Figure 31 shows that the 0.2% proof stress is relatively unaffected by these elements and is more affected by the refining cycle, in this case 1300°C followed by 950°C. However, the major effect of trace elements is on ductility which is dramatically reduced. The improvements in ductility following refinement are consistent and relatively independent of the composition in the range examined. Figure 32 shows the strength reduction and ductility increase expected with tempering time at the post weld heat treatment temperature of 700°C.

For austenitic Type 304 and 316 materials, the effects are broadly similar with the following exceptions. Firstly, the mechanical properties of the HAZ structures do not differ markedly from those of the parent material. Secondly, the effect of PWHT, where the precipitating phase is very dependent on the maximum temperature, can be very great [80,15].Dependent on the temperature attained, the weld metal, which contains δ-ferrite for weldability reasons, can nucleate non-metallic sigma phase, leading to very low ductility. Carbides of this type and Chi and Lavé phase can also develop.

6.3 High Temperature Creep Properties

6.3.1 Creep Life

$\frac{1}{2}$Cr$\frac{1}{2}$Mo$\frac{1}{4}$V:2$\frac{1}{4}$Cr1Mo Weldments

Published rupture data show that 2$\frac{1}{4}$Cr1MoV weld metal has a lower creep strength than the $\frac{1}{2}$CrMoV parent material for lives up to at least 10,000 h [81]. However, a definitive assessment of the rupture properties of all constituents of the weld requires the effect of structure etc., to be examined for one particular cast and welding conditions for a range of simulation treatments. The most comprehensive data set is that prepared by Cane, [76] as part of a collaborative programme within the CEGB to examine the correlation between uniaxial and welded component behaviour [82] and the comparisons below rely very heavily on this work.

Figure 33 compares rupture data for $\frac{1}{2}$Cr$\frac{1}{2}$Mo$\frac{1}{4}$V parent, 2$\frac{1}{4}$Cr1Mo weld metal tempered for 3 h at 700°C, and HAZ simulated coarse grained bainite [83]. The complete data set also included refined parent, double refined parent and intercritically transformed grain coarsened structures. The position of these structures within the HAZ of real welds have been described earlier. The properties can be summarised as:

1. Coarse grained bainitic structures have longer lives than the parent material, ∿ 4 - 5 x at equivalent stresses.

2. The $\frac{1}{2}$Cr$\frac{1}{2}$Mo$\frac{1}{4}$V parent material is stronger in creep than the 2$\frac{1}{4}$Cr1Mo weld metal by factors of up to 4 - 7 x in life.

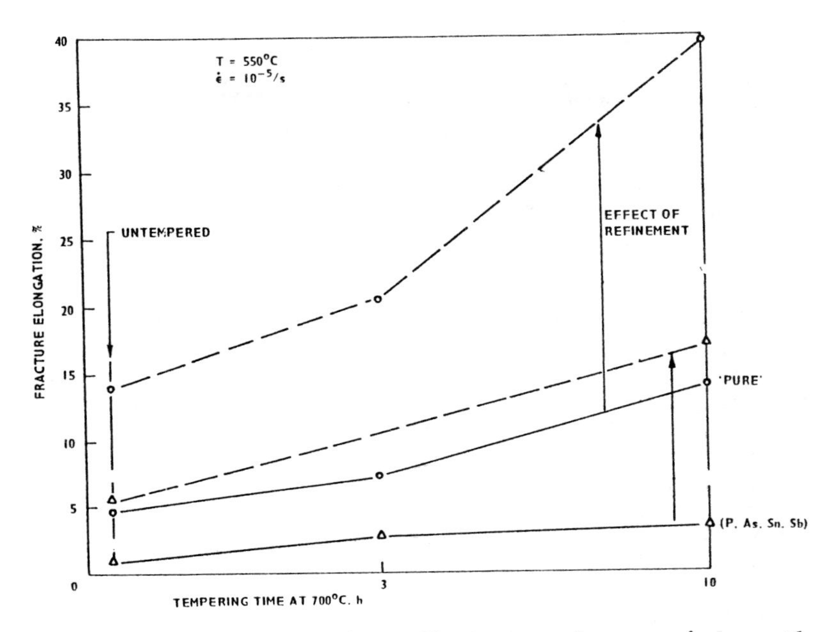

Fig. 32. The effect of grain refinement and tempering on the ductility of $\frac{1}{2}$Cr$\frac{1}{2}$Mo$\frac{1}{4}$V. [79]

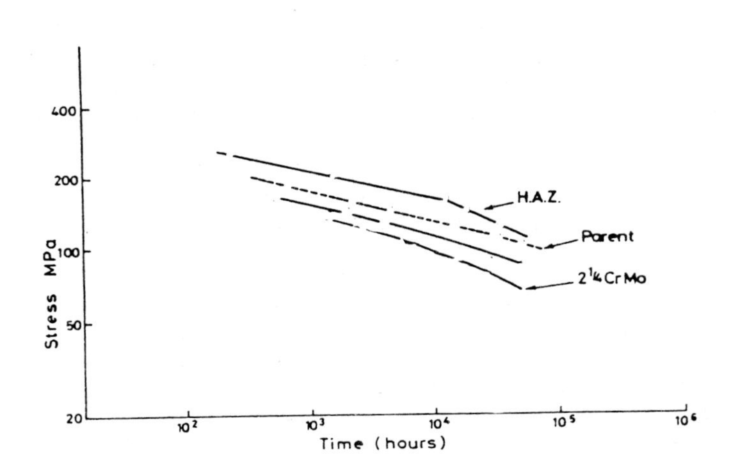

Fig. 33. Stress rupture data for weld metal, parent material and simulated HAZ in $\frac{1}{2}$Cr$\frac{1}{2}$Mo$\frac{1}{4}$V:2CrMo weldments.[83]

3. The refined and doubly refined parent structures are similar to the parent $\frac{1}{2}Cr\frac{1}{2}Mo\frac{1}{4}V$ material.

4. The intercritically reheated coarse grained structures are closer to the coarse grained bainitic structures than the parent material, although limited data suggest a lower life. The creep curves presented in Figure 34 [76] illustrate these characteristics at 565°C and a stress of 160 MNm^{-2}. As is readily seen, the ductilities also follow the rupture data.

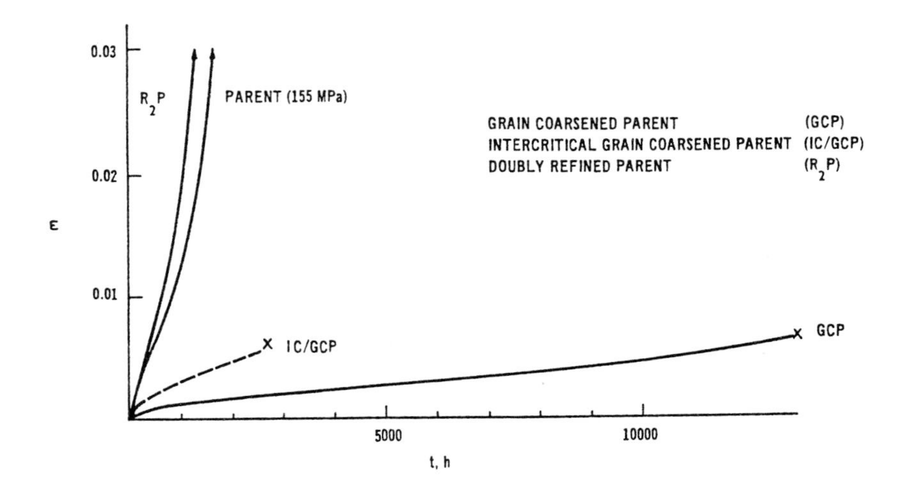

Fig.34. Typical creep curves for HAZ constituents of $\frac{1}{2}Cr\frac{1}{2}Mo\frac{1}{4}V$ weldments [76]

Austenitic Welds

Similar comparisons have been reported for 316 austenitic weldments, the first using samples cut from full size weldments and the second using simulation techniques.In the former, [84] the results in Figure 35 show that at longer times the weld metal is weaker than the parent by ∿ 2 - 3 in life. For the HAZ regions denoted as I and II and corresponding to distances of 5 and 10 mm,respectively, HAZ I has similar properties to the parent, while HAZ II, is somewhat stronger than the parent. Using simulation methods, Ivarsson and Sandstrom [85] found, generally, little difference between HAZ and parent materials strengths. They also incorporated data for different Type 316 weld metals. This is reproduced in Figure 36 and is a good example of the differences that can be found.

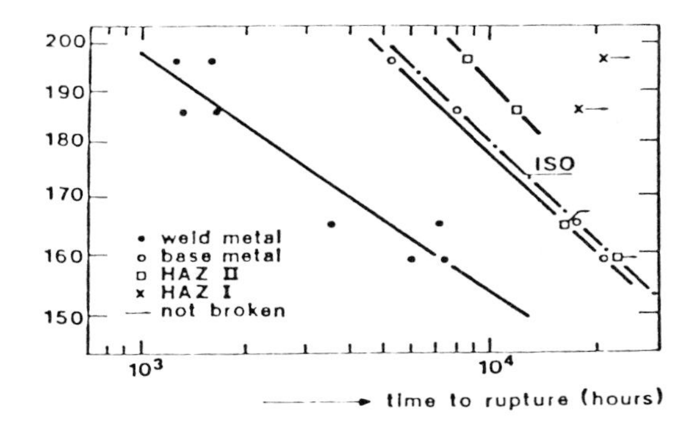

Fig. 35a. Stress rupture curves at 600°C for constituents of austenitic welds. [84]

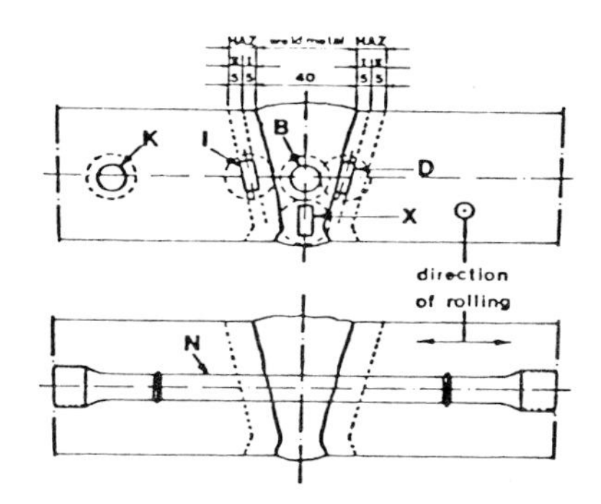

Fig. 35b. Position of test bars used in study.[84]

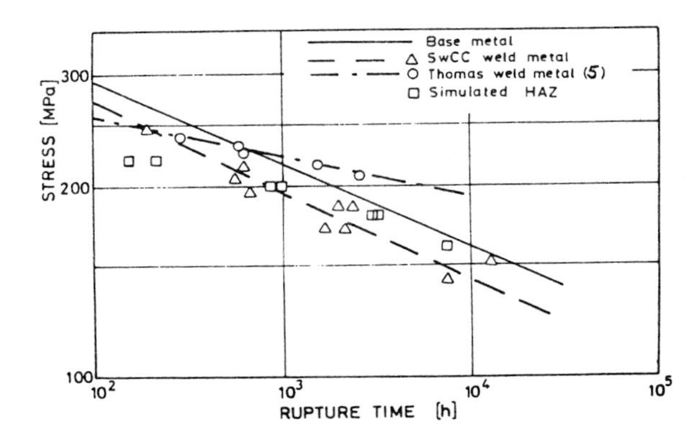

Fig. 36. Stress rupture data for constituents of austenitic welds. [85]

6.3.2 Creep Strain Behaviour

$\frac{1}{2}$Cr$\frac{1}{2}$Mo$\frac{1}{4}$V:2$\frac{1}{4}$Cr1Mo Weldments

Early design codes did not require the use of strain:time data other than to specify strain limits at the design life, and little data of a general nature are available. However, with the commencement of benchmark programmes to validate analytical methods of determining the life of welds and newer codes requiring the use of such data, more information is becoming available in the open literature.

The most accessible data concerns the secondary or steady state creep rate:stress relationships which are normally described using Norton's Law:

$$\dot{\varepsilon} = A \, {}^{n} e^{-Q/RT} \qquad \ldots \qquad (10)$$

where $\dot{\varepsilon}$ is the secondary or steady state creep rate,
σ is the applied stress,
T is the temperature, K,
Q is the activation energy,
and A and n are materials constants which may be stress and temperature dependent.

Many alternatives have been proposed in the literature which produce better correlation between $\dot{\varepsilon}$ and σ, one of the most recent being the σ_o concept proposed by Wilshire and co-workers, [86]. However, as data for σ_o in commercial materials are limited, the creep rate dependence of the various structures is compared here using Norton's Law.

Again, the best documented data for such welds is due to Cane [76]. Figure 37 compares the strain rate dependence of parent $\frac{1}{2}$Cr$\frac{1}{2}$Mo$\frac{1}{4}$V, 2$\frac{1}{4}$Cr1Mo weld metal tempered at 700°C and simulated coarse grained bainite structures [83]. As expected from the rupture data, the weld metal is indeed weaker whereas the coarse grained bainitic structures are stronger than the parent $\frac{1}{2}$Cr$\frac{1}{2}$Mo$\frac{1}{4}$V material. The refined and double refined parent structures are generally similar to the parent material. Finally, the intercritically heated structure has broadly similar properties to the coarse grained bainitic structure.

As described earlier there are overtempered and partially transformed regions near the HAZ:parent interface. In this location, the strain rate relationshops are not as well identified. Obviously, in the overtempered region, the material will be soft with a high ductility whereas the coarse grained, but intercritically heated region, will have properties similar to the coarse grained bainite. Between the coarse grained and intercritically transformed material, properties may change locally but these have not been established experimentally.

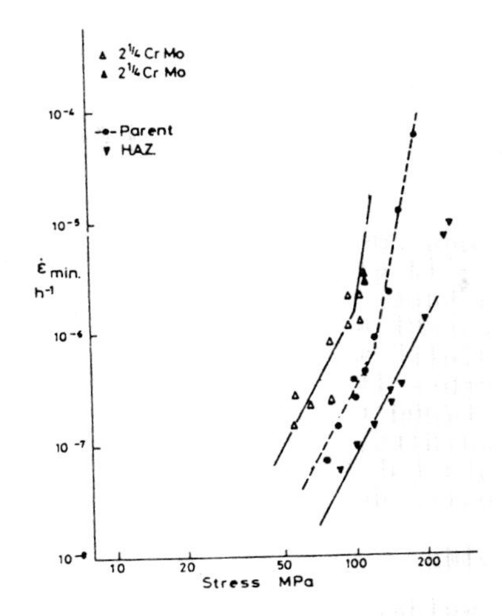

Fig. 37. Stress dependence of the minimum creep rate for 2CrMo weld metal, parent and simulated HAZ $\frac{1}{2}Cr\frac{1}{2}Mo\frac{1}{4}V$.[83]

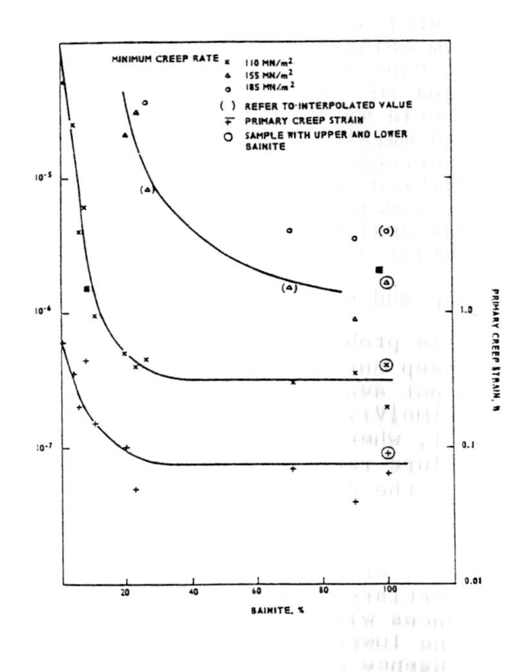

Fig. 38. Minimum secondary creep rate and primary strain as a function of bainite content in as cast CrMoV.[87]

The work above was concerned with homogeneous parent material exposed to weld simulation cycles tailored to provide homogeneous HAZ structures throughout the sample. Earlier work, [87] on a different cast of material examined the properties of ferrite:bainite mixtures. As the percentage of bainite in the structure is increased, the strain rate at constant stress was found to decrease rapidly, reaching a plateau at > ∿ 40% bainite, Figure 38. This is not at variance with the previous data but the reference should be consulted for a more comprehensive description.

Austenitic Welds

For austenitic welds, equivalent although less comprehensive results, are shown in Figure 39 for the Type 316 welds used by Ivarsson and Sandstrom [85]. Little difference was found between the steady state creep behaviour of simulated HAZ structures and the parent 316.

The results of Ettienne and co-workers [84] however, who used samples cut from welds,show generally similar effects to the ferritic data of Cane with the weld metal being weaker, compare Figures 37 and 40. However, as in the case of the rupture properties, weld metal strain behaviour is critically dependent on the weld metal used, there being stronger or approximately equal strength variants available. The effect of weld metal composition and structure has been comprehensively studied, for example, by Thomas [15] and the research team at Oak Ridge National Laboratories in the USA [88] and these should be considered for further information.

6.3.3 Tertiary Creep and Ductility

Tertiary creep is probably the least documented and understood part of the creep curve for commercial materials and wide data bases are not available. The most comprehensive data are for the ½Cr½Mo¼V:2¼Cr1Mo weld combination [76], which is shown in Figure 41, where average values of the time to tertiary:time to failure ratios are shown as a function of stress. In addition, the data of Klueh [89] for a fully bainitic 2¼Cr1Mo, shown to have similar properties to 2¼Cr1Mo weld metal is incorporated. Suffice it to say that as the weld metal becomes more brittle, this ratio increases, indicating very limited tertiary creep. For parent materials, the ratio is low and reduces with decreasing applied stress suggesting that at the lower stress levels, the ratio may be ∿ 0 signifying the absence of a true steady state regime. Obviously, such statements must be qualified by an examination of the strain measurement accuracy and it is found that as the resolution and accuracy increase, the value of the ratio tends to decrease, particularly for the more ductile constituents. Comparable values for austenitic weldments are not generally available although the limited data suggest that reduced

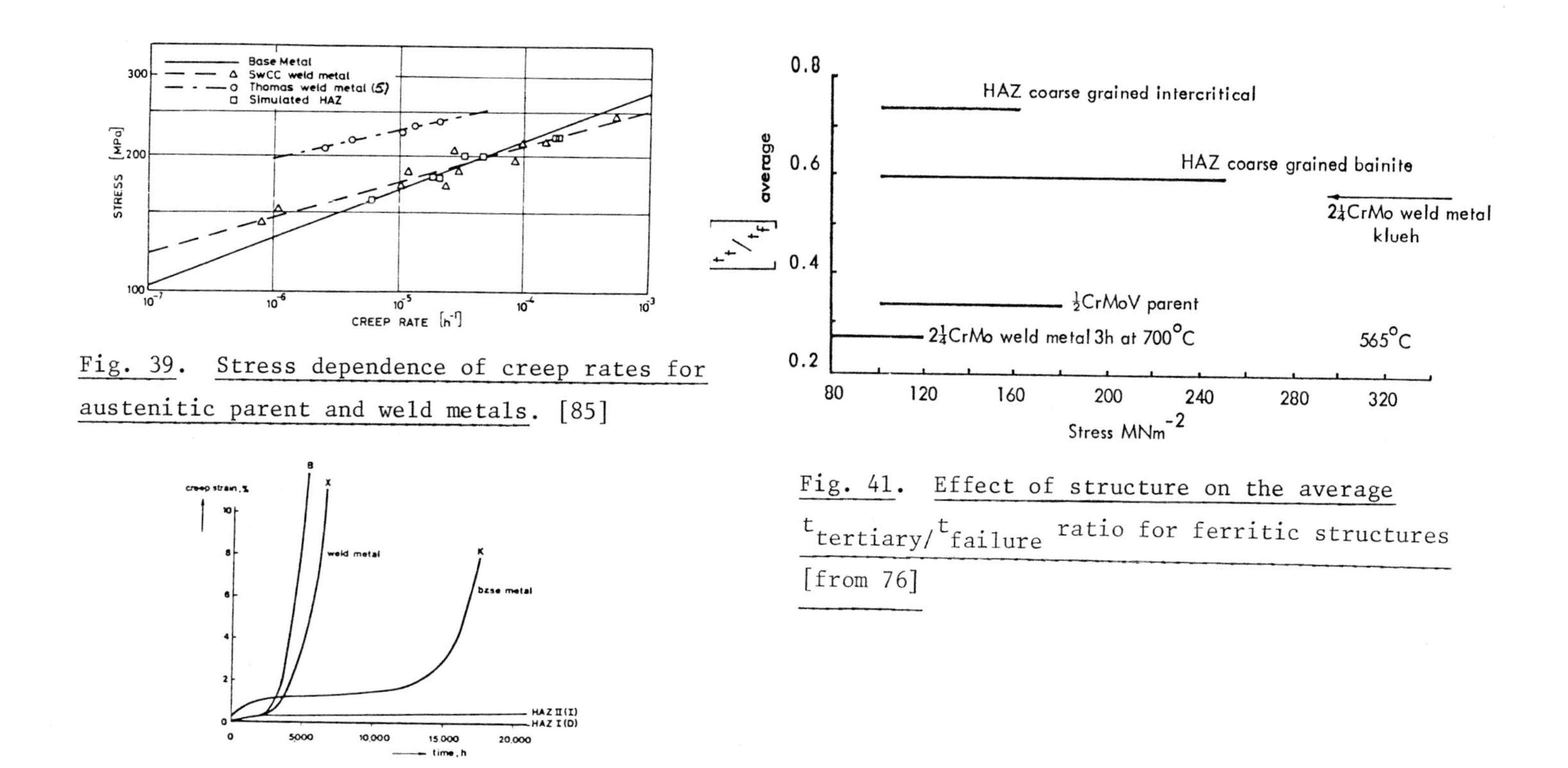

Fig. 39. Stress dependence of creep rates for austenitic parent and weld metals. [85]

Fig. 40. Typical creep curves at 600°C and 159MPa for constituents of austenitic weldments.[86]

Fig. 41. Effect of structure on the average $t_{tertiary}/t_{failure}$ ratio for ferritic structures [from 76]

ductility is characterised by smaller times in tertiary creep.

Ductility itself can be defined in numerous ways but here, only two forms, the $\dot{\varepsilon}t_f$ product and the total ductility are considered. The former has some relevance as it can be referred to as the useable creep ductility whereas the latter is a general parameter. The results for the $\frac{1}{2}Cr\frac{1}{2}Mo\frac{1}{4}V$:$2\frac{1}{4}Cr1Mo$ series suggest values of $\dot{\varepsilon}t_f$ at 565°C of:

Normalised and tempered parent $\frac{1}{2}Cr\frac{1}{2}Mo\frac{1}{4}V$	1-3%
Refined and double refined parent	1-3%
$2\frac{1}{4}Cr1Mo$ weld metal 3 h at 700°C	1-2%
Coarse grained bainitic structures ≡ HAZ	<0.5%

It is emphasised here that PWHT can markedly affect the $\dot{\varepsilon}t_f$ product for $2\frac{1}{4}Cr1Mo$ weld metal as shown by Wolstenholme [90]. He heated uniaxial samples of as-welded $2\frac{1}{4}Cr1Mo$ weld metal under stress, to simulate PWHT and subsequently creep tested them at various temperatures, as shown in Figure 42. It is readily seen that at low PWHT temperatures or in the as-welded state the $\dot{\varepsilon}t_f$ value is < 1% and temperatures in excess of ∿ 660°C are required to produce the necessary ductility. This phenomenon is thought to have led to a particular cracking problem in practice and is considered later.

Finally, the total recorded strains have also been documented for the $\frac{1}{2}Cr\frac{1}{2}Mo\frac{1}{4}V$:$2\frac{1}{4}Cr1Mo$ weld combination and are consistent with the sequence found with the $\dot{\varepsilon}t_f$ product above.

Normalised and tempered parent $\frac{1}{2}Cr\frac{1}{2}Mo\frac{1}{4}V$	6-30%
Refined and double refined parent	7-15%
$2\frac{1}{4}Cr1Mo$ weld metal, 3 h at 700°C	3-14%
Coarse grained bainitic structures ≡ HAZ	0.3-1%

These are quoted as ranges for comparison but the ductility in creep is lowest at low stresses. Exceptions to this rule can occur at very low stress levels.

6.4 High Temperature Creep Crack Growth

The correlation of macroscopic creep crack growth in materials with the parameters of stress, crack length, geometry and temperature has been considered by many workers, [91-95] Various crack growth descriptions based on stress intensity factors, reference stress, crack opening displacement and C* have been strong contenders at various times. At present, there is no generally accepted parameter to characterise crack growth for all materials, although it is clear that the strain accumulation in the vicinity of the crack is a significant factor. There are several excellent reviews available in this field [96,97] which consider the problem in depth. However, for the purposes of this paper, the parameter, stress intensity factor which has been applied in the largest number of cases is

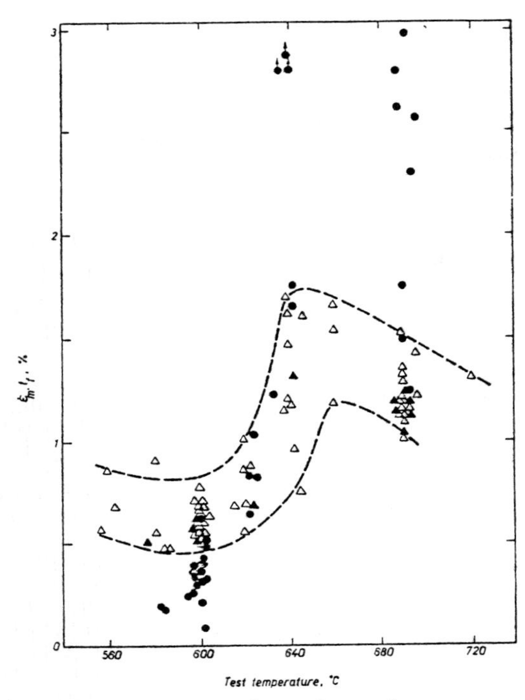

Fig. 42. $\dot{\varepsilon}t_f$ product as a function of temperature for 2CrMo weld metals. Δ-MMA 4mm, Δ-MMA 6mm, O-SA. [90]

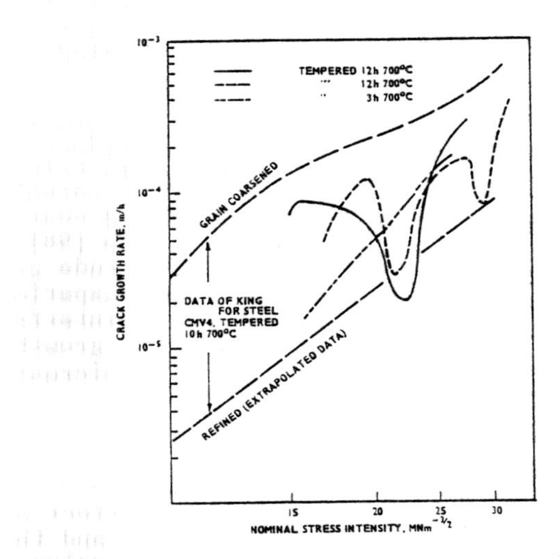

Fig. 43. Variation of crack growth rate with nominal stress intensity factor in ½CrMoV HAZ structures. [98]

used for convenience. It is not suggested that this is the correct correlation but merely a method of comparing the behaviour of defects under identical loading and geometrical conditions as a function of structure. It is therefore a comparative exercise and useful if due account is taken of the crack initiation behaviour.

On loading a macroscopically cracked sample under creep conditions, two extremes of behaviour are found which are dependent on macroscopic ductility. For ductile, and brittle materials, displacement records measured across the crack show the characteristic behaviour found in creep strain:time curves for uniaxial uncracked samples, i.e. primary, secondary and tertiary creep stages. In ductile materials, no crack growth is recorded for a finite time and thus there is an incubation time or displacement required before crack growth can occur. In brittle materials, the incubation time and displacements are much lower and can lead to crack growth from,effectively,a time of zero. These characteristics are associated with the build up of local creep damage at the tip of the crack. The crack growth relationships shown against stress intensity discussed below only refer to crack growth where the incubation time is ignored and thus the effect of the various structures could be underestimated. Where possible data for one geometry and from one cast subjected to different heat treatments or from one laboratory are used. It is emphasised that this is a comparative exercise only and the absolute values of any crack growth rate should not be used in case or failure studies without a more detailed examination of the base data.

The most extreme case of crack growth is normally considered to occur for ferritic $\frac{1}{2}$Cr$\frac{1}{2}$Mo$\frac{1}{4}$V:2$\frac{1}{4}$Cr1Mo weldments in the coarse grained bainitic region of the HAZ, this being very much faster than crack growth rates in the ferritic parent material. Figure 43 compares the growth rates in fine and coarse grained bainitic structures, relevant to the HAZ region [98]. The growth rate is approximately an order of magnitude greater in the coarse grained variant. For comparison, experimental data obtained for a crack growing down a real weld interface comprising regions of both these structures shows growth rates approaching these limits reflecting the local microstructure at the crack tip.

The effect of PWHT time and temperature on the growth rates of a mixed upper and lower bainite structure generated by simulation is shown in Figure 44 [99]. The effect of conventional heat treatment practice is readily seen and the state of tempering is indicated through the hardness value.

Crack growth in 2$\frac{1}{4}$Cr1Mo weld metal has been studied by Gooch and King [100] and here the crack growth rate is related to both the creep rate of the material and ductility. These can be varied in many ways and the extreme condition, worthy of consideration here, is the behaviour of cracks in as-welded or poorly heat treated weld metal. The structure is primarily an untempered or even secondary hardened bainite although initiation occurs primarily in the coarse columnar regions. Under these conditions, crack growth is similar to that found for the coarse grained bainite region of the HAZ.

Crack growth data on austenitic parent materials are readily available [101] but, with limited exceptions [102,103] are lacking for the austenitic weld metals and are unknown for simulated HAZ structures. As a first estimate, in the absence of any data, the behaviour should be related to the creep rates and ductility values, suggesting faster growth rates in some of the austenitic weld metals.

As will be noted, the identification of the properties of the constituent parts of a weldment are critically dependent on the local microstructures. It is for this reason that the general characteristics only have been identified here. In many systems, there is still a need for additional work.

7. DEFORMATION AND FAILURE OF WELDMENTS AT HIGH TEMPERATURES

The previous sections have considered the general principles underlying metallurgical structure variations and residual stresses in weldments and how they are related to welding procedure and the material characteristics. Each microstructural constituent can have different mechanical and creep properties. It is therefore essential to determine the extent to which inhomogeneity in structural or creep properties affects the high temperature performance of the weld.

If a weld is loaded uniaxially, each of its component parts is subjected initially to the same stress. A very simple view suggests that failure will always occur in the weakest material. Practically, this is not the case as evidenced by creep of uniaxial cross weld geometries and the behaviour of welds in service. The reason is that stress and strain redistribution takes place which is related to the distribution and strength differences of the constituents present in the weldment. Finite element and other methods are now available which allow the creep deformation of material combinations to be analysed, see for example, the work of Goodall and Walters [104] and Walters [105]. This has been a very active field in Europe and the UK, particularly in the last five years and the outline that follows relies very heavily on the recently published research.

For simplicity but to conform with the realistic geometries, the case of a circumferential butt weld in a pressurised cylinder is considered; the generalised form of the constituent elements of the weldment being indicated in Figure 45. Also, following the convention of Browne, et al., [83]

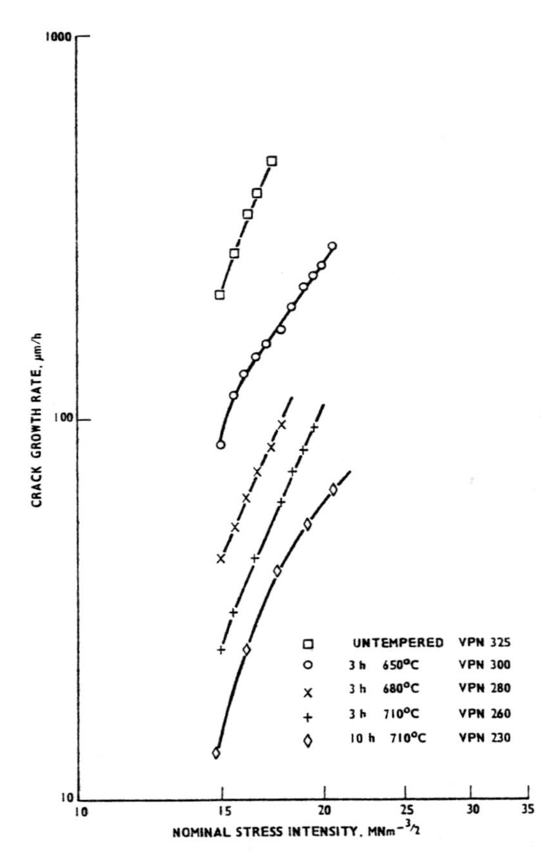

Fig. 44. Variation of crack growth rates at 565°C with tempering of ½CrMoV (Fast oil quench, SENB) [99]

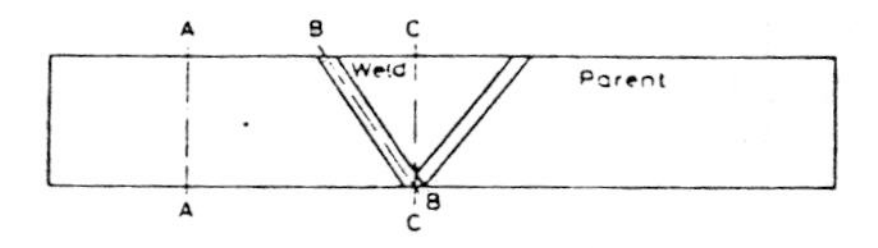

Fig. 45. Schematic representation of weld. [83]

strength differences between weld metal and parent material are defined as the ratio of the uniaxial strain rates for the materials at a constant applied stress. For steady state creep conforming to Norton's Law, this leads to equations of form:

$$\left[\frac{\dot{\varepsilon}_{wm}}{\dot{\varepsilon}_{pm}}\right]_{\sigma=\text{constant}} = \frac{A_{wm}}{A_{pm}} \left[\sigma^{(n_{wm} - n_{pm})}\right] \quad \ldots \quad (11)$$

where the suffixes wm and pm refer to the weld and parent metal, respectively. For the specific case where $n_{wm} = n_{pm}$, this ratio is independent of the applied stress.

7.1 Pressurisation of a Welded Cylinder at Temperature

On pressurisation, the whole section will be elastically loaded and the hoop, axial and radial stresses, σ_h, σ_a and σ_r, respectively, are given by Lamé's equations [106]:

$$\varepsilon_h = \frac{1}{E}\left[\sigma_h - \nu(\sigma_a - \sigma_r)\right] \quad \ldots \quad (12)$$

with equivalent equations for σ_a and σ_r. The elastic stresses are therefore a function of the elastic modulus, E, and Poisson's ratio, ν. Unless compositional extremes are considered, both E and ν will be relatively independent of composition so that the elastic loading is similar in both the weld and parent materials.

At temperature, the elastic state redistributes by creep and in this treatment it is assumed that steady state creep uniaxial data, represented by Norton's Law, can be applied to each of the constituent parts. The creep will result in stress redistribution within the cylinder. Bailey analysed the case of a plain cylindrical pressurised pipe and derived a new series of principal stresses corresponding to the steady state creep condition [107], namely:

$$\sigma_h = P\left[\left(\frac{2-n}{n}\right)\left(\frac{R_o}{r}\right)^{2/n} + 1\right] / \left[\left(\frac{R_o}{R_i}\right)^{2/n} - 1\right] \quad \ldots \quad (13)$$

$$\sigma_a = P\left[\left(\frac{1-n}{n}\right)\left(\frac{R_o}{r}\right)^{2/n} + 1\right] / \left[\left(\frac{R_o}{R_i}\right)^{2/n} - 1\right] \quad \ldots \quad (14)$$

$$\sigma_r = -P\left[\left(\frac{R_o}{r}\right)^{2/n} - 1\right] / \left[\left(\frac{R_o}{R_i}\right)^{2/n} - 1\right] \quad \ldots \quad (15)$$

where R_o/R_i is the outer:inner radius ratio and r is any radius. These stresses are a function of geometry, pressure and the steady state creep rate index, n. Thus, if $n_{wm} = n_{pm}$, and assuming continuity of stress along the welded cylinder, the weld metal should be subjected to a stress identical to that in the stronger parent material. However, considering the particular case of a weld metal which is weaker in creep than the parent, the weld metal cannot sustain these loads and sheds the stress on to the surrounding parent pipe. This is called off loading and the extent to which it occurs will be controlled by the geometry of the weld and the differences in the material properties [105]. At present, there is no analytical solution for thick welded pipe for these parameters and finite element or finite difference solutions must be employed.

7.2 Creep of Cylindrical Butt Welds

The general case of a weld has been considered by Browne, et al., [83] with weld metal, parent material and HAZ properties being separately ascribed, see Figure 45. It was assumed that steady state conditions existed for all materials and that Norton's Law was obeyed. Also it was assumed that $n_{wm} = n_{pm}$. For the experimental data considered by these workers, these assumptions were valid and n = 4. For weld metal:parent:HAZ creep strain rate ratios of 10:1:0.25, the calculated stresses were computed across various sections of the weldment. The stress distributions along the outer surface, the HAZ centre line and the weld metal centre line, respectively, are given in Figures 46a-c [83].

The stress distribution on the outer surface has several important characteristic features. The hoop stresses in the weld metal are reduced and off loading is manifest as an increase in hoop stress in the HAZ region. As the pipe section is traversed away from the weld, the hoop stress reduces to a value compatible with the steady state creep stress for the plain pipe, as calculated from the Bailey equations. The axial stress is relatively unaffected by off-loading, as might be expected, although a small peak is found in the HAZ region.

The distributions along the line through the HAZ and the line through the weld metal centre are also given in Figures 46b and c. The highest axial and hoop stresses are found at the outer surface and it is worth noting that the equivalent stress is a slowly varying function. While the curves are only relevant for the specific data input used in the finite element programme, a parametric study with different weld metal:parent metal creep strain rate ratios was carried out to allow general patterns to be defined for steady state creep. An example is given in Figure 47 which shows the maximum stresses in each region of the weld, normalised by the test pressure. The maximum principal and equivalent stresses, σ_1 and $\bar{\sigma}_s$, respectively, have the following characteristic features:

1. The maximum principal stress in the weld, primarily in the hoop direction, is continuously affected by the creep strain rate ratio although above a ratio of $\sim$ 100, changes are small.

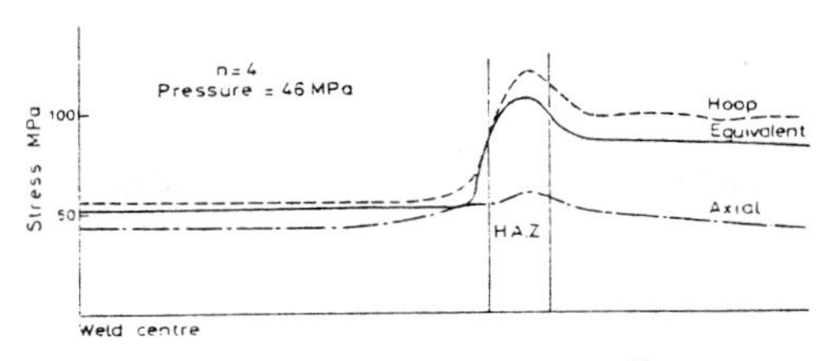

a. Stress Distribution along Outer Surface B.C.

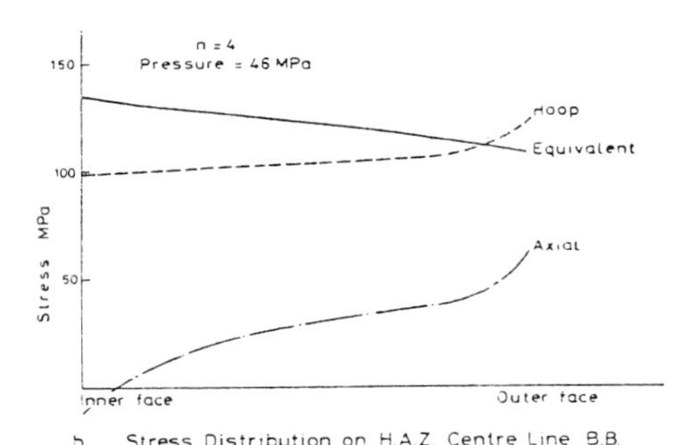

b. Stress Distribution on H.A.Z. Centre Line B.B.

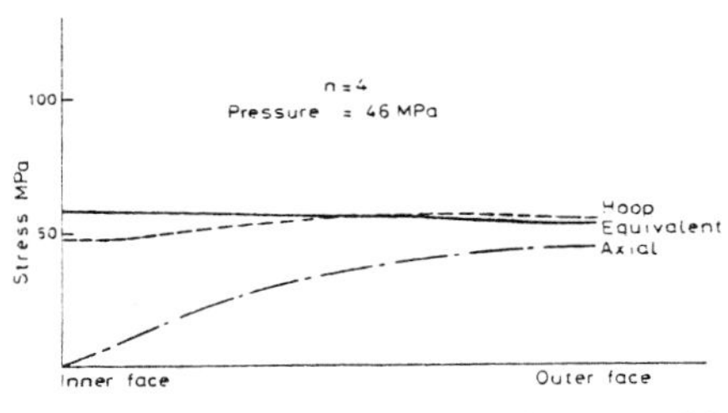

c. Stress Distribution on Weld Centre Line C.C.

Fig. 46. Steady state creep stress distribution in a heavy section ferritic weldment. (See Fig. 45) [83]

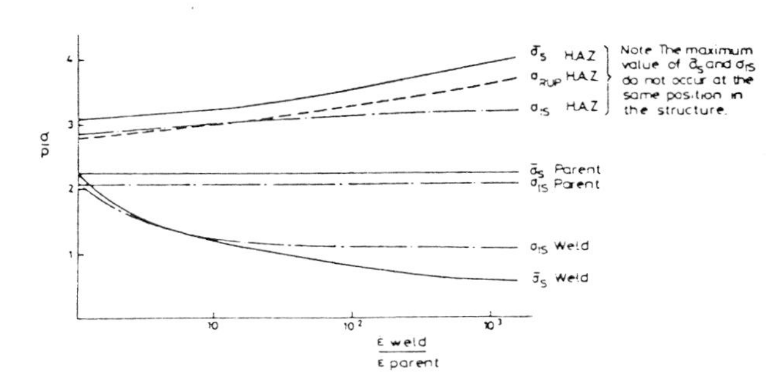

Fig. 47. Variation of peak stresses in weldment with weld metal: parent metal creep rate ratio. [83]

2. The parent pipe stresses, away from the weld are independent of the weld properties, as might be expected.

3. The HAZ stresses, σ_{1s} and $\bar{\sigma}_s$, gradually increase as weld metal strength decreases but apparently still have a high value for a creep strain rate ratio of weld metal:parent material of 1. Thus, the HAZ region has some stiffening properties in its own right.

4. Finally, though not illustrated in the Figure, redistribution through the thickness of the axial stress occurs in an increasing manner at creep strain rate ratios of > 10.

Thus, a fairly complete picture has been obtained for steady state creep by these workers.

In a similar exercise Ivarsson and Sandstrom [85] used finite difference methods to analyse the case of a thin walled welded austenitic tube. Their results, summarised in Figure 48, show similar features to those in the ferritic steel. The study considered only one series of data, specific to their tube tests. Figure 48, however, shows the stress progression as a function of time towards the steady state condition and again illustrates the characteristic off loading of the weld metal stress into the HAZ; the weld metal again being weaker in creep.

The above programmes considered weld metals weaker than parent. If a weld metal:parent metal ratio of < 1 is envisaged, i.e. the weld metal is stronger than the parent material, stress should off load on to the stronger weld metal producing highest stresses in the weld.

Experimentally, strain has to be measured rather than stress and thus, verification of analysis must be based on local strain measurements and detailed comparison of computed and experimental data. The experimental information available to compare with analytical or finite element methods is primarily based on two programmes. Within CEGB a comprehensive research programme concerns the creep behaviour of ferritic $\frac{1}{2}Cr\frac{1}{2}Mo\frac{1}{4}V$ tube and pipes welded with ferritic weld metal, examining a range of weld metal:parent metal strength ratios and post weld heat treatments in real components of pipe and tube sizes [82]. The work includes a comprehensive uniaxial data base [76]. The second programme on Type 316 weldments is based only on tube size welds but has incorporated primary and secondary creep laws [85]. This comparison is shown for austenitic tube welds in Figure 49 which illustrates good agreement for between experiments and prediction using constitutive equations which include both primary and steady state creep in the computations.

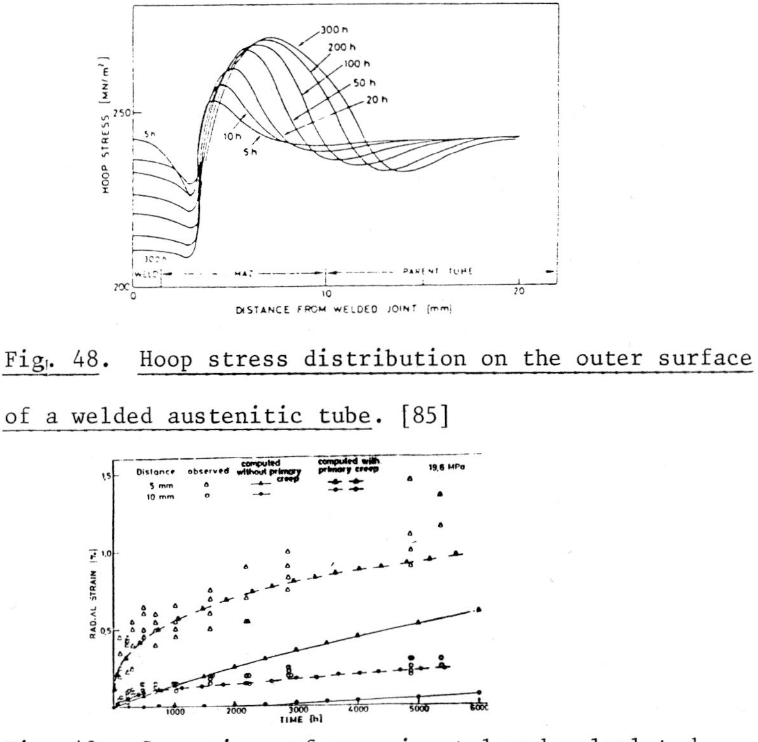

Fig. 48. Hoop stress distribution on the outer surface of a welded austenitic tube. [85]

Fig. 49. Comparison of experimental and calculated strains at 5 and 10mm from the weld centre. [85]

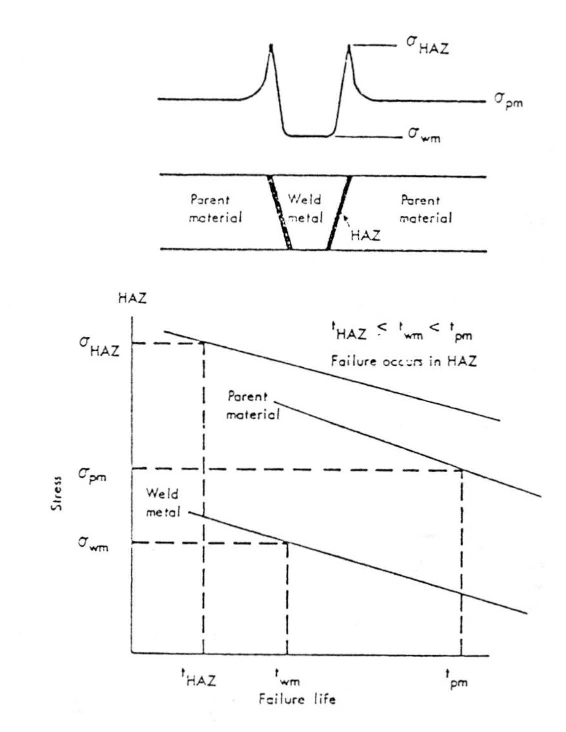

Fig. 50. Schematic description of the simplified life assessment method. [108,114]

A similar comparison has been reported for the thick walled ferritic pipe welds by Coleman, et al., [108]. Their approach involved a very extensive uniaxial data base, and considered only the secondary, true steady state regimes.

Strain rates were measured at the weld metal centre, adjacent to the HAZ and in the parent pipe and very good agreement was found for weld metal:parent material creep strain rate ratios in the range $\sim$ 1 to $\sim$ 1700. The validity of using uniaxial data and finite element analysis to predict steady state deformation rates in real welds has thus been demonstrated for a wide range of creep applications.

A unifying feature in the experiments on both ferritic and austenitic materials is that the strain rates generated at the weld metal and parent pipe centre positions, are virtually identical in steady state creep. This further supports the off loading argument that, in essence, requires that there be strain compatibility across the weld. Simply, this means that the weld metal will be constrained to deform at a rate controlled by the surrounding parent pipe during steady state creep. This is a very useful concept, which may allow relatively complex practical situations to be treated in a simple manner.

The stress analyses, which refer to weld metal weaker than the parent show that the highest stresses are in the hoop direction of the HAZ. As shown earlier this has low ductility in ferritic ½CrMoV materials when a coarse grained bainitic structure is present. It further suggests that the first indications of eventual failure would be small cracks within the HAZ, running parallel to the pipe axis. Such a failure pattern is indeed shown in the thick walled tube tests of Browne, et al., [83] and thick walled pipe tests of Coleman,et al., [108]. No work has yet been reported on the incorporation of tertiary creep within the analytical framework.

Cockroft and Walters [109] first suggested that the creep life of welded joints could be estimated using the maximum values of the stationary state stress. This has been used by by Browne, et al., for defining failure in welded thick wall ½CrMoV tube and by the Swedish workers for welded thin wall Type 316 tube and is shown schematically in Figure 50. It produces a lower bound limit of life by using the maximum principal stress, established under steady state conditions in each zone of the weldment, to define the life from uniaxial stress rupture data. Ivarsson and Sandstrom [85] showed that this gave a good though conservative estimate of the rupture life. Browne, et al., [83] found reasonable agreement using this method. They also investigated the possibility that creep failure in some materials under complex stresses may be controlled not by the principal stress alone but by a representative rupture stress. This was determined experimentally to be $({\sigma_{1s}}^{0.35} . {\bar{\sigma}_s}^{0.65})$ for the coarse grained bainitic material in

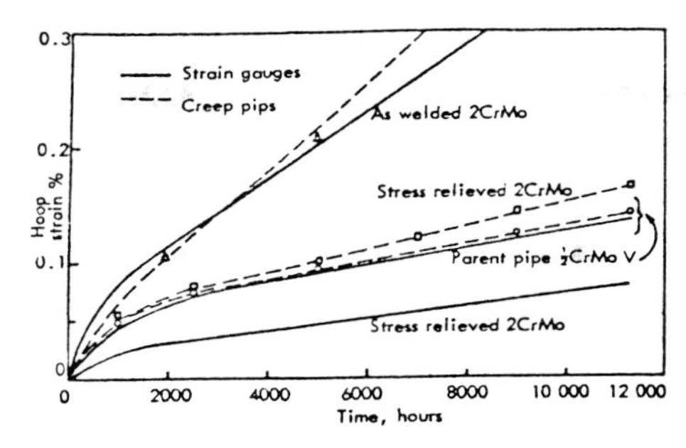

Fig. 51. The variation of hoop strain with time for stress relieved and as welded 2CrMo welds at 455 bar and 838K. [110]

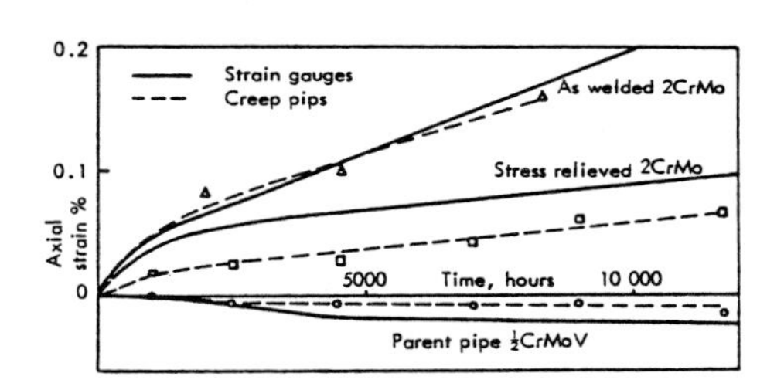

Fig. 52. The variation of axial strain with time for stress relieved and as welded 2CrMo welds at 455 bar and 838K. [110]

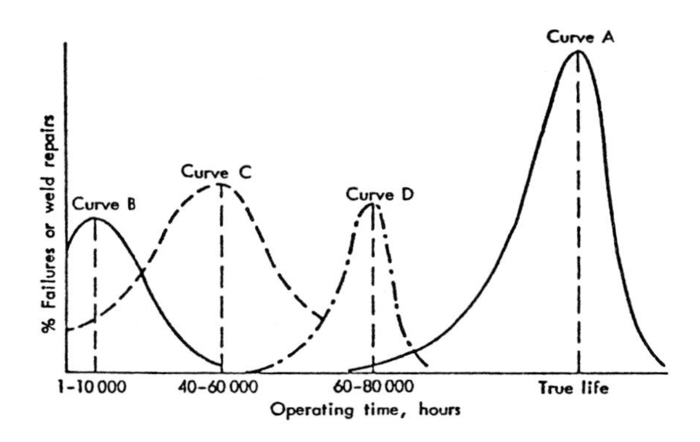

Fig. 53. Schematic representation of weld failure modes in ferritic pipe welds. [114]

the HAZ and agreement between predicted and actual life for these thick wall welded tube was conservative.

In summary for stress relieved welds the predictive capability of the finite element approach is very good for deformation but has yet to be tested for tertiary creep and failure predictions in heavy section welds.

Redistribution of stress is an important feature associated with weldments subjected to creep loading. As stated earlier, welds are normally subjected to PWHT so that the residual stresses are relaxed to a low level and the constituent parts of the weldment are tempered. However, under some circumstances, welds are subjected to service conditions in the un-heat treated state. Limited data are available on the effect of residual stress on the creep of weldments. The salient features of the effect are illustrated in Figures 51 and 52 from the work of Coleman and Parker [110]. Here, the presence of residual stress has increased both the hoop and axial creep strain accumulation in the weld region and such an effect has also been found in other vessel experiments where inadequate or no PWHT was carried out before testing [111]. A finite element analysis which incorporated residual stress and pressure loading in a large diameter pipe weld under steady state creep has been carried out by Hepworth [112]. This analysis fully supports the effects observed.

This section has attempted to highlight how the microstructural components of a weldment operating under creep conditions influence deformation patterns. Comment has concentrated on circumferential butt welds in a cylinder geometry. However, the effects revealed are quite consistent with published data on creep in cross weld samples, for example see the work of Ettiene and co-workers [84,113] who have considered Type 316 and $2\frac{1}{4}$Cr1Mo weldments under uniaxial loading conditions, and the work of Manjoine [103]. Nevertheless, at present the influence of PWHT residual welding stresses on creep deformation and the investigation of the parameters that are important in tertiary creep and failure are still being explored.

8. WELD FAILURE MECHANISMS IN PRACTICE

So far, factors such as the metallurgical structure distributions in welds, the variation of properties associated with these structures, the nature of the residual stresses generated by welding and the effects of post weld heat treatment have been considered. In addition, the general concepts of the analytical methods available for defining weld life under creep conditions have been examined. While the main characteristic feature of any weld is the inherent structural inhomogeneity, which will be present to some extent irrespective of the material combination chosen, the deformation and cracking mechanisms will depend on the interactions between the various parameters. Analysis of service experience is the most relevant method of determining the importance of these factors and identifying the conditions under which they promote failure. It is therefore intended to examine cracking and failure statistics from high temperature steam generating plant operated by the CEGB. The low alloy ferritic ½Cr½Mo¼V type steel welded with 2¼Cr1Mo weld metal has been well documented and is chosen to illustrate the behaviour of complex welded structures.

The ½Cr½Mo¼V material is used within the UK and Europe for plant operating at 565 - 575°C with design lives up to 200,000 h. In the UK, it is welded with 2¼Cr1Mo weld metal although German practice is to use a ½Cr½Mo¼V weld metal nominally of matching composition and creep strength. Standard PWHT practice is to stress relieve at 690 - 710°C for 1 - 2 h per 25 mm thickness.

Were it possible to obtain failure statistics on a single type of weld operating under design conditions, there would be a distribution of failure times around a mean value, defined here as the component life. For a single cast of material failing by creep, as shown earlier, the distribution is Gaussian, although for other failure mechanisms different distributions could arise. In principle, the design life could be set at some proportion of the actual life, say 3 standard deviations below the mean, so that the design life would be inherently safe.

For operating plant, examination of the incidence of the cracking in welds that require repair or where steam leaks arise has identified various premature failure modes:

Heat affected zone cracking ... B.
Transverse weld metal cracking ... C.
Type IV cracking ... D.

These cracking modes are found in the ½Cr½Mo¼V:2¼Cr1Mo combination and they are associated with inherent features of weldments. Their distributions are illustrated schematically in Figure 53, where it can be seen that they occur at different fractions of the design life [114,115]. Each failure mode will now be treated separately to examine how the various factors arising from welding influence creep life.

8.1 Heat Affected Zone Cracking

This occurs in the circumferential direction around the heat affected zone. It is primarily associated with the coarse grained bainite in the HAZ, Figure 54, and can occur during PWHT although crack initiation and growth during short term service operation are also found [116]. It is generally

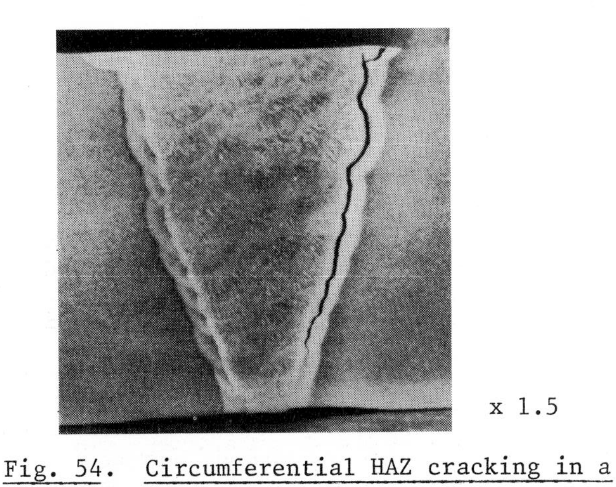

x 1.5

Fig. 54. Circumferential HAZ cracking in a weldment [115]

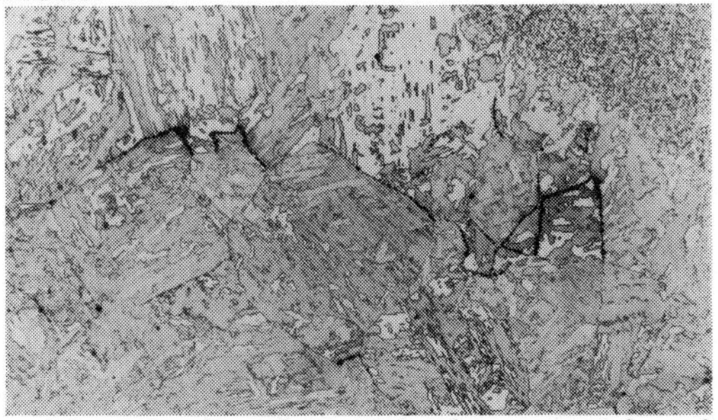

Fig. 55. Cavitation and microcracking in a HAZ coarse grained bainitic region.[115]

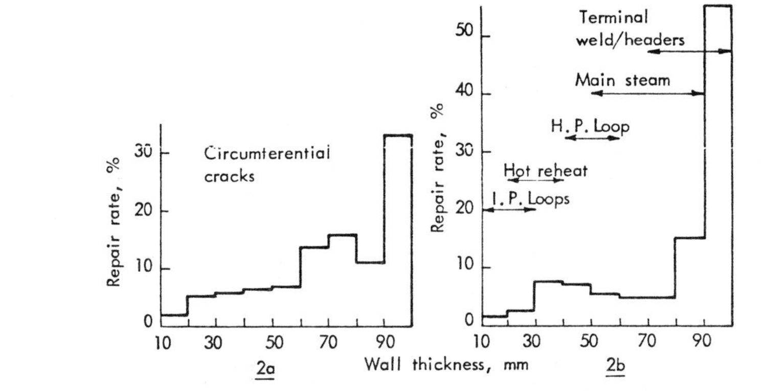

Fig. 56. Repair rate in ferritic welds as a function of size. [2]

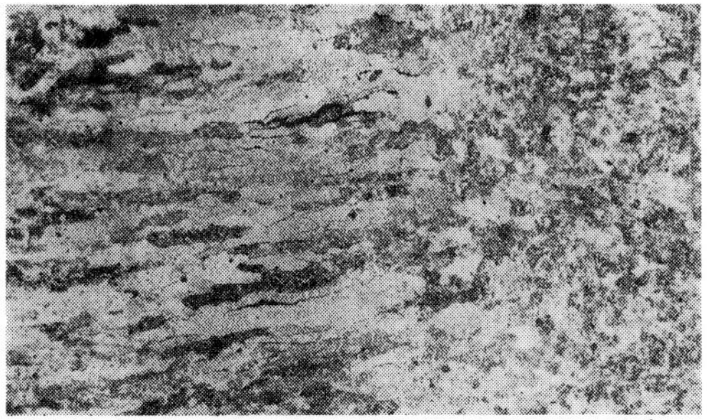

x 20

Fig. 57. Transverse weld metal cracking in a longitudinal cross section of a weld.[115]

detected in the 0 - 20,000 h period and although found in pipe size welds of most sizes, it is particularly prevalent at terminal welds, which generally are the thickest.

Microscopically, the cracks initiate as cavities on prior austenite grain boundaries in the coarse grained bainitic region, Figure 55. While cracking is confined to this HAZ band, the orientation of the microcracks to the main pipe axis is generally random showing the effect of stress multiaxiality in the early stages. These cracks or cavities may link up to form macropscopic cracks which can, under some circumstances, grow completely through the HAZ to the pipe bore, leading to steam leakage.

Mechanistically, the cracks are initiated by creep occurring under the action of the residual welding stresses and it is now known that, in many cases where cracking occurred in service, significant residual stresses remained because PWHT was carried out at temperatures that were too low. As the locked-in elastic strains corresponding to the welding residual stresses are only of the order of 0.1 - 0.2%, total recovery of this amount of creep strain will lead to cracking only if local low ductility materials are present in the HAZ region during PWHT and early service life.

The detailed structure of the HAZ depends on the welding procedure used and the manner in which weld beads are laid down. Historically, before this was understood, certain procedures were used which under specific circumstances would generate regions of coarse grained bainite that were almost continuous completely across the pipe HAZ resulting in easy crack nucleation and growth. This occurred primarily when low weld bead overlap and high heat input procedures were used. With the improved understanding of the factors affecting the formation of coarse grained bainite during welding, it is now possible to specify controlled welding procedures that, in practice, either fully refine or greatly diminish the extent of the coarse grained bainitic structures. The fine grained bainitic structures with which they are replaced are more ductile and the cracking problem can be overcome [78,79,98]. Based on this simple model the effect of other factors which affect the HAZ cracking resistance can now be considered.

Numerous laboratory and plant studies have shown that cracking sensitivity is markedly affected by the vanadium content [117], by the presence of trace elements in the parent pipe material [79], by post weld heat treatment conditions [116], and, apparently, by weld thickness, Figure 56 [2], although the latter may simply be a reflection of welding procedure and residual stress factors.

Firstly, trace element effects such as Sn, P, As and Sb do not generally affect the kinetics and sequence of the transformations occurring although they do segregate during PWHT and service. Their presence at prior austenite grain boundaries markedly affects the local ductility and consequently the overall ductility is reduced. It is usually claimed that these segregates lead to reduced surface energy requirements for crack initiation and growth. In addition MnS has been suggested as a major crack initiator [118]. Obviously any low energy site will be a candidate for initiation and it is probably true that any combination of factors leading to low ductility locally will favour HAZ cracking.

The effect of increasing the amounts of the major alloying element, vanadium, is more straightforward. Vanadium levels greater than 0.25% markedly increase the matrix strength, thereby limiting any deformation during stress relief to regions near grain boundaries [117]. This is associated with a lower fracture stress in the HAZ and increased cracking sensitivity.

In summary, it is now recognised that HAZ cracking can be eliminated by imposing limits on the main alloy and trace element contents, by ensuring that PWHT is carried out at a temperature which achieves both residual stress relaxation and tempering of the weld materials and, also, by using welding procedures that restrict the formation of potentially weak and brittle coarse grained bainitic HAZs.

8.2 Transverse Weld Metal Cracking

In the early 1970's, inspections of operating ½CrMoV welds fabricated with 2¼Cr1Mo weld metal, revealed multiple cracking, parallel to the pipe axis, on the outer weld metal surface, Figure 57. The cracks were small and many in number and were referred to as transverse weld metal cracking (TWMC). Very occasionally, they were found after PWHT but more usually were detected after 10,000 - 30,000 h service in plant. They were found in $\sim$ 5 - 10% of 25 mm thick reheat pipe welds and $\sim$ 60% of thicker 75 mm thick sections, Figure 56 [2]. There were, however, very few of these latter welds.

Metallographically, the cracks initiate in the coarse columnar regions of the weld metal in the beads deposited immediately at the fusion boundary with the parent material. Initially, they occur as cavitation and link to form large numbers of small microcracks, Figure 58 [115]. In the early stages of development, they are constrained to the weld bead size by the refined microstructure in the weld metal associated with adjacent weld beads. However, in service further growth occurs and can, although rarely, continue into the more ductile parent pipe material. Hardness measurements on the weld metal were $\sim$ 300 - 350 VPN, corresponding to PWHT temperatures of less than 650°C. Residual stress measurements on selected welds confirmed that the stresses were high compared with those resulting from PWHT at $\sim 700^oC$ as specified in current codes. Finally, in some cases, the trace elements such as Cu were somewhat high though within the contemporary specification. Plant monitoring showed that in some weldments the numbers of these cracks increased with time as did crack depth.

Laboratory investigation of welds taken from service and experimental welds showed that these cracks initiated within the weld metal near the weld fusion boundary at the top of the weld [90]. In addition, the compositional factors and PWHT conditions under which such cracks are generated were identified [119]. The key parameter was found to be the PWHT temperature. This was shown conclusively by Wolstenholme[90] who identified a creep ductility minimum in the range 600 - 650°C in all $2\frac{1}{4}$Cr1Mo weld metals, although the effect was more marked in the presence of trace elements. Post weld heat treatment at 700°C did not result in the same weld metal embrittlement.

Mechanistically, the low PWHT temperatures have two main effects. Firstly, the residual stresses are not reduced very much and, secondly, the weld metal structure is not tempered sufficiently. The result is that during PWHT, any creep relaxtion takes place in a brittle bainitic columnar solidification structure and microcracks can initiate at columnar grain boundaries. In service, two stress systems are present, the residual stress and the pressure stress from the design conditions. These predominate in the hoop direction on the outer surface and the result is to increase hoop strain accumulation and the propensity to form axial cracks in the weld metal. Little or no cracking is found near the bore, as might be expected, because the residual stresses distribution is compressive in this region [42].

With longer service times, the residual stresses relax and, in addition, tempering of the weld metal is completed. This should lead to the cessation of crack growth because of the reduction in the driving force due to the combined residual and pressure stress fields and, indeed, this is the case with operating plant. In practice it was possible to remove the majority of the cracks by surface grinding; weld repairs being carried out where necessary.

This cracking mode is similar mechanistically to HAZ cracking, albeit in a different location and material and over a more extended timescale. It is a good example of the interactions between parameters arising from materials, welding procedure and engineering considerations even in conventional weldments. This cracking can be eliminated by measures similar in concept to those used for the control of HAZ cracking. In future, difficulties with TWMC should not arise although, like HAZ cracking, it is possible that more rapid creep degradation may take place late in life as a result of any incipient damage remaining in the structure from the early life cracking propensity.

8.3 Type IV Cracking

This is named after a German classification of weld cracking modes and was reported in Germany in $\frac{1}{2}Cr\frac{1}{2}Mo\frac{1}{4}V$ parent material welded with matching $\frac{1}{2}Cr\frac{1}{2}Mo\frac{1}{4}V$ weld metal [120]. In more recent years, it has also been reported in Denmark [121] and in the UK and probably exists on a much wider scale.

It is normally detected in the 40,000 - 80,000 h operating regime and occurs in a circumferential mode just outside the fully transformed HAZ region. It has only been detected in weldments subjected to additional axial or bending system loading and is thus associated with section changes, terminal welds, multiple pipe intersections and errors in setting pipe hanger loading systems [121].

Microstructurally, the cracking occurs initially as creep cavitation or microcracking in the region of the HAZ which was partially transformed or overtempered, Figure 59 [122]. As shown earlier, the partially transformed structures have low creep ductility and creep rates slightly higher than the coarse grained bainitic structures, whereas the soft overtempered structures are of low creep strength but high ductility. These zones occur in all ferritic welds to some degree and their extent and characteristics are affected by the welding procedures and by the parent material properties, the latter being the more important. Such a microstructural configuration is weak when subjected to axial or bending loads and may lead to failure.

The failure mechanism is therefore by creep of locally constrained regions under the action of some additional operating stresses, which are not always allowed for within the design. Plant experience suggests that the development of creep cavitation and cracking is relatively rapid, very localised and can lead to steam leakage [121].

The mode of failure re-emphasises that quite specific interactions between metallurgical structure and applied stress can be important when welds are operated at elevated temperatures.

8.4 Failure Corresponding to the Design Life

Obviously, components subjected to stress under creep conditions would fail given sufficient time and would have a failure distribution curve, as suggested in Figure 53. At present, there is no information regarding the failure mode that will predominate, although as the CEGB in line with other

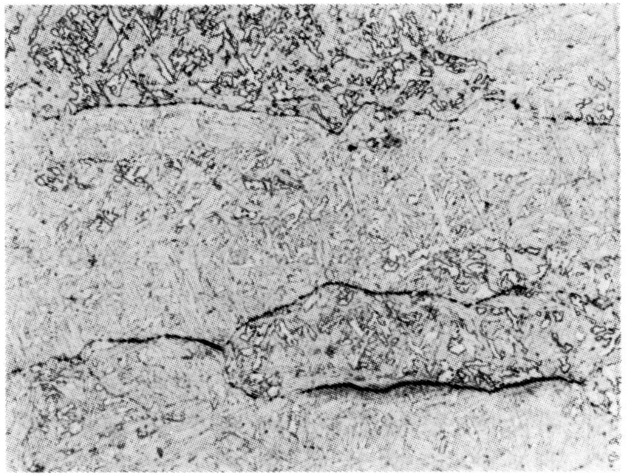

x 200

Fig. 58. Cavitation and cracking on the weld metal columnar grain boundaries in 2CrMo. [115]

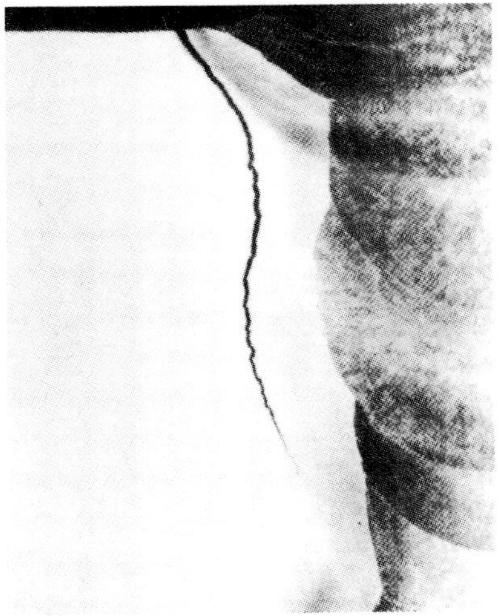

x 5

a. Circumferential crack in the transition HAZ/base metal region.

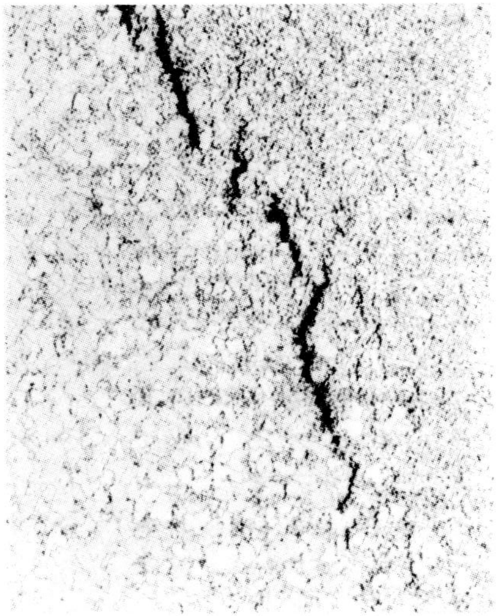

x 100

b. Detail at crack tip showing cavitation and structure.

Fig. 59. "Type IV" Cracking in ferritic components. [122]

organisations, intends to operate plant to 2000,000 h and possibly longer in future, evidence may be obtained. However, it is interesting to hypothesise on the failure modes suggested by the analytical models and the results of accelerated tests.

If it is assumed that the premature failure modes discussed earlier have been overcome, then a failure mode based only on applied pressure stresses must be examined. For discussion this is taken as the $\frac{1}{2}$Cr$\frac{1}{2}$Mo$\frac{1}{4}$V cylindrical pipe butt welded using 2$\frac{1}{4}$Cr1Mo initially reported by Goodall and Walters [104] and Walters [105].

The majority of welds in service are fabricated with weld metal of slightly lower creep strength than the parent and will contain HAZ regions which are of higher creep strength and lower ductility. In steady state creep, the hoop stresses will offload from the weaker weld metal into the stronger HAZ. Thus the highest stresses will be in the HAZ with the parent and weld metal stresses being determined by the relative strengths of these materials. In steady state there is evidence that the hoop strain rates will be similar completely across the weldment and thus, crack initiation time and position will be controlled by the relative strains to failure of the constituent microstructures of the weld. As pointed out by Walters [105], cracking would be expected to occur parallel to the pipe axis, in the HAZ region. Such behaviour has been reported from a series of full size experimental welded vessel tests on welds in both tube and pipe geometries [83,108]. Probably the next most susceptible region will be the weld metal and then the parent material. Obviously, as a region cracks, the relative load bearing capacities of the constituents change locally and tertiary rather than steady state creep will ensue possibly with modification to the stress and strain distributions. At this stage, there will be a balance between macroscopic crack growth and microcrack initiation in other parts of the weldment. The former should dominate, so that the orientation of the major crack should remain parallel to the pipe axis having been nucleated originally in the HAZ region.

It is also possible that a Type IV cracking mode may predominate at very long lives, but there is no creep data available to give guidance.

Certain weld designs, between austenitic and ferritic steels have not been considered. In these, there are differences in the coefficients of thermal expansion and thermal cycling, creep stresses and local structures combine to alter the failure mode and position [123-127]. These dissimilar metal welds undergo a complex loading pattern, although failure is still directly influenced by the interaction of the pressure and thermal expansion stresses with microstructure in the inhomogeous austenitic:ferritic interface region. Such failures are being subjected to considerable research and the problems have been reviewed in the literature.

9. SUMMARY AND CONCLUSIONS

This paper has attempted to rationalise the high temperature creep failure of weldments through a detailed consideration of what constitutes a weld, in terms of micro-structural, property and residual stress variation. Using the specific case of the ferritic ½CrMoV:2Cr1Mo weld combination, the main parameters controlling failure have been outlined and identified. There is no reason to believe that the view taken here should not be applicable to other material combinations if the loading systems and the relevant properties are known.

In conclusion, this paper is an attempt to lead physical metallurgists, engineers and managers through the minefield of black art that is normally associated with welding and the performance of welds at elevated temperatures, by providing a logical framework for consideration of this multi-disciplinary field. Within the next 2 - 4 years, it is expected that definitive experiments currently being carried out in UK, Sweden and USA will allow improved design rules for weldments to be evolved.

10. ACKNOWLEDGEMENTS

The authors wish to thank numerous colleagues both within and outside the CEGB for helpful discussions over many years and particularly those at MEL who have helped with this paper. This paper is published by permission of the Central Electricity Generating Board.

11. REFERENCES

1. TOFT, L.H. & YELDHAM, D.E. Weld Performance in High Pressure Steam Generating Plant in Midlands Region CEGB. International Conference on Welding Research Related to Power Plant, CEGB, London, 1972, 5.

2. CHEETHAM, D., FIDLER, R., JAGGER, M. & WILLIAMS, J.A. 'Relationship between Laboratory Data and Service Experience in Cracking in CrMoV Weldments'. Conf. on Residual Stresses in Welded Constructions and their Effects. London, 1977, The Welding Institute.

3. HOULDCROFT, P.T. Welding Process Technology, Cambridge University Press, 1977.

4. LANCASTER, J.F., The Metallurgy of Welding, Brazing and Soldering. Inst. of Metallurgists. Modern Metallurgical Texts, vol.3, Allen and Unwin Ltd., London 2nd Ed. 1970.

5. SMITH, D.C.,'Development, Properties and Usability of Low Hydrogen Electrodes', Welding J.Res.Supp., 1959, 38, p377s.

6. BONISZEWSKI, T., 'Formulation of a basic flux coating for an experimental 2CrMo electrode', Metal Construction and British Welding Journal, 1971, 3, 18.

7. WOLSTENHOLME,D.A., 'Reactions involving oxygen between gas, metal and slag during welding with basic electrodes', Conference 'Trends in steels and consumables for welding', The Welding Institute, 1978, London 123-134.

8. DAVIS, Miss L., An introduction to welding fluxes for mild and low alloy steels, The Welding Institute, Cambridge 1981.

9. DAVIES, G.J. & GARLAND, J.G., 'Solidification Structures and Properties of Fusion Welds', International Metallurgical Reviews, 1975, 20, 83.

10. SAVAGE, W.F., 'Solidification Segregation and Weld Imperfections', 1980, Houdremont Lecture, Welding in the World, 1980, 18, 90.

11. BAEKIE, B.L. & YAPP, D., 'Solidification and Casting of Metals, Book 192, Metals Society, London, 1979, 438.

12. HEMSWORTH, B., BONISZEWSKI, T. & EATON, N.F., 'Classification and definition of high temperature welding cracks in alloys', Metal Construction, 1969, 1, (2s), 5.

13. BROOKS, J.A. & LAMBERT, Jn.,F.J., 'Welding J.Res.Supp., 1978, 57, 139s.

14. SUUTALA, N., TAKALO, T. & MUISIO, T., 'The Relationship between Solidification and Microstructure in Austenitic and Austenitic-Ferritic Stainless Steel Welds', Met.Trans.A., 1979, 10A, 512.

15. THOMAS, R.G., 'The effect of δ-ferrite on the creep properties of austenitic weld metals', Welding J. Research Suppl., 1978, 57, 81.

16. CHRISTENSEN, N., DAVIES, V.de L. & GJERMUNDSEN, K., 'Distribution of Temperatures in Arc Welding', British Welding Journal, 1965, 12, 54.

17. LANCASTER, J.F., 'Energy distribution in argon shielded welding arcs', British Welding J., 1954, 1, 412.

18. GLICKSTEIN, S.A. & YENISCAVICH, W., 'A review of element effects on the welding arc and weld penetration', WRC Bulletin No.266, 1977.

19. GHENT, H.W., ROBERTS, D.W., HERMANCE, C.E., KERR,H.W. & STRONG, A.R., 'Arc efficiency in TIG welds', Welding Institute Conference on Arc Physics and Weld Pool Behaviour, 1979.

20. ROSENTHAL, D., 'Mathematical theory of heat distribution during welding and cutting', Welding J. Res. Supp., 1941, 220, 220s.

21. HESS W.F., MERRILL,L.L., NIPPES, E.F., & BUNK, A.P., 'The measurement of cooling rates associated with arc welding and their application to the selection of optimum welding conditions', Welding J. Res. Supp., 1943, 22, 370.

22. GLADMAN, T., 'On the theory of the effect of precipitate particles on grain growth in metals', Proc.Roy.Soc., 1966, 294A, 298.

23. ALBERRY,P.J., & JONES, W.K.C., 'Structure and hardness of 0.5CrMoV and 2CrMo simulated heat affected zones', Metals Technology, 1977, 4, 557.

24. ALBERRY,P.J., CHEW,B., & JONES, W.K.C., 'Prior austenite grain growth in heat affected zone of a 0.5CrMoV steel', Metals Technology, 4, 317.

25. IKAWA, H., OSHIGE,H., NOI,S., DATE,H.,& UCHIKAWA,K., 'Relation between welding conditions and grain size in weld heat affected zone', Transactions of the Japan Welding Society, 1978, 1, 47.

26. ALBERRY,P.J., & LAMBERT,J.A., 'The welding metallurgy of SA508 class 2 heat affected zones', Int.Conf.on Welding Technology for Energy Applications, AWS-ASM, 1982.

27. ALBERRY,P.J., & JONES, W.K.C., 'Diagram for the prediction of weld heat affected zone microstructure', Metals Technology, 1977, 4, 360.

28. CONSTANT,A. & MURRY, G., 'Contribution à L'étude des transformations rapides en relation avec les problemes de soudage des aciers. Dispositifs d'essai et examples d'application'., Soudages et Techniques Connexes, 1963, 17,405.

29. BERNARD,G., 'A viewpoint on the weldability of carbon manganese and microalloyed structural steels', Microalloying 75, 1975, Washington, 52.

30. MULLERY,F. & CADMAN,R.O.L., 'Cracking of welded joints in ferritic heat resisting steels', Brit.Weld.J.,1962,9,(4).

31. BAILEY,N. & JONES,S.B., 'Solidification cracking of ferritic steels during submerged arc welding', The Welding Institute, 1977.

32. COE,F.R., Welding Steels without hydrogen cracking, The Welding Institute, Cambridge, 1973.

33. ALBERRY, P.J. & JONES, W.K.C., 'A computer model for the prediction of heat affected zone microstructures in multipass weldments', 1979, CEGB Report R/M/R282.

34. ALBERRY,P.J., BRUNNSTROM,R.R.L. & JONES, K.E., 'A computer model for the prediction of heat affected zone structures in mechanised tungsten inert gas multipass weld deposits', 1981, CEGB Report RD/M/1158R81.

35. CHUBB,E., FIDLER,R. & WALLACE,D., 'Development and relaxation of welding stresses', Int.Conf.,"Welding Research Related to Power Plant", CEGB, London 1972, 143.

36. JONES,W.K.C. & ALBERRY,P.J., 'Model for stress accumulation in steels during welding', Conf., "Residual Stresses in Welded Construction and their Effects", London, 1977, The Welding Institute, UK.

37. JONES, W.K.C. & ALBERRY,P.J., 'The role of phase transformation in the development of residual stresses ...', Conf. "Ferritic Steels for Fast Reactor Steam Generators", BNES, London, 1977.

38. FIDLER,R. & JERRAM,K., 'Residual stresses in $2\frac{1}{4}$Cr1Mo welds', Conf. "Ferritic Steels for Fast Reactor Steam Generators', BNES, London, 1977.

39. FIDLER,R., 'Residual stresses associated with austenitic welds', Conf. "Residual Stresses in Welded Construction and their Effects", London, 1977, The Welding Institute, UK.

40. FIDLER, R., 'The effect of time and temperature on residual stress in austenitic welds', Conf. "Pressure Vessels and Piping", 1982, Orlando, Florida, USA., ASME.

41. UEDA,Y., FUKUDA,J. & NAKACHO,K., 'Basic procedures in analysis and measurement of welding residual stresses by the finite element method', Conf. "Residual Stresses in Welded Construction and their Effects", London, 1977, The Welding Institute, UK.

42. FIDLER,R., 'The complete distribution of residual stress in a $\frac{1}{2}$Cr$\frac{1}{2}$Mo$\frac{1}{4}$V/2CrMo main steam pipe weld in the as-welded condition', CEGB Report R/M/R261, 1978.

43. DOWNE,B.G. & FIDLER,R., 'A manual for the measurement of residual stress by trepanning using EDM', CEGB Report R/M/N1120, 1981, CEGB, UK.

44. BEANEY,E.M. & PROCTOR,E., 'A critical evaluation of the centre hole technique for measurement of residual stresses', 1972, CEGB Report RD/B/N2492.

45. FIDLER,R. Unpublished work on centre hole drilling using a stationary air abrasive jet technique.

46. FIDLER,R., 1978, Private Communication.

47. HEPWORTH,J.K., 'Residual stresses in submerged arc bead-on-plate welds', CEGB Report R/M/N1090, 1981.

48. HARDY,A.K., 'Circumferential distribution of residual stress in a single bead MMA weld on reheat pipe', 1975, CEGB Report R/M/N833.

49. MASUBUCHI,K., 'Models of stresses and deformation due to welding. A Review', Journal of Metals, 1981, Dec. 19-23.

50. RYBICKI, E.F. & STONESIFIER,R.B., 'Computation of residual stresses due to multipass welds in piping systems', J.Press. Vessel Technol., 1979, 101, 139-154.

51. UEDA,Y. & YAMAKAWA,T., 'Analysis of thermal elastic/ plastic stress and strain during welding', Document X-616-71, Commission X of the International Institute of Welding, 1971

52. HEPWORTH, J.K., 'Finite element calculation of residual stresses in welds', Conf. "Numerical Methods for Non-Linear Problems', Swansea, 1980, Pineridge Press, Swansea, UK.

53. RYBICKI,E.F., SCHMUESER,D.W., STONESIFER,R.B., GROOM,J.J. & MISHLER,H.W., 'A finite element model for residual stresses and deflections in girth butt welded pipes', J.Press.Vessel Technol., 1978, 100, 256-262.

54. RYBICKI,E.F. & McGUIRE,P.A., 'A computational model for improving weld residual stresses by induction heating', J. Press.Vessel Technol., 1981, 103, 294-299.

55. KING,B.L., 'Some effects of composition and microstructure on the high temperature ductility of CrMoV steels', 1976, CEGB Report RD/L/R1945.

56. WOLSTENHOLME,D.A., 'Transverse cracking and creep ductility of 2CrMo weld metal', 1976, CEGB Report R/M/R266.

57. THOMAS,R.G. & KEOWN,S.R., 'Phase transformations in Duplex type 316 weld metals', 1981, CEGB Report RD/M/R312.

58. HORTON,C.A.P., MARSHALL,P. & THOMAS,R.G., 'Time dependent changes in microstructures and mechanical properties of AISI 316 steel and weld metal', 1981, CEGB Report RD/L/2045N81.

59. COLE, N.C., GOODWIN,G.M. & SLAUGHTER,G.M., 'Effect of heat treatments on the microstructure of stainless steel weld metal', "Microstructural Science, Vol.3", 1975, American Elsevier Publishing Co.Inc., NY.

60. FINNIE,I. & HELLER,W.R., Creep of Engineering Materials, Magraw Hill, London 1959.

61. MACKENZIE,A., Unpublished work,1973, University of Glasgow, Department of Mechanical Engineering, UK.

62. TOYOOKA,T. & TERAI,K., 'On the effects of PWHT', Welding J. Res. Supp., 1973, June, 247-254.

63. FERRILL,D.A., JUHL,P.B. & MILLER,D.R., 'Measurement of residual stresses in a heavy weldment', Welding J. Res. Supp., 1966, Nov., 504-515.

64. FIDLER,R., 'A finite element analysis for the stress relief of a 2CrMo-CrMoV main steam pipe weld', CEGB Report R/M/R270, 1980.

65. SMITH,E. & NUTTING,J., 'The tempering of low alloy creep resistant steels containing Cr, Mo and V', JISI, 1957, Dec., 314-329.

66. HIPPSLEY,C.A., 'Precipitation sequences in HAZ of $2\frac{1}{4}$Cr1Mo steel during PWHT', Met.Sci., 1981, 15, 137-147.

67. WOLSTENHOLME,D.A., 'Deoxidation and mechanical properties of 2Cr-Mo weld metals', 1975, Ph.D., Thesis, University of London, UK.

68. ARGENT,B.B., NIEKERK,M.H., van and REDFERN, G.A., 'Creep of ferritic steels', JISI, 1970, 208, 830-843.

69. CONFERENCE "High Temperature Properties of Steels", Eastbourne, 1966, BISRA and ISI, England, ISI Publication P.97.

70. ARONSSON,B. & HEDE, A., 'Some observations on the reproducibility of creep rate determinations', ibid. 69. p41.

71. RUTTMAN,W., KRAUSE,M. & KREMER,K.J., 'International Community Tests on long term behaviour of $2\frac{1}{4}$Cr1Mo steel',ibid.69.23.

72. TAIRA,S. & SAKUI,S., 'Standardisation of the creep testing machine', ibid. 69., p31.

73. MURRAY,G.,'A new formula for the extrapolation of creep test results: An example of its application to a CrMoV steel'. ibid. 69., p17.

74. CONFERENCE 'The Presentation of Creep Strain Data', Sheffield, 1971. BISRA, Sheffield, UK.

75. JOHNSON,R.F., MAY,M.J., TRUMAN,R.J. & MICKLERAITH,J., 'Elevated temperature tensile, creep and rupture properties of 1Cr½Mo, 2¼Cr1Mo and ½Cr½Mo¼V steels', ibid. 69., p229.

76. CANE,B.J., 'Collaborative programme on the correlation of test data for high temperature design of welded steam pipes', 1981, CEGB Report RD/L/2101/N81.

77. WILLIAMS,J.A. Unpublished work.

78. ALBERRY,P.J. & JONES, W.K.C., 'A comparison of the mechanical properties of 2CrMo and ½CrMoV simulated HAZ's', 1977, Metals Tech., Jan., 45.

79. KING,B.L., 'Some effects of composition and microstructure on the high temperature ductility of CrMoV steels', 1976, CEGB Report RD/L/R1945., CEGB,UK.

80. LAI,J.K.L., CHASTELL,D.J. & FLEWITT,P.E.J., 'Precipitate phases in 316 stainless steel resulting from long term, high temperature service', 1980, CEGB Report RD/L/N97/80.

81. BATTE,A.D. & MURPHY,M.C., 'Creep rupture properties of 2¼Cr1Mo weld deposits', Welding J. Res. Supp., 1973, June, 261-267.

82. ROWLEY,T. & COLEMAN,M.C., 'Collaborative programme on the correlation of test data for the design of welded steam pipes', 1973, CEGB, Report RD/M/N710.

83. BROWNE,R.J., CANE,B.J., PARKER,J.D. & WALTERS,D.J., 'Creep Failure analysis of butt welded tubes', Conf. 'Creep and Fracture of Engineering Materials and Structures', Swansea, 1981, Pineridge Press, Swansea, UK.

84. ROODE,F., ETIENNE,C.F. & VAN ROSSIM,O.,'Stress and strain analyses for creep and plasticity of welded joints in AISI 316', Conf.'Engineering Aspects of Creep', Sheffield, 1980, Inst.Mech.Engrs., London,UK.

85. IVARSSON,B.G. & SANDSTROM,R., 'Creep of butt welded tubes', Conf. 'Creep and Fracture of Engineering Materials and Structures', Swansea, 1981, Pineridge Press, Swansea,UK.

86. DAVIES,P.W. & WILSHIRE,B., 'On internal stress measurement and the mechanism of high temperature creep', Scripta Met., 1971, 5, 475.

87. TOWNSEND,R.D., 'The effect of structure on the creep properties of low alloy ferritic steels', 1971, CEGB Report RD/L/R1740.

88. COLE, N.C., KING,R.T., GOODWIN,G.M. & BERGGREN,R.G., 'The effect of ferrite content on the properties of austenitic welds', AWS Annual Meeting, Houston, Texas, May 6th-10th, 1974.

89. KLUEH,R.K., 'Creep and creep rupture behaviour of a bainitic 2¼Cr1Mo steel', Int.J.Press.Vessels and Piping, 1980, 8, 165-185.

90. WOLSTENHOLME,D.A., 'Transverse cracking and creep ductility of 2CrMo weld metals', Conf. 'Trends in Steels and Consumables for Welding', London, 1978, The Welding Institute, UK.

91. SIVERNS, M.J. & PRICE,A.T., 'Crack propagation under creep conditions in a quenched 2¼Cr1Mo steel'. Int.J.Fracture, 1973, 9, 199-207.

92. NEATE,G.J., 'Creep crack growth in ½Cr½Mo¼V steel at 565°C'., Eng.Fr.Mechs., 1977, 9, 297-306.

93. HAIGH,J.R., 'Some aspects of creep crack growth in a coarse grained 1CrMoV steel (simulation of a weld HAZ)', 1977, CEGB Report RD/L/N217/76.

94. JAMES,L.A., 'Preliminary observations on the extension of cracks under static load at elevated temperature', Int.J. Fr.Mechs., 1972, 8, 347.

95. GOOCH,D.J., HAIGH,J.R. & KING,B.L., 'Relationship between engineering and metallurgical factors in creep crack growth', Met.Sci.J., 1977, Nov., 545.

96. ELLISON,E.G. & HARPER,M.P., 'Creep behaviour of components containing cracks', J.Strain Analysis, 1978, 13, 35.

97. FU,L.S., 'Creep crack growth in alloys at elevated temperature; a review', Eng.Fr.Mechs., 1980, 13, 307.

98. GOOCH,D.J. & KING,B.L., 'Creep crack growth in controlled microstructure CrMoV HAZ's', 1979, CEGB Report RD/L/N178/78.

99. GOOCH,D.J., 'Creep crack growth in variously tempered coarse grained bainitic and martensitic $\frac{1}{2}$Cr$\frac{1}{2}$Mo$\frac{1}{4}$V steel', 1976, CEGB Report RD/L/N80/76.

100. GOOCH,D.J. & KING,B.L., 'High temperature crack propagation in 2$\frac{1}{4}$Cr1Mo MMA weld metals', Conf.,'Weldments, Physical Metallurgy and Failure Phenomena', 5th Bolton Landing Conf., Bolton Landing, NY., 1978, GEC, Schenectady, USA.

101. SADANANDA,K. & SHAHINIAN,P., 'Crack growth under creep and fatigue conditions', Symposium, 1978, Fall Meeting AIME, Milwaukee, USA.

102. HAIGH,J.R., 'Creep crack growth in Type 316 weld metal', Welding J., 1977, 56, 149-153.

103. MANJOINE,M.J., 'Crack initiation and growth in welds at 593°C', 1977, WARD-HT-3045-28, Westinghouse Co., Pittsburg, USA.

104. GOODALL,I.W. & WALTERS,D.J., 'Creep of butt welded tubes', Int.Conf. on 'Creep and Fatigue in Elevated Temperature Application', 1973, Sheffield, Inst.of Mech.Engrs,London.

105. WALTERS,D.J., 'The stress analysis of cylindrical butt welds under creep conditions', 1976, CEGB Report RD/B/N3716.

106. LAMÉ,G., 'Lecons sur la theoré ... de l'elasticité', 1852, Gauthier-Villars, Paris.

107. BAILEY,R.W., 'The utilisation of creep test data in engineering design', Proc.Int.Mech.Engrs, 1935, 131, 131.

108. COLEMAN,M.C., PARKER,J.D. & WALTERS,D.J., 'The behaviour of ferritic weldments in thick section $\frac{1}{2}$Cr$\frac{1}{2}$Mo$\frac{1}{4}$V pipe at elevated temperature', CEGB Report RD/M/1204R81.

109. WALTERS,D.J. & COCKROFT,R.D.H., 'A stress analysis and failure criteria for high temperature butt welds', 1972, International Institute of Welding Symposium, 1972, Toronto, Canada.

110. COLEMAN,M.C. & PARKER,J.D., 'Deformation in 2CrMo-$\frac{1}{2}$CrMoV pressure vessel weldments at elevated temperature', Conf. 'Creep and Fracture of Engineering Materials and Structures', Swansea, 1981, Pineridge Press, Swansea, UK.

111. WILLIAMS,J.A. & HEPWORTH,J.K., 1982, Research in progress.

112. HEPWORTH,J.K., 'The effect of residual stress on the creep deformation of welded pipe',CEGB Report RD/M/1202R81.

113. ETIENNE,C.F., VAN ROSSUM,O. & ROODE,F., 'Creep of welded joints in AISI 316', Conf.,'Engineering Aspects of Creep', 1980, Sheffield, Inst.Mech.Engrs, London, UK.

114. WILLIAMS,J.A., 'Methodology for high temperature failure analysis', 1981, Symposium 'The Behaviour of Joints in High Temperature Materials', Commission of European Communities, Petten, The Netherlands, North Holland Publishers.

115. COLEMAN,M.C., 'The structure of weldments and its relevance to high temperature fracture', Conf."Weldments, Physical Metallurgy and Failure Phenomena', 5th Bolton Landing Conf., Bolton Landing, NY 1978, GEC, Schenectady, USA.

116. COLEMAN,M.C., 'The testing, monitoring and metallurgical examination of a cracked throttle pressure vessel, 5th Int. Conf. on Fracture, ICF5, 1981, Cannes, France.

117. MYERS,J., 'Influence of alloy and impurity content on stress relief cracking on CrMoV steels'. 1976, CEGB Report RD/M/R254, CEGB,UK.

118. MIDDLETON,C., 'Reheat cavity nucleation in bainitic creep resisting low alloy steels', 1979, CEGB Report RD/L/N56/79, CEGB,UK.

119. WOLSTENHOLME,D.A., 'Transverse cracking on reheating of a 2CrMo **submerged arc weld metal**', 1976, CEGB Report R/M/N868, CEGB,UK.

120. SCHULLER,H.J., HAIGH,L. & WOITSCHECK,A., 'Cracking in the weld region of shaped components in hot steam lines', Der Maschinenschaden, k974, 47, 1-13.

121. ARNSWALD,V.W., BLUM,R., NEUBAUER,B. & POULSEN,K E., 'Einsatz von Oberflachengefügeuntersuchungen für die Prüfung.. Kraftwerksbauteile', Kraftwerkstechnik, 1979, 59, 581-593.

122. FRANK,R., HAGN,L. & SCHÜLLER,H.J., 'Creep damage and intention for prevention', Conf. 'Prediction of Residual Life Time of Constructions Operating at High Temperature', 1977, The Hague, The Netherlands, NIL.

123. NICHOLSON,R.D. & PRICE,A.T., 'Service experience of nickel based transition joints', 1981, Conf. 'Dissimilar Joint Welding', Leicester 1981, The Welding Institute, UK.

124. LUNDIN,C.D., 'Dissimilar metal welds - transition joints literature review', Welding Research Supp., 1982, Feb.,58-63.

125. NICHOLSON, R.D., 'The creep rupture properties of austenitic and nickel based transition joints', 1981, CEGB Report RD/M/1176N81, CEGB,UK.

126. DOOLEY,R.D., STEPHENSON,G.G., TINKLER,M.J., MOLES,M.D.C. & WESTWOOD,H.J., 'Ontario Hydro experience with dissimilar metal welds in boiler tubing', Welding Research Supplement, 1982, Feb., 45-49.

127. BAGNALL,B.I., ROWBERRY,T.R. & WILLIAMS,J.A., 'Service experience with heavy section dissimilar joints operating at elevated temperature', 1981, Conf. 'Dissimilar Joint Welding', Leicester,June 1981, The Welding Institute, UK.